THE BEST OF
Food&Wine
VEGETABLES, SALADS & GRAINS

THE BEST OF
Food&Wine
VEGETABLES, SALADS & GRAINS

American Express
Publishing Corporation
New York

The Best of Food & Wine / VEGETABLES, SALADS & GRAINS
Designer: Loretta M. Sala
Assistant Editor: Martha Crow
Illustrations: Joanna Roy

American Express Publishing Corporation
Editor in Chief/Food & Wine: Mary Simons
Art Director/Food & Wine: Elizabeth Woodson
Managing Editor/Books: Kate Slate
Marketing Director: Elizabeth Petrecca
Production Manager: Joanne Maio Canizaro

© 1993 American Express Publishing Corporation

All rights reserved. No part of this book may be reproduced or transmitted in any form or by any means, electronic or mechanical, including photocopying, recording, or by any information storage and retrieval system, without permission in writing from the publisher.

Published by American Express Publishing Corporation
1120 Avenue of the Americas, New York, New York 10036

Manufactured in the United States of America
ISBN 0-916103-20-X

TABLE OF CONTENTS

FOREWORD • 7

FINDING THE RIGHT WINE MATCH • 8

APPETIZERS • 16

FIRST COURSES • 32

SOUPS & CHOWDERS • 60

MAIN COURSES: VEGETABLES & GRAINS • 84

MAIN COURSES: SALADS • 114

SIDE DISHES: VEGETABLES & GRAINS • 148

SIDE DISHES: SALADS •200

DESSERTS• 240

DRESSINGS, CONDIMENTS & PICKLES • 258

INDEX • 270

CONTRIBUTORS • 283

FOREWORD

BY MARY SIMONS, EDITOR IN CHIEF

Not so very long ago, vegetables were, on the American table at least, pretty much an afterthought. Often they were overcooked. Most were certainly not painstakingly prepared. Many were eaten out of a sense of duty...especially by children and adult males.

But that's all changed. These days, some of the very best cooking in America focuses on vegetables. Now we treat them with the same care that we use in preparing all our foods and have become truly imaginative in combining them with other ingredients.

Across the nation, farmers' markets and truck gardens whet our new vegetarian appetites by providing us with seasonal produce that rivals the best in the world. Suddenly those servings of vegetables that we're told should be part of our daily diet are easy to come by.

Combining them with grains of various kinds is what this book is all about. Grains and vegetables...for every course in the meal...from sophisticated first courses to inventive side dishes, stir-fries, main-course salads, stews and desserts are all here. It's a fresh way to think about food that is nutritious and delicious.

FINDING THE RIGHT WINE MATCH

BY ELIN McCOY & JOHN FREDERICK WALKER

If you're choosing just one wine for dinner, it makes sense to try to find a wine that will complement the main dish. Unfortunately, traditional advice on wine and food matching assumes that meals are almost always built around a main dish of meat, fowl or fish. But meat, fowl and fish don't appear on American tables only as a separate dish—in today's recipes they often turn up as one ingredient among many, if, indeed, they appear at all. Vegetables and grains have come to the fore, and salads are no longer relegated to side dish status; they've all become main courses in their own right. But while there's no dearth of advice on what wines to serve with lamb, for example, there's precious little on what to serve with dishes where lamb is little more than a flavoring or garnish.

Today's cooks have questions like these: What do you serve with a lentil-based stew or a hearty thirteen-bean soup? What wine complements a salad composed of arugula, sun-dried tomatoes, sliced duck breast and walnut oil dressing? What goes with an asparagus omelet, stuffed eggplant or a rich risotto?

That's where this chapter comes in. In the following pages you'll find everything you need to know to choose appropriate wines for dishes whose principal flavors derive from vegetables, grains and greens, whether they're simple or sophisticated, side dish or main dish.

COMBINING TASTES

Here's the bottom line: Choosing a bottle of wine that will make an appealing match for a particular dish is no more difficult than choosing compatible dishes to serve at the same meal. The idea, in both cases, is to arrive at attractive combinations of flavors and textures.

There are two main ways to do that. One is to put similar tastes and textures together to create a harmonious taste combination—say, the sweet lusciousness of peaches with fresh cream or the ripe fruitiness of late-harvest Riesling with fresh strawberries—and it is one of the simplest and most effective ways to create attractive wine/food pairings.

The second way is to use contrasting flavors and textures to accent and heighten each other. The sharp taste of caper sauce on broiled salmon, for example, points up the richness of the fish. Similarly, a crisp, tart white wine such as Muscadet acts as a palate-cleansing foil to the briny flavor and succulent texture of raw oysters.

But regardless of whether the wine echoes the taste of the food or contrasts with it, the flavor *intensity*—the impact—of the food and the wine should be roughly similar. It's easy to understand why: A very light, delicate wine makes very little impression at all if served with hearty, rich food; conversely, a powerful, deep wine overwhelms the mild flavor of a delicate dish. It's not just the lack of flavor affinities that makes a fruity, simple Chenin Blanc a poor choice to pair with a lamb chop or makes a big, mouth-filling Barolo an unsatisfactory partner to boiled shrimp—there's simply too much disparity in *weight* of flavor. In brief, substantial fare calls for hearty wines; light, subtle dishes call for delicate wines—whether red or white.

If matching food and wine is just a matter of these simple

principles, then why do so many people find it daunting? Often, they have the mistaken notion that the world of wine is enormously complex, with endless variations from thousands of appellations and producers, with each and every vintage. They forget that in terms of taste—the key thing, after all—wine presents a far simpler picture. There are fresh, light wines (mostly, but not all white) and bigger, richer wines (mostly, but not all, red). There are dry and sweet wines within these groupings, and of course there are the the subtle differences in flavor that come from the variety or varieties of grapes used (see "Wine Flavors," page 11).

The characteristic taste of each basic wine type—light, semisweet, tart white; rich, tannic, mouth-filling red; and so on—provides all the clues you need to find suitable matches. How? It's a process of elimination. Let's follow a train of thought that will be familiar to anyone who has prepared a menu: Assume you're serving sole in lemon sauce and want a side dish that will enhance it. You would consider the possibilities and make mental comparisons. You might discard, for example, mashed potatoes (too bland, too soft, too white) or marinated red peppers (too strong). Let's say you finally decide wild rice would add just the right contrast of texture and subtle, nutty flavor that will enhance the fish.

Choosing a wine for the dish is a parallel process that draws on your experience with a few basic wine types. For example, as you think about it, a big, flavorful Chardonnay would be likely to overpower the delicate fish. Gewürztraminer? A spicy, fruity wine doesn't strike you as likely to work with the lemon sauce. A crisp Sauvignon Blanc? The herby nuances and clean taste should underscore the sole, sauce and wild rice nicely...Of course at the table you'll discover if your choices of side dish and wine accompaniment are merely compatible with the fish or a brilliant marriage of tastes. But fine-tuning recipes, orchestrating courses and selecting just the right bottle is all part of the fun of menu planning.

The wine suggestions given in this chapter for various vegetable- and grain-based dishes are just that: suggestions. Armed with the information in this section you should be able to readily choose attractive alternatives and find wines to enhance dishes not specifically mentioned here.

VEGETABLES, LEGUMES AND GRAINS

In dishes where meat or fish is just one ingredient among many—say, a rice pilaf garnished with a few shrimp—the shrimp don't dominate the flavor of the dish or provide the key taste element that affects the choice of wine. In such cases, it makes more sense to pay attention to the underlying vegetable or grain base—here, the rice pilaf. Of course, sometimes the underlying flavor is so bland (as with noodles, boiled potatoes or plain white rice) no clear indication of suitable wine type emerges—it all depends on what's added. With other vegetables and grains, however, such as peas, beans, bulgur and the like, there's usually a taste and a texture that suggest a basic wine type; then the other ingredients help narrow the possibilities within that category to a specific choice.

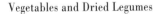

By themselves, for example, all types of cooked beans have a texture that suggests red wine. Pork, duck, lamb, sausage, goose, etc.—ingredients typical of such hearty bean casseroles as cassoulet—add considerably to the richness and fattiness of the basic dish and tilt the wine choice toward a hearty red with enough taste and tannin to match it: a traditional French country red such as Madiran or Cahors or a vigorous Côtes du Rhône, red Zinfandel from California or Australian Shiraz.

The following lists cover many vegetables, legumes and grains often served on their own or as the prime component of many dishes.

Vegetables and Dried Legumes

As vegetables and legumes have become primary ingredients in more of our meals, we have discovered just how rich a world of flavors and textures they constitute, ranging from crunchy to smooth and soft, bland or subtle to pleasantly bitter, delicate to rich. Most have an underlying character and flavor that is not wholly changed by the type of cooking or the sauce with which they are served.

Artichokes: The long-standing injunction against serving wine with artichokes whether whole or just the hearts is based on the fact that after tasting artichokes, wine tastes oddly sweeter. This makes artichokes an unflattering companion to fine, subtle wines, but the effect is less noticeable on simple, assertive dry reds and whites such as Bordeaux Blanc, Sauvignon Blanc, Côtes du Rhône and the like. Egg-, oil- or butter-based sauces help make artichokes more compatible with wine.

Asparagus: Another vegetable traditionally regarded as a poor match for wines, asparagus works very well (especially when simply steamed and served with butter) with light, crisp, fruity whites, especially those that are off-dry to semisweet, such as Chenin Blanc, Vouvray, Riesling or Sylvaner. The common green variety of asparagus is less intense than the white European type and, especially when topped with a butter- or cheese-based sauce, is quite companionable with a variety of reds.

Avocado: This tropical fruit is so frequently added to salads it might as well be regarded—like the tomato—as an honorary vegetable. Its bland, creamy-textured flesh works well with a number of round-textured whites, especially fruity California or Australian Chardonnay.

Beans and Other Dried Legumes: There are over 25 types of beans, peas and lentils available in this country, among them strongly flavored black beans; lighter kidney beans; rich fava beans; delicate, elegant flageolets; earthy yellow, red and brown lentils; mildly nutty chickpeas; green and yellow split peas. What they all share when cooked is considerable texture, and it's that characteristic that provides an important clue to finding a wine match. All legumes are partial to reds, notably those with some tannin and texture. A thick lentil or bean soup, for example, blends harmoniously with a young California Merlot with soft tannins. A spicy black bean chili needs a California red Zinfandel. A mild vinaigrette dressing, on the other hand, will dominate a bean salad, making it into a dish best paired with a tart, simple white—Soave, Vernaccia di San Gimignano, Sauvignon Blanc. Off-dry whites, such as Riesling and Chenin Blanc, complement the savory but sweet flavors of split pea soup with ham.

Green Beans: Crisp green beans are one of the most versatile vegetables and lend themselves well to a wide variety of dishes. When young and tender, boiled quickly and coated with butter, a dash of lemon juice and a sprinkling of parsley, green beans are delicious with light whites. But they can also be cooked with slivered almonds or mushrooms or in a rich casserole with bacon, onions and green pepper, in which case they complement light reds.

Beets: The strong, earthy taste of these root vegetables works best with simple vigorous reds such as Châteauneuf-du-Pape, Zinfandel or Australian Shiraz.

Broccoli: Steamed broccoli, particularly if enhanced with a rich, mild sauce, pairs very well with a variety of reds, including Cabernet Sauvignon and Bordeaux.

Brussels Sprouts: The distinctive flavor of this seasonal vegetable is especially supportive of fine reds, including Bordeaux and Burgundy. Brussels sprouts are a top choice for a side dish at a meal at which better bottles will be broached, especially when they are cooked with walnuts or chestnuts.

Cabbage: Depending on the way it is cooked, green cabbage can have an assertive, strong flavor (when served with a sweet-and-sour sauce, as a hearty soup or stuffed with vegetables and spices) to mild and almost sweet (when steamed or boiled). With strong-flavored green cabbage dishes, choose Chardonnay or other big whites; with mild-flavored cabbage dishes, choose light, tart whites such as Pinot Grigio. Red cabbage (especially when cooked with red wine) is a fine choice for a side dish at a meal where Cabernet Sauvignon, Merlot or Bordeaux will be poured. The mild, slightly earthy flavors of red cabbage seem to find a flavor echo in the characteristic herbaceousness of such reds.

Carrots: While there is nothing problematic about matching wine and carrots, there is nothing particularly compelling about the combination, either. At best the two will coexist peacefully at the same meal.

Cauliflower: The mild flavor of cauliflower is not particularly enhanced by wine—or vice versa. It's best with wine when served with cheese sauce or buttered toasted bread crumbs, in which case choose a medium-bodied red such as Chianti or Rioja.

Celery Root (Celeriac): Raw celery root dishes, such as *céleri rémoulade*, call for a crisp, assertive white, such as Alsace Pinot Blanc. When cooked, celery root has a more delicate, even elegant flavor that is particularly good with subtle, aged reds.

Chestnuts: The flavor of chestnuts and their rich, smooth texture have a strong affinity for deep, fruity or earthy reds—especially Syrah, Cabernet Sauvignon, Merlot, Brunello di Montalcino and Barolo—making chestnut puree a classic side dish when these reds are served.

Corn: The sweetness of fresh corn on the cob makes a perfect foil for light, fresh, fruity Chenin Blanc. Similar light, fruity wines will complement creamy corn pudding and even a spicy southwestern corn casserole with tomatoes and chiles.

Eggplant: The meaty quality of this vegetable makes it a prime partner for medium-bodied reds, especially Chianti, Rioja and St-Emilion, whether the eggplant is sautéed and then baked with a sauce, combined with other vegetables in a casserole or brushed with oil and grilled.

Fennel: The mildly sweet anise flavor of this delicate crunchy bulb is particularly harmonious with Italian reds with a licorice nuance, such as Barolo and Barbaresco.

Leeks: The mild, sweet, subtle but distinctive flavor of

continued on page 12

WINE FLAVORS

For convenience in food matching, the following checklist of commonly available wines is organized into broad taste categories and descriptive terms are given for key flavors of each wine (remember that such descriptions aren't to be taken too literally—oak-aged Chardonnay often has a faint vanillan nuance; it does not taste like vanilla extract).

Within each category wines are listed from lightest to weightiest in taste, but bear in mind that this ranking is necessarily approximate; individual examples can vary greatly.

LIGHT, REFRESHING WINES

Dry, Fruity Reds
Gamay/Gamay Beaujolais/Beaujolais: grapy, simple
Oregon Pinot Noir: cherry-like, delicate
Valpolicella: lean, scented
Dolcetto: soft, fruity

Crisp, Dry Whites
Soave, Orvieto and Galestro: fresh, clean, neutral, simple
Muscadet: savory, clean and lean
Bordeaux Blanc, Entre-Deux-Mers: simple, mineral-like
Pinot Grigio: simple, soft, fruity-herbal
Sauvignon Blanc/Sancerre: pleasingly sharp, assertive; melony, herbaceous; simple to complex
Mâcon: simple, apple-like Chardonnay flavors without oak character
Champagne/brut sparkling wine: palate-cleansing bubbles and tartness

Fruity, Fragrant Dry to Semisweet Whites
Riesling: fragrant ripe fruitiness enhanced by lively acidity; citrusy; simple to complex
Sylvaner: simple, fruity, clean
Chenin Blanc/Vouvray: soft, somewhat peachy fruitiness, bright acidity
California Gewürztraminer: spicy, soft-textured, fragrant

BIG RICH WINES

Dry, Characterful Reds
Côtes du Rhône: straightforward, assertive
Burgundy: ripe cherries, smoky-earthy, silky
California Pinot Noir: raspberry-cherry flavors
Rioja: lean, pleasantly woody
Chianti: scented, piquant
Bordeaux: complex, lean, tannic, olive-ish, cedary
Merlot: soft, plummy, Cabernet Sauvignon-like flavors
Cabernet Sauvignon: black-curranty, complex, tannic
Zinfandel: bright, spicy, blackberry
Barbera: tart and intense
Barbaresco/Barolo: tannic, earthy, licorice-like nuances
Brunello di Montalcino: powerful, earthy, tannic
Rhône, Syrah, Shiraz: earthy, intense, peppery, simple to complex

Dry, Full-Flavored Whites
Pinot Blanc: lean, austere, assertive
Sauvignon Blanc-Sémillon blends/White Graves: round, firm, herby, savory, complex
Chardonnay/White Burgundy: fat-textured and fleshy with oaky-vanillan overtones; complex
Gavi: firm, subtle
White Rhône: powerful, perfumed, mineral-like
Alsace Gewürztraminer: dramatic, clove-like richness

Sweet, Concentrated Dessert Wines
Late-Harvest Riesling: apricot-like
Sauternes: lemony, honied
Vintage Port: sweet black plums; concentrated and thick

continued from page 10

leeks finds flavor echoes in the pungency of Pinot Noir and red Burgundy. Alternatively, rich oak-aged Chardonnay works well.

Mushrooms: The earthy, complex tastes and dense, almost meaty texture of various mushrooms (especially cèpes or porcini and shiitake) make them noteworthy accompaniments to wines, particularly fine dry weighty whites such as Chardonnay and white Burgundy, white Graves, white Rhônes, Alsace Pinot Gris and virtually the entire range of fine reds, especially aged examples.

Onions: The redolent pungency and sweetness of cooked onions is best matched by light, relatively fruity reds such as Gamay, Beaujolais and Oregon Pinot Noir or a heavier white such as an oak-aged Chardonnay. Red onions and scallions—as a salad accent—are considerably more assertive and require a simple neutral white, such as Soave or Gavi.

Peas (Green Peas): Fresh peas—at least first-of-the-season baby peas—are relatively sweet and can be accompanied successfully by tart, fruity whites like Pinot Grigio or off-dry whites such as Riesling and Chenin Blanc, which also pair very well with edible-pod peas such as sugar snaps and snow peas.

Peppers: Bell peppers actually share a similar organic compound with Cabernet Sauvignon, which is one reason why they are often good partners. Another good match is Sauvignon Blanc or Sancerre, whose grassy, herby nuances are very similar in taste.

Potatoes: Plain potatoes, like plain rice or plain pasta, are quite bland and mellow tasting, but depending on other typical additions—butter, spices, gravies, sauces, sour cream and the like—can play a significant supporting role to a wide variety of wine flavors, both white and red. The sauce or spice is the key to a wine choice.

Spinach: Spinach and similar leafy green vegetables such as Swiss chard, despite a slightly sharp flavor, are particularly wine-enhancing when cooked, complementing dry crisp whites such as Alsace Pinot Blanc or a Mâcon, lighter reds such as Rioja and Chianti, and even a rich Merlot.

Sweet Potatoes: Semisweet white wines, fruity reds such as Beaujolais and Dolcetto, and uncomplicated Pinot Noirs and Zinfandels work best with the strong, sweet flavors of this vegetable.

Tomatoes: The high acidity of this fruit/vegetable adds sharpness to salads when it is served raw and considerable bite to cooked tomato sauce, making many dry wines taste too sweet, flabby, even tinny. Sauvignon Blanc or Sancerre work best with salads based on raw tomatoes. A straightforward not-too-rich red works best with tomato sauce—a simple Chianti or California Zinfandel. Another good choice is Barbera from Italy's Piedmont district. Its natural tartness makes it a fine partner to many tomato-sauced dishes as well as dishes that include rich, almost meaty sun-dried tomatoes.

Winter Squash: Richly flavored, smooth textured, with a warm sweetness, yellow-fleshed winter squashes such as acorn, butternut and hubbard are best with the same kinds of wines that accompany sweet potatoes.

Zucchini: This ubiquitous vegetable is as versatile as the potato. Steamed zucchini—as well as other summer squashes such as yellow zucchini and pattypan squash—are light and refreshing and work well with simple whites such as Mâcon or Sauvignon Blanc. With creamy or cheesy toppings, zucchini pairs well with a variety of reds.

Grains

The most common grains used in main and side dishes are forms of wheat, corn, rice and wild rice. They have a narrower range of flavors than vegetables although they can be delicate to nutty to toasty to earthy. Their differences in texture are subtle too: mostly soft and fluffy, but they can be thick, heavy and almost chewy.

Bulgur: Bulgur is the dried and cracked whole grain of precooked wheat and is used for a variety of Middle Eastern pilafs and, of course, tabbouleh. Its distinct nutty quality and chewy texture work well with simple reds and whites.

Cornmeal: In Italy, coarse cornmeal is used to make polenta, which is often combined with cheese, tomato sauce or mushrooms. In its various forms it can underscore a variety of whites and reds, as can other dishes made from cornmeal, such as southern-style spoonbread and corn bread.

Couscous: Couscous, a North African grain, is fairly delicate in flavor—much like pasta—and is particularly good with Chardonnay, other ingredients permitting.

Kasha: Kasha consists of toasted whole buckwheat groats, and its strong, rustic flavor is best matched to hearty simple reds such as Côtes du Rhône and Zinfandel.

Rice: By itself, long-grain white rice is nearly neutral in flavor, although basmati is notably fragrant and pecan rice attractively and delicately nutty-sweet. They are good with both light whites and reds. Rich short-grain arborio rice is used in risotto, which, depending on the recipe, can pair marvelously with a wide range of fine whites and reds. *Risotto al Barolo*, for example, is an ideal foil for that powerful Piedmontese red, while *risotto primavera* is delicious with Pinot Grigio. Brown rice has a heavier, more flavorful character, with almost the nutty quality of bulgur, and is best with a heavier white or red. But in most dishes with rice, it is the other ingredients that will tune your wine choice.

Wild Rice: Wild rice (which is actually a grass) is a peerless, wonderfully nutty grain that makes a superb addition to grilled fish, lamb and game and is particularly good with dry white Graves, red Bordeaux and Cabernet Sauvignon.

THE SALAD ISSUE

Although traditional wine and food matching advice largely ignored the vegetable kingdom, it did draw attention to salads—by putting them in the category of things wine shouldn't be served with. This blanket prohibition was based largely on the acidic sharpness of salad dressings based on vinegar. The reason? The intense acidity of vinegar easily overpowers the relatively mild acidity found in any wine. After a bite of salad drenched in strong vinaigrette, any wine tastes flabby, flat or stripped of flavor. This effect makes many winelovers shy away from salads entirely, particularly at meals where fine wines are served. At formal dinners where each course is paired with a special wine, it's typical to find salad offered as a palate-refresher after the main dish and before dessert—although unaccompanied by wine.

But salads and wines aren't incompatible. It all depends on what kind of salad it is, the choice of dressing, the types of greens used and the presence of other ingredients. Let's consider these issues one by one:

Salad Type

These days, the term "salad" covers a wide culinary territory, from a simple plate of tossed greens to main-course composed salads. Some dishes—seafood salads, for example—may be salads only in the sense of a mix of ingredients: Greens (if included at all) may amount to little more than a decorative bed of lettuces that probably won't be eaten anyway and the dressing may be omitted entirely. Depending on how wine-friendly the ingredients are, complex contemporary salads may be relatively easy to match with wine.

Greens

Salad greens vary greatly in intensity and flavor from mild, delicate types such as butter or Boston lettuce to more flavorful types such as oak leaf, red leaf, romaine and endive to assertive, peppery-sharp types, such as watercress, escarole, arugula and mustard greens. The bitterness of some of the strongly flavored greens is less flattering to wines than milder greens, which can taste almost sweet in comparison. Arugula and the like are best used as accents (if at all) in salads paired with wine. For a simple salad of mild tossed greens, Sauvignon Blanc copes best.

Dressing

The simplest way to make a salad (whether plain green or elaborate) more flattering to wine is to tone down the sharpness of the dressing by decreasing the amount of vinegar used or even by using wine or a small amount of lemon juice as a substitute. Mayonnaise, which is oil based, is not very wine-friendly for the opposite reason: It tends to accentuate the acidity of wines; whites, in general, seem less affected by it than reds.

Other Ingredients

Bits of chèvre, blue cheese, Parmesan and other cheeses; nuts (especially pine nuts and walnuts); and strips of meat, poultry or fish will add wine-flattering flavors to any salad. The higher proportion of wine-friendly ingredients, the subtler and more sophisticated the wine can be. That's because protein and fats diminish the astringency of red wines and ameliorate the acidity of both reds and whites, rounding out their flavor. Salads composed of a wide mix of ingredients—such as *salade niçoise* or antipasto—are best matched to wines with broad, straightforward flavors (Soave, Pinot Grigio and dry rosés, for example), which can tie together the welter of individual tastes.

On the other hand, anchovies, capers, peppercorns and citrus slices are as hard on wine as the strongest vinaigrette, although they can be used sparingly without clashing unpleasantly with an accompanying wine. But if the salad is full of sharp, assertive, aggressively spicy flavors, there's little point in serving a fine, complex wine alongside—its subtleties won't be noticed. Save those bottles for more delicately flavored salads and serve a modest, simple wine to provide a refreshing backdrop of flavor for the dish.

1
APPETIZERS

SPICY PUMPKIN SEEDS

These seeds are deliciously addictive! Make a double or triple batch to have on hand for nibbling.
Makes About ½ Cup

1 garlic clove, minced
1 tablespoon olive oil
2 teaspoons cumin
1 teaspoon sweet paprika
½ teaspoon cayenne pepper
½ cup shelled unsalted pumpkin seeds (pepitas)*
1 teaspoon coarse (kosher) salt
*Available at Latin American markets and health food stores

1. Preheat the oven to 400°. In a small skillet, cook the garlic in the oil over low heat for about 5 minutes without browning. Strain the oil through a fine sieve and discard the garlic. Return the garlic-flavored oil to the skillet and add the cumin, paprika and cayenne. Cook over low heat, stirring occasionally, until fragrant, about 1 minute.

2. Place the pumpkin seeds on a baking sheet. Scrape the spice mixture over the seeds, sprinkle with the salt and toss to coat evenly. Spread the seeds in a single layer and bake in the oven for about 5 minutes, or until they turn light brown. The seeds will pop and dance while cooking.
—Marcia Kiesel

DOUBLE RADISH SANDWICHES

These Double Radish Sandwiches make excellent economical hors d'oeuvres and appetizing foils to creamy cold summer soups. Although freshly grated horseradish is best for this recipe, you may substitute the bottled variety.
Makes 32 Small Sandwiches

1 stick (4 ounces) unsalted butter, softened
4 tablespoons finely shredded fresh horseradish, or 2 tablespoons bottled horseradish, drained
½ teaspoon fresh lemon juice (omit if using bottled horseradish)
16 slices thinly sliced white bread, crusts removed
1 bunch radishes

1. Beat the butter with a wooden spoon until it is light and fluffy. Continue beating as you gradually add the horseradish and lemon juice (if using). Keep the butter at room temperature until you are ready to assemble the sandwiches.

2. Arrange all the bread slices in front of you on a work surface. Spread the horseradish butter evenly over each slice, using approximately ½ tablespoon for each slice of bread.

3. Thinly slice the radishes and place them close together in a single layer over half the bread slices. Top with the remaining bread slices to make sandwiches, pressing them gently so the butter adheres to the radish slices. Cut the sandwiches in half on the diagonal (and, if desired, in half again to form triangles).

4. Arrange the sandwiches on a serving platter and cover loosely with a tea towel that has been moistened and then very thoroughly wrung out. Refrigerate for 30 to 60 minutes, or until chilled.
—F&W

TRIANGLES WITH TWO-MUSHROOM FILLING

To make larger triangles, which make a nice first course, make the strips of phyllo (see Step 5) 4 inches wide and use 3 tablespoons of filling for each.
Makes About 64 Small Triangles

1 ounce dried shiitake mushrooms
2 tablespoons unsalted butter
1 tablespoon vegetable oil
¼ cup chopped shallots
1 pound fresh mushrooms, chopped
½ cup dry Madeira
1 tablespoon sherry vinegar
1 teaspoon fresh lemon juice
1 teaspoon salt
½ teaspoon freshly ground pepper
2 tablespoons dry bread crumbs
1 package (1 pound) phyllo dough (see Note)
About ¼ cup clarified butter

1. Place the shiitake mushrooms in a small bowl. Cover with boiling water and let soak for 30 minutes.

2. Meanwhile, in a large skillet, melt the butter in the oil over moderate heat. When the foam subsides, add the shallots and sauté until softened but not browned, about 2 minutes. Add the chopped fresh mushrooms, increase the heat to moderately high and cook, stirring, until the juices evaporate and the pieces separate, about 5 minutes.

3. Preheat the oven to 375°. Drain the shiitake mushrooms and squeeze dry. Cut off and discard the woody stems. Chop the caps and add them to the skillet.

4. Add the Madeira and cook until reduced to a glaze, about 5 minutes. Add the vinegar, lemon juice, salt and pepper and cook for 2 minutes. Add the bread crumbs and cook, stirring frequently, for 2 minutes longer. Remove the skillet from the heat and let cool to room temperature.

5. Lay one sheet of phyllo on a flat work surface. Brush lightly with clarified butter, cover with a second sheet of phyllo and butter the top sheet. Cut the layered phyllo crosswise into 2-inch strips. Working quickly with one strip at a time, spoon a heaping teaspoon (1½ teaspoons) of the filling onto the bottom of the strip. Fold up using the classic flag-folding technique. Place on a buttered baking sheet and brush the tops with clarified butter.

6. Bake in the top third of the oven for 10 minutes. Brush again with butter. Bake for 10 to 15 minutes longer, until the triangles are golden brown.

NOTE: There will be leftover phyllo dough. It can be wrapped tightly and refrigerated for another use for up to 3 days if it is not dried out.
—*F&W*

BRUSSELS SPROUTS HORS D'OEUVRES
6 to 8 Servings

- 2 pints (10 ounces each) brussels sprouts
- 4 ounces blue cheese, at room temperature
- 4 ounces cream cheese, at room temperature
- 6 scallions (white and tender green), minced
- 1 teaspoon salt

1. In a large pot of rapidly boiling water, cook the brussels sprouts until crisp-tender, about 8 minutes. Drain in a colander and refresh under cold running water. Set aside to drain.

2. Meanwhile, in a medium bowl, combine the blue cheese, cream cheese, scallions and salt; mix until well blended.

3. Cut off ⅛ inch from the stem end of each brussels sprout to give it a flat base. Using a small melon baller, hollow out a deep well in the top of each sprout.

4. Pack the cheese mixture into a pastry bag fitted with a ⅜-inch plain tip (#5) and pipe into the hollowed sprouts. Arrange the hors d'oeuvres on a platter and serve at room temperature.
—*W. Peter Prestcott*

SPINACH-STUFFED MUSHROOMS WITH CHEDDAR AND CHERVIL
These chervil-flavored mushrooms make wonderful appetizers.
4 to 6 Servings

- 2 tablespoons unsalted butter
- 1 small onion, chopped
- 1½ cups chopped cooked fresh spinach or 2 packages (10 ounces each) frozen chopped leaf spinach, thawed
- ½ cup heavy cream
- ½ cup plus 1 tablespoon chopped fresh chervil
- ¾ teaspoon salt
- ¼ teaspoon freshly ground pepper
- ½ teaspoon fresh lemon juice
- ¾ cup grated mild Cheddar cheese
- 1 tablespoon olive oil
- 2 pounds medium to large fresh mushroom caps

1. Preheat the oven to 500°. In a medium saucepan, melt the butter over moderate heat. Add the onion and cook over low heat until softened, about 10 minutes.

2. Add the spinach. Increase the heat to high and stir in the cream. Cook, stirring, until the cream is very thick, about 3 minutes. Remove the mixture from the heat and stir in ¼ cup of the chervil, the salt, pepper, lemon juice and ⅓ cup of the grated Cheddar cheese. Set aside.

3. In a large skillet, heat the olive oil over high heat. Add the mushroom caps and 1 tablespoon of water. Cover and cook over high heat until the mushrooms are tender and glazed, about 4 minutes.

4. Using a slotted spoon, transfer the mushrooms to a baking sheet, hollow-side up. Divide the spinach mixture evenly among the mushroom caps, filling them generously. Top with the remaining grated Cheddar and bake until the mushrooms are heated through and the cheese is melted, about 5 minutes. Arrange the stuffed mushrooms on a serving dish. Sprinkle with the remaining chopped chervil and serve at once.
—*Marcia Kiesel*

EGGPLANT COOKIES WITH GOAT CHEESE AND TOMATO-BASIL SAUCE
These savory eggplant rounds can also be served as an accompaniment to a main course.
4 Servings

- 6 tablespoons plus 2 teaspoons olive oil
- 1 large garlic clove, minced
- 1 small shallot, minced
- 1 tablespoon tequila, preferably golden
- 1½ tablespoons dry red wine
- 2 large tomatoes—peeled, seeded and chopped
- Pinch of salt and freshly ground pepper
- 2 tablespoons shredded fresh basil
- ½ cup all-purpose flour, sifted
- 1 egg, lightly beaten
- ½ cup fresh bread crumbs

2 long, narrow Japanese eggplants, peeled and cut into twenty-four ½-inch rounds
4 ounces cylindrical goat cheese, cut into 12 rounds

1. In a medium saucepan, heat 2 teaspoons of the oil. Add the garlic and shallot and cook over moderate heat, stirring, until softened but not browned, about 1 minute. Add the tequila and red wine and cook until almost completely absorbed, about 1 minute. Stir in the tomatoes, salt and pepper and cook until slightly thickened, about 5 minutes. Stir in the basil and keep warm.

2. Put the flour, egg and bread crumbs in 3 separate shallow bowls. Dip the eggplant rounds in the flour until completely coated. Next, coat with the egg and then roll in the bread crumbs.

3. In a large heavy skillet, heat 2 tablespoons of the olive oil over moderately high heat. Add one-third of the eggplant slices and sauté, turning once, until crisp and golden brown, about 4 minutes on each side. Drain on paper towels. Repeat 2 more times with the remaining oil and eggplant.

4. Sandwich 1 slice of the goat cheese in between 2 slices of hot fried eggplant. Spoon the tomato-basil sauce onto 4 heated plates and arrange 3 eggplant "cookies" on each one. Serve warm.
—Robert McGrath, Four Seasons Hotel, Dallas

RED POTATOES WITH GORGONZOLA CREAM, BACON AND WALNUTS
12 Servings

¼ cup walnut halves
½ cup sour cream
3 ounces Gorgonzola cheese, at room temperature
½ teaspoon freshly ground black pepper
⅛ teaspoon hot pepper sauce
6 slices of bacon, cut into ¼-inch dice
12 small red potatoes (about 1¼ pounds)
1 tablespoon olive oil
½ teaspoon coarse (kosher) salt
2 tablespoons minced chives

1. Preheat the oven to 400°. Place the walnuts on a baking sheet and roast for about 5 minutes, until fragrant and lightly toasted. Let cool, then chop coarsely and set aside.

2. In a small bowl, using a fork, combine the sour cream and Gorgonzola until well blended. Stir in ¼ teaspoon of the black pepper and all of the hot pepper sauce. *(The recipe can be made to this point up to 1 day ahead. Store the walnuts in an airtight container at room temperature. Cover and refrigerate the Gorgonzola cream until about 30 minutes before serving.)*

3. Heat a medium skillet over moderately high heat until hot. Add the bacon and cook, stirring often, until crisp and brown, 4 to 5 minutes. Using a slotted spoon, transfer the bacon to paper towels to drain. Set aside.

4. In a roasting pan, toss the potatoes with the oil, salt and the remaining ¼ teaspoon black pepper until well coated. Bake for 25 to 30 minutes, until tender. Keep warm.

5. To serve, stir the reserved bacon and toasted walnuts into the Gorgonzola cream. Halve the potatoes lengthwise. If necessary, cut a small slice from the rounded (skin) side of the potato halves so that they can stand. Place the potato halves on a platter and dollop each one with a teaspoon of the filling. Sprinkle the chives on top and serve.
—Kathy Casey

FRIED POLENTA WITH GORGONZOLA
Polenta is the perfect foil for the sharp, distinctive flavor of Gorgonzola cheese in this tempting hors d'oeuvre.
Makes About 4 Dozen

1½ teaspoons salt
1 cup instant polenta
½ pound Gorgonzola cheese, at room temperature
2 eggs
2 cups fine dry bread crumbs
Vegetable oil, for deep-frying

1. In a medium saucepan, bring 3 cups of water to a boil over high heat. Add the salt. Gradually stir in the polenta and cook, stirring constantly, for 10 minutes. Pour the polenta onto a lightly greased baking sheet and shape it into a square about 1 inch high, smoothing the surface with a wet spatula. Cover the polenta with a dampened kitchen towel and let cool to room temperature, about 30 minutes.

2. Cut the cooled polenta into even slices, about ¼ inch wide. Cut these slices into 2-inch lengths.

3. Take half of the polenta slices and spread each with 1 teaspoon of Gorgonzola. Cover with the remaining polenta slices to make small sandwiches.

4. Beat the eggs with a splash of water. Dip each sandwich into the egg wash and coat evenly with the bread crumbs; set aside. *(The recipe can be prepared to this point up to 12 hours ahead. Cover loosely and refrigerate.)*

5. Heat the vegetable oil in a deep-fryer to 375° to 400°. One at a time, drop the sandwiches into the hot oil in small batches so they are not touching and fry until golden brown, about 1 minute. Drain on paper towels. Repeat with the remaining sandwiches and serve hot.
—*John Robert Massie*

VEGETARIAN SPRING ROLL

The spring roll is a delicate preparation, quite different from the egg roll that evolved from it. In Canton and Shanghai, it is traditionally made with vegetables combined either with pork or shrimp. This all-vegetable roll is Chinese in feeling, but with a fresh and different taste. (These can be frozen after they have been lightly browned. To cook: Defrost, let come to room temperature and dry off any moisture with paper towels. Deep-fry until golden brown.)
Makes 18

10 dried Chinese black mushrooms
1½ tablespoons oyster sauce*
1½ teaspoons mushroom soy sauce*
1 teaspoon dry sherry
1 teaspoon distilled white vinegar
1 teaspoon Oriental sesame oil
1 teaspoon sugar
½ teaspoon salt
Pinch of freshly ground white pepper
1 tablespoon cornstarch
5 cups peanut oil
1 bunch of scallions, quartered lengthwise and cut into 1½-inch pieces
1 large Chinese celery cabbage* (about 1½ pounds), stalks quartered lengthwise and cut into ¼-inch slices
18 egg roll skins (in plastic bags)*
2 eggs, lightly beaten
Sweet and Sour Sauce or Chili Soy Sauce (recipes follow)
*Available at Asian markets

1. Put the mushrooms in a small bowl, add hot water to cover and soak until softened, about 30 minutes. Rinse the mushrooms and squeeze dry. Discard the stems; slice the caps thinly.

2. In a small bowl, stir together the oyster sauce, soy sauce, sherry, vinegar, sesame oil, sugar, salt, white pepper, cornstarch and 1½ tablespoons of water. Set the sauce aside.

3. Heat a wok over high heat for 30 seconds. Add 2½ tablespoons of the peanut oil, stirring with a Chinese metal spatula to coat the sides of the wok. As soon as a wisp of white smoke appears, add the mushrooms and scallions and stir-fry for 45 seconds. Add the celery cabbage and stir-fry until wilted, about 2 minutes.

4. Make a well in the center of the vegetables. Stir the reserved sauce and pour it into the well. Toss and stir to thoroughly combine all the ingredients. Cook over moderate heat, stirring until the sauce thickens, about 1 minute. Remove from the heat and transfer the filling to a shallow dish. Let cool to room temperature, then cover and refrigerate for at least 8 hours or overnight.

5. One at a time, lay an egg roll skin on a flat surface with a corner facing you. Place 2 tablespoons of the filling across the bottom third of the wrapper. Brush the beaten egg around the edges of the entire wrapper. Fold up the bottom corner and roll once. Then fold in both sides and roll up all the way. Press the end to seal tightly. Repeat with the remaining wrappings and filling.

6. In a large saucepan or deep-fryer, heat the remaining peanut oil to 350°. Deep-fry the spring rolls in batches without crowding, turning frequently, until golden brown, 3 to 5 minutes. Drain on paper towels and serve with Sweet and Sour Sauce and/or Chili Soy Sauce for dipping.
—*Eileen Yin-Fei Lo*

SWEET AND SOUR SAUCE

Dip egg rolls or spring rolls, such as Vegetarian Spring Roll (at left), in this tart-sweet sauce.
Makes About 1¾ Cups

½ cup red wine vinegar
⅓ cup sugar
1 can (8 ounces) tomato sauce
⅛ teaspoon salt
2 tablespoons cornstarch mixed with 2 teaspoons cold water
2 tablespoons minced green bell pepper
2 teaspoons minced fresh coriander (cilantro)

In a small nonreactive saucepan, combine the vinegar, sugar, tomato sauce, salt and dissolved cornstarch. Bring to a boil over moderately high heat, stirring frequently. Boil for 1 minute. Add the bell pepper and coriander. Pour the sauce into a bowl and serve warm or at room temperature.
—*Eileen Yin-Fei Lo*

CHILI SOY SAUCE

Along with Sweet and Sour Sauce (above), this makes a nice dipping sauce for Vegetarian Spring Roll.

Makes About ½ Cup

1 tablespoon dark soy sauce
1 tablespoon light soy sauce
1¼ teaspoons minced fresh hot red chiles, or ½ teaspoon crushed hot pepper
1 tablespoon distilled white vinegar
3 tablespoons chicken broth or water
½ teaspoon sugar
2 teaspoons Oriental sesame oil
1½ teaspoons minced garlic
1½ teaspoons minced fresh ginger
1 tablespoon thinly sliced scallions

In a small bowl, combine all the ingredients. Let stand for 30 minutes and serve at room temperature.
—Eileen Yin-Fei Lo

AVOCADO TEMPURA WITH CORIANDER AND LIME

6 Servings

1 cup all-purpose flour
½ cup cornstarch
2 teaspoons salt
1 cup ice water
1 teaspoon baking soda
⅓ cup chopped fresh coriander (cilantro)
Vegetable oil, for deep-frying
2 avocados, preferably Hass, cut lengthwise into 12 slices
Lime wedges, for serving

1. Into a large bowl, sift the flour, cornstarch and 1 teaspoon of the salt. Form a well in the center and pour in the ice water all at once. Whisk briskly to form a smooth batter.

2. Cover the bowl with plastic wrap and refrigerate for at least 30 minutes. When ready to use, whisk in the baking soda and coriander.

3. In a deep-fryer or large heavy saucepan, heat about 1 inch of oil to 375°. Sprinkle the avocado slices with the remaining 1 teaspoon salt. Let sit for a minute.

4. In batches, dip the avocado slices in the batter to coat lightly; fry without crowding in the hot oil, turning once, for about 30 seconds on each side, until lightly browned. Remove and drain on paper towels. Serve hot (see Note) with wedges of lime.

NOTE: These are best served right away, but they can be held for a short time in a warm (200°) oven.
—Anne Disrude

10-INGREDIENT VEGETABLE FRITTERS

6 Servings

¼ pound bean sprouts (about 2 cups)
1 red bell pepper, cut into thin slivers
8 scallions, thinly sliced into rounds
3 mushrooms, slivered
2 teaspoons capers, rinsed and patted dry
1 tablespoon slivered black olives
1 teaspoon minced garlic
1 teaspoon minced fresh tarragon or ½ teaspoon dried
½ teaspoon minced fresh rosemary or ¼ teaspoon dried
2 tablespoons chopped (oil-packed) sun-dried tomatoes
Vegetable oil, for deep-frying
1 cup all-purpose flour
1 teaspoon baking powder
1 cup ice water
Coarse salt and lemon wedges, for serving

1. In a bowl, toss together the bean sprouts, bell pepper, scallions, mushrooms, capers, olives, garlic, tarragon, rosemary and sun-dried tomatoes.

2. Preheat the oven to 250°. Place a baking sheet with a rack on it in the oven. Fill a deep-fryer with 4 inches of oil and heat to 390°.

3. Meanwhile, sift the flour and baking powder into a medium bowl. Make a well in the center, pour in the ice water and whisk until smooth.

4. Add the batter to the vegetables and stir to coat well. Using a large flat spoon, slide about ¼ cup of the fritter mixture into the hot oil, trying not to make the fritters too thick in the center. Add as many as will fit without crowding and fry, turning once, until browned and crisp, 3 to 4 minutes. Drain on paper towels; then transfer to the rack in the oven to keep warm until all the batter is used up.

5. Serve the fritters in a napkin-lined basket and pass coarse salt and wedges of lemon on the side.
—Anne Disrude

ARTICHOKE FRITTERS WITH REMOULADE SAUCE

Makes 40

REMOULADE SAUCE

1 cup mayonnaise
1 teaspoon Dijon mustard
½ teaspoon anchovy paste
1 tablespoon minced capers
1 tablespoon minced cornichons (French gherkin pickles)
1 tablespoon minced parsley
1 tablespoon minced fresh chives
2 teaspoons fresh lemon juice

FRITTERS

¼ cup fresh lemon juice
5 large artichokes
1 cup all-purpose flour
1 teaspoon baking powder

1 cup ice water
1 quart vegetable oil, for deep-frying

1. Make the remoulade sauce: In a bowl, combine the mayonnaise, mustard, anchovy paste, capers, cornichons, parsley, chives and lemon juice with ½ cup of water. Whisk to blend well. Cover and refrigerate until serving time.

2. Make the fritters: Pour the lemon juice into a large nonreactive saucepan with about 2 quarts of water.

3. Trim each artichoke by snapping off the stem; bend back and pull off the tough outer dark green leaves. Using a serrated stainless steel knife, cut off the top two-thirds of the artichoke, leaving about 1½ inches of the heart. Pare away the bases of the dark green outer leaves. As you trim each artichoke, drop immediately into the acidulated water to prevent discoloration.

4. When all the artichokes are in the saucepan, bring to a boil over high heat. Reduce the heat to moderate and boil until the artichoke hearts are tender, 20 to 25 minutes. Let the hearts cool in the cooking liquid to retain their color. (*The artichokes can be cooked 1 day ahead and refrigerated in their cooking liquid overnight.*)

5. When cool, remove the artichokes from the liquid and cut each heart into eighths. Remove the hairy choke and drain well.

6. In a medium bowl, combine the flour and baking powder with the ice water. Blend thoroughly.

7. In a deep-fryer or large deep saucepan, heat the oil to 400°. One by one, dip each piece of artichoke into the batter and add to the hot oil. Fry in batches without crowding until crisp and brown, about 3 minutes. Drain on paper towels. Serve the fritters hot with the remoulade sauce for dipping.
—*John Robert Massie*

MARINATED GREEN BEANS

Serve these beans as one of several antipasti. They would also make a nice side salad or warm-weather vegetable side dish.
6 Servings

1½ pounds green beans
2 hard-cooked eggs
Red Wine Vinaigrette (p. 258)
1 pint cherry tomatoes

1. Cook the beans in a large pot of boiling salted water until just crisp-tender, 6 to 10 minutes, depending on their size and age. Drain in a colander, then refresh under cold running water to stop the cooking and set the color.

2. Peel the eggs, then chop the yolks and whites separately.

3. Pat the cooled beans completely dry on paper towels, then toss them in a bowl with enough of the Red Wine Vinaigrette to coat them. Let the beans marinate at least 1 hour. In a separate bowl, marinate the cherry tomatoes for an hour in enough of the vinaigrette to coat them.

4. When you are ready to serve, arrange the beans in a shallow bowl or on a platter, with the cherry tomatoes around the edge, and garnish with the yolks and whites of the hard-cooked eggs. Drizzle a little more vinaigrette over all.
—*F&W*

MARINATED RED AND GREEN PEPPERS

A piquant dish, excellent with crusty French or Italian bread as an hors d'oeuvre, or as a salad on a buffet or with dinner.
6 Servings

3 green bell peppers
3 red bell peppers
Salt and freshly ground black pepper
2 tablespoons red wine vinegar
2 cloves garlic, cut into fine slivers
Anchovy fillets (optional)
Olive oil
3 tablespoons drained capers

1. Preheat the broiler. Broil the peppers as close to the heat as possible until the peppers are charred on all sides. Place the peppers in a paper bag, then close the bag tightly. Allow the peppers to steam in the bag for a few minutes to loosen the skins.

2. Remove the skins of the peppers, then stem, seed and derib them. Cut the peppers into strips about 1½ inches wide. Gently pat the strips dry.

3. Arrange a layer of the red and green pepper strips in a shallow serving dish. Sprinkle on a small amount of salt and pepper, a few drops of vinegar, and some of the garlic slivers and anchovy fillets. Repeat with another layer of peppers, salt, pepper, vinegar, garlic, and anchovies, repeating the layering until the peppers are all used.

4. Drizzle enough olive oil over all so that the peppers are almost covered. Marinate the peppers for at least two hours at room temperature. Scatter the capers over the top and serve.
—*F&W*

ROASTED PEPPERS AND CELERY ON TOASTED CROUTONS

Without the bread, this delicious vegetable makes an excellent accompaniment to roasted or grilled meats.

8 Servings

- 8 bell peppers—halved lengthwise, seeded and deribbed
- 10 large celery ribs, peeled and cut into 2-by-¼-inch strips
- 1 teaspoon thyme
- ⅔ cup extra-virgin olive oil
- 1 teaspoon grated orange zest
- ½ teaspoon salt
- ¼ teaspoon freshly ground black pepper
- 16 slices Italian bread, cut 1 inch thick

1. Preheat the oven to 350°. Place the peppers in a large roasting pan in a single layer, cut-side down. Distribute the celery around the peppers and sprinkle with the thyme. Drizzle the oil over the vegetables.

2. Roast in the oven until the pepper skins are wrinkled and loose, about 45 minutes. Remove from the oven and let stand until cool enough to handle. (Leave the oven on.) Peel the peppers. Tip the pan and pour the oil and juices into a bowl; reserve.

3. Cut the peppers into 2-by-¼-inch strips. Place the peppers and celery in a bowl. Add the orange zest, salt and black pepper and toss to mix. *(The recipe can be prepared to this point up to 5 days ahead. Refrigerate, covered. Let return to room temperature before proceeding.)*

4. Brush both sides of the bread with the reserved oil and place in the roasting pan. Bake, turning once or twice, until the croutons are golden and crisp, about 20 minutes.

5. Spoon the pepper mixture over the croutons and serve.

—Anne Disrude

GRILLED EGGPLANT BRUSCHETTA WITH PROSCIUTTO

12 Servings

- 1 large eggplant (1¼ pounds), peeled
- 1 tablespoon table salt
- About ⅔ cup extra-virgin olive oil
- Six ¾-inch slices from a round loaf of country bread
- 1 large garlic clove, halved
- Sea salt
- Freshly ground pepper
- 6 ounces very thinly sliced prosciutto, cut crosswise into ½-inch-wide strips
- 2 tablespoons finely chopped parsley

1. Trim and slice the eggplant into twelve ½-inch rounds. Place the rounds in a colander, sprinkling them with the table salt. Place a plate on top to weight them; set the colander over a bowl and set aside for 1 hour.

2. Prepare a medium charcoal fire or preheat the broiler. Remove the eggplant from the colander and blot off excess moisture and salt with paper towels. Brush both sides of the eggplant slices with some of the oil. Grill for 10 to 15 minutes, turning once, until brown and crusty on the outside and soft on the inside. Alternatively, broil the eggplant slices on a large baking sheet.

3. Cut the bread slices in half and brush very lightly on both sides with some of the remaining olive oil. Grill the bread, turning once, for about 2 minutes, until toasted. Alternatively, broil the bread. Rub each slice with the cut garlic clove and brush with more olive oil.

4. Place a slice of eggplant on a slice of bread. Top with a little sea salt and pepper and the prosciutto. Sprinkle with the parsley.

—David Rosengarten

TOMATO BREAD CRISP

6 to 8 Servings

- 4 tablespoons unsalted butter, softened
- 1 long loaf (about ½ pound) Italian bread, sliced ¼ inch thick or 1 Italian peasant bread, sliced into ¼-inch rounds
- 2½ pounds large beefsteak tomatoes (about 4), thinly sliced crosswise
- ¼ cup olive oil
- ½ teaspoon salt
- ¼ cup minced parsley
- ¼ teaspoon sugar
- 1 teaspoon freshly ground pepper
- 2 English muffins, split and lightly toasted
- ¼ cup plus 1 tablespoon freshly grated Parmesan cheese

1. Preheat the oven to 350°. Generously butter a 9-by-13-inch glass baking dish with 2 tablespoons of the butter.

2. Arrange the bread slices in a jelly-roll pan and bake until lightly toasted, 10 to 12 minutes.

3. Line the bottom of the baking dish with half the toasted bread slices in a single layer. Patch any empty spaces with small bits of toast. Arrange half of the tomato slices on top, overlapping if necessary.

4. Drizzle 2 tablespoons of the olive oil over the layer of tomatoes, then season with ¼ teaspoon of the salt, 2 tablespoons of the parsley, ⅛ teaspoon of the sugar and ½ teaspoon of the pepper.

5. Cover the tomatoes with the remaining toast and layer the remaining tomato slices on top. Drizzle with the remaining 2 tablespoons olive oil and sprinkle the remaining ¼ teaspoon salt, 2 tablespoons parsley, ⅛ teaspoon sugar and ½ teaspoon pepper on top. Dot with the remaining 2 tablespoons butter.

6. Tear the toasted English muffins into large pieces and process in a food processor to form coarse crumbs. In a small bowl, toss the crumbs with the Parmesan cheese. Sprinkle the crumbs evenly over the tomatoes and bake, uncovered, for 45 minutes. Cover with foil and bake for 15 minutes longer, or until the crisp is soft and lightly browned. Serve hot.
—Lee Bailey

HERBED PITA CRISPS

This recipe can easily be multiplied to yield larger batches. These triangles can be made in advance and stored in airtight containers.
Makes 40 Crisps

3 tablespoons olive oil
1 teaspoon basil
½ teaspoon coarse (kosher) salt
5 large whole wheat pita bread pockets (12-ounce bag), each cut into eighths

1. Preheat the oven to 450°. In a large bowl, combine the olive oil, basil and salt. Add the pita pieces and toss to coat well.

2. On 1 or 2 baking sheets, spread out the pita pieces in a single layer. Bake for 4 minutes. Using tongs, turn the pita pieces over and continue baking for 4 more minutes, until golden and crisp. *(The recipe can be made up to 2 days ahead. Let cool, then store in an airtight container. If necessary, recrisp in a 400° oven for 2 minutes.)*
—Kathy Casey

SALAD NACHOS VINAIGRETTE

These cold and crunchy nachos are a reconstructed version of those I enjoyed years ago at a now-defunct Boulder, Colorado, restaurant called Tico's. Although they take more fussing to assemble than hot nachos, the oven—and the cook—stays cool. Arrange them on big white platters and, instead of beer or Margaritas, drink white or rosé wine.
Makes 36

¾ cup canned refried beans
3 pickled jalapeño peppers, seeded and minced
2 tablespoons fresh lime juice
1 teaspoon Dijon mustard
½ teaspoon salt
Freshly ground black pepper
⅓ cup corn oil
1½ avocados, preferably Hass
1 medium Spanish or red onion, finely diced
¼ cup minced fresh coriander (cilantro)
36 unsalted corn tortilla chips
4 ounces medium-sharp Cheddar cheese, grated
1 cup finely shredded inner leaves of romaine lettuce
4 medium radishes, very thinly sliced

1. In a bowl, stir together the refried beans and jalapeño peppers.

2. In another bowl, whisk together the lime juice and mustard. Season with ¼ teaspoon of the salt and black pepper to taste. Gradually whisk in the corn oil. Mix until incorporated and set aside.

3. Halve and pit the avocados. With a spoon, scoop the flesh into a bowl and mash with a fork. Stir in the onion, coriander and remaining ¼ teaspoon salt.

4. Evenly spread 1 teaspoon of the bean mixture on each tortilla chip. Top the bean layer with 1 teaspoon of the avocado mixture. Arrange the nachos on two serving platters as you go along.

5. Sprinkle the cheese evenly over the nachos. Sprinkle the lettuce evenly over the cheese. Sprinkle the radishes evenly over the lettuce. Drizzle the reserved vinaigrette dressing over the nachos and serve at once.
—Michael McLaughlin

CLASSIC NACHOS

These are nachos just like Mom used to make—if Mom ran a Tex-Mex bar. Baked in a single layer and topped with the minimum amounts of refried beans and cheese, these nachos are thin, crisp and addictive. (If you're from the goopy school of nacho making, just use more beans and cheese.)
Makes 36

36 unsalted corn tortilla chips
¾ cup canned refried beans
4 pickled jalapeño peppers, each sliced into 9 rounds
3 to 4 plum tomatoes, diced
2 scallions, thinly sliced
3 ounces medium-sharp Cheddar cheese, grated
3 ounces Monterey Jack cheese, grated

1. Preheat the oven to 475°. Dollop each tortilla chip with 1 teaspoon of the refried beans. Arrange the chips, bean-side up, in a single layer, on two large jelly-roll pans or edged baking sheets.

2. Top each chip with a slice of jalapeño pepper. Sprinkle the tomatoes and scallions evenly over the nachos. In a bowl, combine the grated cheeses and sprinkle evenly on top.

3. Bake on the top rack of the oven, one pan at a time, until the cheese is melted and the nachos are sizzling slightly, about 5 minutes. With a metal spatula, transfer the nachos to a platter and serve at once.
—Michael McLaughlin

CARROT CHIPS
Chef Leslie Revsin's bright, chewy chips also make convivial partners for poultry dishes.
8 to 10 Servings

1½ quarts peanut oil
3 to 4 small, firm carrots (about 1 pound total), trimmed and peeled
Salt (optional)

1. In a deep-fat fryer, heat the oil to 350°.

2. Meanwhile, using a small sharp knife, trim off the thin ends to make the carrots fairly uniform in length. Cut off a thin lengthwise strip from each of the carrots. Using a vegetable peeler or a mandoline with the cutting blade set at 1/16 inch, starting with the cut side, slice the carrots lengthwise into thin strips. Stack the carrot slices into manageable piles and cut into rectangles about 1½ inches long. Alternatively, cut the carrots to fit the feed tube horizontally and slice in a food processor fitted with an extra-thin blade.

3. Divide the carrot slices into 4 batches. Place 1 batch in the fryer basket, lower the basket into the hot oil and fry until the chips are light brown and curled, about 2 minutes. Lift the basket out of the oil; shake gently to remove any excess oil. Transfer the chips to several layers of paper towels to drain. Repeat with the remaining batches, checking the temperature of the oil and adjusting the heat, if necessary, to maintain 350°.

4. Season the chips lightly with salt, if desired, and toss gently. Transfer to a bowl and serve.
—Doris Tobias

CELERY ROOT CHIPS
Chef Gérard Pangaud serves these for nibbles.
4 to 6 Servings

2 quarts peanut oil
3 tablespoons lemon juice or distilled white vinegar
1 firm celery root (¾ to 1 pound total), trimmed and peeled
Salt

1. In a deep-fat fryer, heat the oil to 360°.

2. In a medium bowl, combine 4 cups of cold water with the lemon juice.

3. Using a large heavy knife, cut a thin slice from each end of the celery root. Using a mandoline with the cutting blade set at 1/16 inch, starting with a cut end, slice the celery root crosswise into disks. Alternatively, cut the celery root to fit into the feed tube and slice in a food processor fitted with an extra-thin blade. As you cut them, transfer the slices to the bowl of acidulated water. Let soak for 5 minutes, then drain and pat dry with paper towels.

4. Divide the celery root slices into 3 batches. Place 1 batch in the fryer basket, lower the basket into the hot oil and fry just until the chips are lightly browned, about 2 minutes. Lift the basket out of the oil; shake gently to remove any excess oil. Transfer the chips to several layers of paper towels to drain. Repeat with the remaining batches, checking the temperature of the oil and adjusting the heat, if necessary, to maintain 360°.

5. Season the chips with salt to taste and toss gently.
—Doris Tobias

PARSNIP CHIPS
Chef Seppi Renggli recommends serving these chips with drinks or as an accompaniment to broiled meats, roast loin of pork, veal or pork chops, and roast chicken or duckling.
8 to 10 Servings

1½ pounds even-size parsnips, trimmed and peeled
About 2 quarts vegetable oil
Coarse (kosher) salt

1. Using a large heavy knife, cut off a little of the thin ends so that the parsnips are fairly uniform in length. Cut off a thin lengthwise strip from each of the parsnips. Using a mandoline with the cutting blade set at 1/16 inch, starting with the cut side, slice the parsnips lengthwise into thin strips. Alternatively, cut the parsnips to fit into the feed tube horizontally and slice in a food processor fitted with an extra-thin blade. Immediately transfer the parsnip strips to a large bowl, cover with cold water and let soak for 1 hour.

2. Drain well and rinse the parsnip strips under cold running water; pat dry with paper towels.

3. Pour 3 inches of vegetable oil into a deep-fat fryer and heat to 425°. (The amount of oil you use will depend on the size of your fryer.)

4. Divide the parsnip strips into 4 batches. Place 1 batch in the fryer basket, lower the basket into the hot oil and fry just until the chips are golden brown, 2 to 3 minutes. Lift the basket out of the oil; shake gently to remove any excess oil. Transfer the chips to several layers of paper towels to drain. Repeat with the remaining batches, checking the temperature of the oil and adjusting the heat, if necessary, to maintain 425°.

5. Season the chips with coarse salt to taste. Transfer to a bowl and serve.
—Doris Tobias

RUTABAGA CHIPS

Rutabagas, or meaty yellow turnips, are preferred by chef Anne Rosenzweig for both their deep color and robust flavor.
8 to 10 Servings

1½ quarts soybean oil
1 very large or 2 medium rutabagas (1½ to 2 pounds total), trimmed and peeled
Salt

1. In a deep-fat fryer, heat the oil to 360°.

2. Meanwhile, using a large heavy knife, cut a thin slice from the end of each rutabaga. Using a mandoline with the cutting blade set at 1/16 inch, starting with the cut side, slice the rutabagas crosswise. Alternatively, cut the rutabagas to fit into the feed tube and slice in a food processor fitted with an extra-thin blade. Pat the rutabaga slices dry with paper towels.

3. Divide the rutabaga slices into 6 batches. Place 1 batch in the fryer basket, lower the basket into the hot oil and fry until the chips are light golden brown, about 2 minutes. Lift the basket out of the oil; shake gently to remove any excess oil. Transfer the chips to several layers of paper towels to drain. Repeat with the remaining batches, checking the temperature of the oil and adjusting the heat, if necessary, to maintain 360°.

4. Season the chips with salt to taste and toss gently. Pile in a bowl and serve.
—Doris Tobias

BEET FRITES

These sweet frizzy chips were created by chef Barry Wine as a garnish for grilled salmon with hot mustard and for lobster with orange. They also make excellent munchies with aperitifs.
6 to 8 Servings

About 1½ quarts corn oil
1 pound even-size medium beets without tops, trimmed and peeled
¼ cup cornstarch
Coarse (kosher) salt (optional)

1. Pour 3 inches of vegetable oil into a deep-fat fryer and heat to 375°. (The amount you use will depend on the size of your fryer.)

2. Using a large heavy knife or a mandoline with the cutting blade set at ⅛ inch, slice the beets crosswise ⅛ inch thick. Alternatively, cut the beets to fit the feed tube and slice in a food processor fitted with an extra-thin blade. Stack the beet slices a few at a time and cut into very thin julienne strips.

3. Place the beet strips in a large bowl. Sprinkle the cornstarch on top and shake the bowl to coat the beets evenly.

4. Divide the beets into 3 batches. Place 1 batch in the fryer basket, lower the basket into the hot oil and fry, stirring constantly at first with tongs or a long-handled metal spoon and then frequently, until the beets darken slightly, 3 to 4 minutes. Lift the basket out of the oil; shake gently to remove any excess oil. Transfer the beet *frites* to several layers of paper towels to drain. Repeat with the remaining batches, checking the temperature of the oil and adjusting the heat, if necessary, to maintain 375°.

5. Season the beet *frites* lightly with coarse salt and toss gently. Pile into a bowl and serve. (If using as a garnish, don't season with salt at all.)
—Doris Tobias

MIDDLE-EASTERN EGGPLANT DIP

This spicy dip should be chilled overnight to blend the flavors, so plan accordingly.
Makes About 2½ Cups

1 large or 2 medium eggplants (2 to 2½ pounds)
6 tablespoons olive oil
1 cup finely chopped onion
½ cup finely chopped green bell pepper
1 garlic clove, minced
2 large tomatoes (about 1 pound)— peeled, seeded and chopped
½ teaspoon sugar
2½ teaspoons salt
½ teaspoon freshly ground black pepper
⅛ teaspoon cayenne pepper
2 tablespoons fresh lemon juice
Cherry tomato halves, for garnish
Triangles of pita bread, for serving

1. Preheat the oven to 425°. Place the eggplant on a rack over a baking sheet and set it in the center of the oven. Turning every 15 minutes, bake 50 to 60 minutes, or until the skin is soft and

wrinkled. Remove from the oven with the rack and set aside to cool.

2. Cut the eggplant in half and scoop out the pulp and then finely chop it with a knife or process it briefly in a food processor. Turn into a colander and let drain for 15 minutes.

3. Meanwhile, in a large skillet, warm 4 tablespoons of the oil over moderate heat. Add the onion and sauté until translucent, 6 to 8 minutes. Add the bell pepper and garlic and cook 5 minutes longer, stirring occasionally. Remove from the heat and combine in a bowl with the eggplant. Stir in the tomatoes, sugar, salt, black pepper and cayenne.

4. In a large skillet over moderate heat, heat the remaining 2 tablespoons olive oil. Add the eggplant mixture and bring it to a simmer, stirring constantly. Cover the skillet, reduce the heat to low and continue simmering for 45 minutes, stirring occasionally.

5. Remove the cover and cook, stirring frequently, for about 15 minutes, or until the moisture has evaporated. The mixture should hold its shape when stirred with a spoon. Remove the skillet from the heat and stir in the lemon juice. Transfer to a bowl, cover and refrigerate several hours or overnight.

6. To serve, turn the mixture into a bowl and decorate with the cherry tomato halves. Serve it with triangles of pita bread.
—F&W

WARM MEDITERRANEAN EGGPLANT DIP WITH HERBED PITA CRISPS
12 Servings

1 medium eggplant (about 14 ounces), trimmed and sliced crosswise ½ inch thick
3 tablespoons olive oil
2 teaspoons coarse (kosher) salt
1 medium yellow squash
1 medium zucchini
5 medium white mushrooms, cut into ¼-inch dice
½ of a medium red onion, finely chopped
1 tablespoon minced garlic
2 tablespoons dry red wine
2 teaspoons balsamic vinegar
1 jar (6 ounces) marinated artichoke hearts, chopped, liquid reserved
1 large tomato, cut into ¼-inch dice
¼ teaspoon crushed red pepper
⅛ teaspoon freshly ground black pepper
¼ cup tomato sauce, preferably homemade
¼ cup finely diced pimiento
1 tablespoon drained capers
1 tablespoon minced Calamata olives
2 tablespoons minced fresh basil
1 tablespoon minced flat-leaf parsley
Herbed Pita Crisps (p. 23)

1. Preheat the oven to 450°. Spread the eggplant slices on a baking sheet and drizzle 1½ tablespoons of the oil on top. Season with 1 teaspoon of the salt. Bake for 25 minutes. Let cool slightly, then chop finely and set aside.

2. Meanwhile, using a sharp knife, cut a ¼-inch-thick lengthwise slice of the skin of the yellow squash. Place the squash cut-side down on a work surface and cut off the remaining skin in ¼-inch-thick lengthwise slices all around. Cut the skin slices into ¼-inch dice. Repeat with the zucchini. (Save the squash pulp for another use.) Set aside.

3. In a large, heavy, nonreactive saucepan, heat the remaining 1½ tablespoons olive oil over moderately high heat until shimmering, 1 to 2 minutes. Add the mushrooms, onion, yellow squash and zucchini. Cook, stirring occasionally, until the vegetables begin to brown lightly, about 4 minutes. Add the garlic and cook, stirring, for 1 minute.

4. Stir in the red wine, vinegar and artichoke liquid. Add the tomato, crushed red pepper, black pepper and the remaining 1 teaspoon salt. Cook, stirring, until the tomato softens and the mixture is thick, about 5 minutes.

5. Stir in the artichoke hearts, tomato sauce, pimiento, capers, olives, basil, parsley and the reserved eggplant. Reduce the heat to moderately low and cook until heated through, about 3 minutes. Serve warm with Herbed Pita Crisps. *(This dip can be made up to 2 days ahead; reheat before serving.)*
—Kathy Casey

EGGPLANT CAVIAR

Set this garlicky eggplant dip out with crudités, a bowl of tomato salsa and some chips.
Makes About 2 Cups

3 medium-large eggplants (about 1¼ pounds each)
1 large head of garlic, unpeeled
⅓ cup olive oil
1 medium-large onion, chopped
2 tablespoons fresh lemon juice
1 tablespoon anchovy paste
Salt and freshly ground pepper

1. Preheat the oven to 400°. Prick the eggplants in several places. Place the eggplants and the whole head of garlic in a roasting pan and bake for about 45 minutes, or until the eggplants and garlic feel soft when pressed. (The garlic may take longer than the eggplant.)
2. Deeply score the eggplants in several places. Place in a colander to drain off the bitter liquid, about 1 hour.
3. Meanwhile, cut off the base of the garlic head to expose the flesh. Squeeze the garlic into a medium bowl.
4. When the eggplants have drained, peel them. Add the pulp to the garlic and mash with a fork to combine.
5. In a large heavy skillet, heat the olive oil. Add the onion and cook over moderate heat until translucent, about 6 minutes. Stir in the eggplant and garlic mixture, 1½ tablespoons of the lemon juice and the anchovy paste. Simmer the mixture, stirring frequently, until very thick and dark, about 30 minutes.
6. Season the eggplant caviar with the remaining ½ tablespoon lemon juice and salt and pepper to taste. Transfer to a bowl and refrigerate until chilled. Serve chilled or at room temperature.
—*John Martin Taylor*

PROVENÇALE SPINACH DIP WITH HERB TOASTS

This health-minded dip is chunky with spinach and olives and gutsy with garlic and goat cheese. If you don't have time to make the herb toasts, use slices of a good sourdough bread.
Makes About 3½ Cups

¼ cup olive oil
1 large red onion, finely chopped
1½ tablespoons chopped fresh thyme or 1½ teaspoons dried
½ cup pine nuts
1 long, thin loaf of French bread, sliced ¼ inch thick
1 teaspoon dried Greek oregano or herbes de Provence
11 ounces mild goat cheese, such as Bucheron, at room temperature
1½ cups milk
1 package (10 ounces) frozen spinach—thawed, squeezed dry and finely chopped
10 olives, preferably Calamata, chopped
2 garlic cloves, minced
½ teaspoon fresh lemon juice
Salt and freshly ground pepper

1. Preheat the oven to 400°. In a large skillet, warm 2 tablespoons of the olive oil over moderately high heat. Add the onion and thyme; reduce the heat to low. Cook the onion, stirring occasionally, until softened but not browned, about 10 minutes. Set aside to cool.
2. Put the pine nuts on a baking sheet and toast in the oven until golden brown, about 4 minutes. Set aside to cool. Leave the oven on.
3. Lightly brush 1 side of each slice of bread with the remaining 2 tablespoons olive oil. Arrange the slices, oiled-side up, on a baking sheet and toast in the oven until golden brown, about 5 minutes. Sprinkle the oregano over the toasts.
4. In a medium bowl, mash the goat cheese with a fork until creamy. Gradually blend in the milk. Fold in the spinach, olives, garlic and the cooled onion mixture and toasted pine nuts. Add the lemon juice and season with salt and pepper to taste. Serve with the herb toasts.
—*Marcia Kiesel*

FRIJOLEMOLE

Similar to guacamole but without the avocado, this dip can (and will) be gobbled up—a real crowd pleaser.
Makes About 6 Cups

1 tablespoon corn oil
1 large Spanish onion, coarsely chopped
4 medium garlic cloves, coarsely chopped
2 cans (16 ounces each) black beans, drained and rinsed
1 tablespoon fresh lemon juice
1 tablespoon fresh lime juice
1 large tomato, finely chopped
2 to 3 medium jalapeño peppers, seeded and minced, or 1 can (4 ounces) chopped mild green chiles
6 medium scallions, minced
¼ teaspoon salt
½ teaspoon freshly ground black pepper
4 to 6 dashes of hot pepper sauce, to taste
⅓ cup (loosely packed) fresh coriander leaves (cilantro), minced (optional)
1 cup sour cream
Tortilla chips, for serving

1. In a large saucepan, heat the oil over moderate heat. Add the onion and cook until translucent, about 5 minutes. Add the garlic and cook until fragrant, about 2 minutes. Add the black beans and ⅔ cup water. Cook until the beans begin to break apart, about 15 minutes. Transfer the beans to a food processor and puree until smooth. Transfer to a bowl and let cool to room temperature.

2. Meanwhile, in a medium bowl, stir together the lemon juice, lime juice, tomato, jalapeños, scallions, salt and black pepper. When the beans are cool, add them to the bowl and stir well to combine. *(The recipe can be prepared to this point up to 8 hours ahead; cover and refrigerate.)*

3. Season the dip to taste with hot pepper sauce and stir in the minced coriander. Pass the sour cream separately and serve with tortilla chips.
—Tracey Seaman

CRISPY SHALLOT
DIP
It takes very few ingredients—though a whole lot of shallots—to produce this ultimate onion dip.
Makes About 3½ Cups

⅔ cup olive oil
2 pounds shallots, thinly sliced
2 sprigs of fresh thyme or ½ teaspoon dried
1 package (8 ounces) cream cheese, at room temperature
1 pint sour cream, at room temperature
Salt and freshly ground pepper
Potato chips, for serving

1. In a large skillet, warm the olive oil over moderately high heat. Add the shallots and stir in the thyme. Reduce the heat to low and cook, stirring occasionally, until the shallots are browned and crisp, about 30 minutes.

2. Using a slotted spoon, transfer the shallots to paper towels to drain. Discard the thyme sprigs, if used.

3. In a medium bowl, mash together the cream cheese and sour cream with a wooden spoon. Stir in the shallots. Season with salt and pepper to taste. *(The recipe can be made up to 1 day ahead; cover and refrigerate.)* Serve with potato chips.
—Marcia Kiesel

GOLDEN MADRAS
DIP
For the complex flavor of the curry powder to fully develop with the other seasonings and ingredients, this dip should sit at least an hour before serving. It can be made up to one day ahead; keep covered in the refrigerator.
Makes About 3 Cups

2 tablespoons olive oil
1 medium green bell pepper, coarsely chopped
1 large onion, coarsely chopped
1 large tart green apple, such as Granny Smith, peeled and coarsely chopped
2 tablespoons curry powder
1 teaspoon cumin seeds
1 teaspoon finely grated fresh ginger
1 cup sour cream
½ cup plain yogurt
1 tablespoon chopped fresh dill or 1½ teaspoons dried
Salt
Papadums,* fried according to package directions, or toasted pita bread, for serving
*Available at Indian markets and specialty food stores

1. In a large skillet, warm the olive oil over moderate heat. Add the green pepper, onion and apple and cook until slightly soft, about 10 minutes.

2. Stir in the curry powder, cumin seeds and ginger. Reduce the heat to low and cook, stirring occasionally, until the vegetables and apple are very soft and the spices are fragrant, about 10 minutes longer. Set aside to cool.

3. Transfer the mixture to a food processor and puree until smooth; stop the machine occasionally to scrape down the sides. Add the sour cream and yogurt and process until thoroughly blended. Pour the dip into a medium bowl and fold in the dill. Season to taste with salt. Let sit at room temperature for at least an hour before serving or refrigerate, covered, for up to 1 day. Serve with fried papadums or toasted pita.
—Marcia Kiesel

GINGERED
EGGPLANT MOUSSE
As this delicately flavored mousse drains overnight in the refrigerator, it takes on the pattern of the basket in which it was molded.
8 to 10 Servings

2 medium eggplants (about 1¼ pounds total)
½ pound farmer cheese
½ cup sour cream
2 tablespoons tomato paste
1 medium garlic clove, minced
1 cup minced scallions (about 1 bunch)
1 teaspoon ground ginger
1 teaspoon salt
2½ teaspoons freshly ground pepper
1 cup heavy cream
1 tablespoon olive oil

1. Preheat the oven to 400°. Place the eggplants on a baking sheet, prick them with a fork and roast for 40 min-

utes, or until very tender when pierced. Let cool completely. Peel the eggplants and puree the pulp in a food processor. Scrape the puree into a large bowl.

2. Combine the farmer cheese, sour cream, tomato paste and garlic in the food processor and puree until smooth. Scrape this mixture into the bowl of eggplant puree and add the scallions, ginger, salt and 2 teaspoons of the pepper; mix well.

3. In a medium bowl, whip the heavy cream until it forms stiff peaks. Fold the cream into the eggplant mixture. Season to taste with additional salt, pepper and ginger.

4. Line a 6- to 8-cup wicker basket with dampened cheesecloth. Gently pour in the mousse mixture and cover with any overhanging cheesecloth. Cover with plastic wrap, place on a large plate and refrigerate overnight.

5. About 30 minutes before serving, remove the mousse from the refrigerator and set aside. Just before serving, unwrap the mousse and unmold onto a decorative platter. Drizzle with the olive oil and sprinkle with the remaining ½ teaspoon pepper.
—W. Peter Prestcott

AVOCADO AND ROASTED CORN GUACAMOLE WITH TOASTED CORN TORTILLAS

You can make a larger batch of the toasted tortillas and store them in an airtight container for future nibbling.
8 Servings

1 cup fresh or thawed frozen corn kernels
¼ cup plus 3 tablespoons corn oil
2 large avocados, preferably Hass, cut into ½-inch dice
1 large tomato, cut into ¼-inch dice
¼ cup chopped fresh coriander (cilantro)
2 tablespoons minced red onion
About 1 teaspoon minced fresh or pickled jalapeño pepper
1 teaspoon minced garlic
2 tablespoons fresh lime juice
1 teaspoon cider vinegar
1½ teaspoons coarse (kosher) salt
¼ teaspoon cumin
2 packages (7 ounces each) 5-inch corn tortillas, quartered
Table salt

1. Preheat the oven to 450°. On a baking sheet, toss the corn with 1 tablespoon of the oil. Roast, tossing often, for 7 to 8 minutes, until golden. Let cool, then transfer to a medium bowl.

2. Fold in the avocado, tomato, coriander, onion, jalapeño and garlic. Stir in the lime juice, vinegar, coarse salt, cumin and 2 more tablespoons of the corn oil. Cover and refrigerate for up to 6 hours.

3. Meanwhile, in a large bowl, toss the tortilla pieces with the remaining ¼ cup oil. Arrange half of the tortillas in a single layer on 2 large baking sheets. Bake for 5 to 6 minutes, or until crisp. Remove from the oven, transfer to paper towels and season with table salt. Repeat with the remaining tortillas and more salt. Serve the tortilla chips warm with the cold guacamole.
—Kathy Casey

GUACAMOLE

The best guacamole is the simplest and the freshest, made with flavorful avocados (the wrinkled, black-skinned Hass variety or a near relative), ripe tomatoes, fresh jalapeños or serrano chiles, a bit of onion, salt and—in my house at least—a generous amount of fresh coriander. For the version below, I also added some fresh marjoram, since its pungent taste is almost as southwestern to my palate as coriander. If you don't have any fresh marjoram or oregano on hand, don't use dried; just increase the coriander if you like.
6 to 8 Servings

¼ cup minced yellow onion
¾ cup coarsely chopped fresh coriander (cilantro), stems included
2 tablespoons chopped fresh marjoram or oregano
1 to 1½ jalapeño peppers, to taste, stemmed
1 teaspoon salt
4 ripe but firm Hass avocados
2 large tomatoes, seeded and cut into ¾-inch chunks
⅓ cup finely diced red onion

1. In a food processor or blender, combine the yellow onion, coriander, marjoram, jalapeños and salt. Process until smooth, about 30 seconds.

2. Halve and pit the avocados. Using the dull side of a knife, deeply score the flesh into ¾-inch chunks. Scoop the chunks out of the skins into a medium bowl.

3. Add the pureed herb mixture, the tomatoes and the red onion to the bowl and toss gently until just combined. *(The recipe can be prepared up to 1 hour ahead. Cover tightly and let stand at room temperature.)*
—Michael McLaughlin

2 FIRST COURSES

FIRST COURSES

MIXED GREENS WITH POLENTA CROUTONS AND GORGONZOLA
6 Servings

1½ teaspoons salt
1 cup instant polenta*
1 head of radicchio
1 small head of chicory (curly endive)
1 small head of romaine lettuce
1 large Belgian endive
1 tablespoon red wine vinegar
3 tablespoons plus ⅓ cup olive oil, preferably extra-virgin
1 tablespoon walnut oil
½ teaspoon Dijon mustard
⅛ teaspoon freshly ground pepper
4 ounces Gorgonzola cheese, at room temperature
*Available at Italian markets and specialty food shops

1. In a heavy medium saucepan, bring 2¼ cups of water to a boil over high heat. Add 1 teaspoon of the salt and stir in the polenta. Cook, stirring constantly with a wooden spoon, until the polenta thickens. Reduce the heat to a simmer and cook, stirring occasionally, until the polenta pulls away from the sides of the pan when stirred, about 15 minutes. (Note that the cooking time, longer than specified on the box, is needed to produce a tight, relatively dry polenta that will hold together when fried.)

2. Scrape the thickened polenta out of the pan onto a nonreactive baking sheet. Using a lightly oiled wooden or plastic spatula, spread the polenta into an even rectangle about 12 by 6 by ¼ inch. Let cool to room temperature, then cover and refrigerate until well chilled.

3. Meanwhile, prepare the salad greens. Cut the core from the radicchio and separate the head into individual leaves. Separate the chicory and romaine into individual leaves, remove the tough ribs and tear the leaves into bite-size pieces. Separate the Belgian endive into separate leaves. Combine the greens, rinse thoroughly and dry. Wrap the greens in paper towels, place in a plastic bag and refrigerate if not serving at once. *(The recipe can be made 1 day ahead to this point.)*

4. In a small bowl, whisk together the vinegar, 3 tablespoons of the olive oil, the walnut oil, mustard, remaining ½ teaspoon salt and the pepper. Set the dressing aside.

5. Preheat the broiler. Remove the polenta from the refrigerator. Cut the polenta into 6 rectangles 6 by 2 inches each. Heat the remaining ⅓ cup olive oil in a large skillet. Add the polenta and sauté over moderately high heat, turning once, until crisp and golden, about 4 minutes. Remove with a slotted spatula and drain on paper towels. *(The recipe can be prepared to this point up to 1 hour ahead.)*

6. Spread the Gorgonzola over the fried polenta.

7. To assemble the salad, toss the greens with the vinaigrette until coated. Divide among 6 plates.

8. Run the polenta croutons under the broiler, until the cheese melts, about 1 minute. Slice each rectangle crosswise into 6 smaller rectangles or cut into triangles and, while still warm, arrange on top of the salad.
—*John Robert Massie*

AVOCADO SALAD WITH GRAPEFRUIT AND BACON
2 Servings

½ cup diced (¼ inch) slab bacon (2 to 3 ounces)
1 pink grapefruit, peeled
1 avocado, preferably Hass, cut lengthwise into 12 slices
½ cup fresh grapefruit juice
¼ teaspoon salt
1 teaspoon olive oil
1 teaspoon sugar
¼ teaspoon freshly ground pepper
2 tablespoons finely chopped red onion
Sprigs of watercress, for garnish

1. Blanch the bacon in a small pan of simmering water until the fat is translucent, about 5 minutes. Drain and pat dry.

2. With a sharp knife, remove all the white pith and the outside membrane from the grapefruit. Remove the sections by cutting between the dividing membranes.

3. Toss the avocado in the grapefruit juice to coat; then remove the avocado and reserve the juice.

4. On two warm plates, arrange alternating slices of avocado and grapefruit sections. Sprinkle the avocado with the salt.

5. In a small skillet, fry the bacon in the oil over moderate heat until lightly browned, about 3 minutes. Add the sugar, pepper and reserved grapefruit juice. Cook for 1 minute to warm through.

6. Pour the bacon dressing over the grapefruit and avocado. Sprinkle with the red onion. Garnish with a sprig of watercress.
—*Anne Disrude*

Guacamole (p. 29).

Wild Mushroom Risotto (p. 109).

Au Pied de Cochon's Onion Soup (p. 68).

SWISS CHEESE BEIGNETS

I found this recipe in my grandmother's papers labeled "how to use leftover egg whites"; it dates from the beginning of the century. What caught my attention was its simplicity and use of only egg whites, rather than whole eggs, which makes the beignets very light. If at all possible, beat the egg whites by hand in a copper bowl to get the most volume.

I love the combination of these crisp cheese tidbits with the tart-bitter salad greens, but they are marvelous all by themselves with cocktails or Champagne. Grate the cheese at the last minute. If grated ahead of time, the cheese tends to pack and the beignets are not as light.
Makes About 1½ Dozen

½ pound French Gruyère cheese
1 quart vegetable oil, for
   deep-frying
3 egg whites
Salt (optional)

1. With a fine shredding grater, flat or rotary, grate the cheese, keeping it as loosely packed as possible. In a large heavy saucepan or deep-fat fryer, heat the oil to 325°.

2. Meanwhile, beat the egg whites until they are stiff but not dry. Sprinkle about one-third of the grated cheese over the beaten egg whites and fold in. Sprinkle on half of the remaining cheese and fold in. Lightly fold in the remaining cheese just until it is evenly distributed. The mixture will remain lumpy.

3. Drop heaping tablespoons of the cheese-egg white mixture into the hot oil, frying in batches without crowding, until golden brown, about 5 minutes. As the beignets are added to the oil, they may stick together; separate carefully with 2 forks.

4. Drain the beignets on paper towels and sprinkle lightly with salt if desired. Serve hot.
—*Lydie Marshall*

ARUGULA AND WATERCRESS SALAD WITH SWISS CHEESE BEIGNETS

An unusual, refreshing change from the standard tossed salad with baked goat cheese, this one tops bitter greens, tossed with an herbal vinaigrette, with exceptionally light, hot cheese fritters.
6 to 8 Servings

2 bunches of arugula, tough stems
   removed
1 bunch of watercress, tough
   stems removed
½ cup Vinaigrette Dressing with
   Garlic and Fresh Herbs (p. 261)
Swiss Cheese Beignets (at left)

1. Wash the salad greens. Dry in a salad spinner and then pat dry in a towel. Refrigerate until ready to toss the salad.

2. In a large bowl, toss the arugula and watercress with the Vinaigrette Dressing with Garlic and Fresh Herbs. Divide among 6 or 8 plates and top with the hot Swiss Cheese Beignets.
—*Lydie Marshall*

SALAD OF PEARS AND MIXED GREENS WITH CHEVRE

This refreshing salad from cheesemaker Laura Chenel would also work as a cheese course.
8 Servings

½ cup coarsely chopped walnuts
1 tablespoon red wine vinegar
½ cup walnut oil
1 garlic clove, lightly crushed
¼ teaspoon salt
⅛ teaspoon freshly ground white
   pepper
2 heads of Bibb lettuce, torn into
   bite-size pieces
2 bunches of arugula, large
   stems removed, torn into
   bite-size pieces
3 pears, cored and sliced lengthwise
½ pound mild, creamy goat cheese,
   cut into 8 rounds

1. Preheat the oven to 350°. Spread the walnuts on a baking sheet and bake until lightly toasted, 10 to 12 minutes.

2. In a small bowl, whisk together the vinegar and oil. Add the garlic and set aside to steep at room temperature for at least 1 hour, or refrigerate overnight. Discard the garlic clove and season the dressing with the salt and pepper.

3. In a large bowl, toss the lettuce and arugula with two-thirds of the dressing. Divide the greens among 8 salad plates.

4. Toss the pears with the remaining dressing. Arrange the slices and a round of goat cheese alongside each salad and scatter the toasted walnuts over the top.
—*Laura Chenel*

FIRST COURSES

FIRST COURSES

SALAD OF PEARS AND STILTON WITH SAGE LEAVES

Sage blossoms are used here (although they are optional), so keep this recipe in mind the next time your sage plants bloom.

❡ With its fresh fruit, raw greens and salty, tangy blue cheese, this dish provides a challenge that can be met by a deep-flavored wine, such as a Sémillon-Chardonnay from Australia—Rosemount's—or a white Hermitage from the Rhône, such as Chapoutier Hermitage Chante-Alouette.

4 Servings

1 teaspoon Dijon mustard
1½ tablespoons Champagne vinegar or white wine vinegar
⅓ cup light olive oil
¼ teaspoon salt
¼ teaspoon freshly ground pepper
4 tablespoons unsalted butter
24 large fresh sage leaves plus 2 tablespoons chopped fresh sage
2 cups French bread cubes (½ inch)
2 pears, preferably Bosc—peeled, halved, cored and sliced into ¼-inch-thick wedges
½ head of Boston lettuce, torn into bite-size pieces
½ head of escarole, torn into bite-size pieces
½ head of romaine lettuce, torn into bite-size pieces
½ head of red leaf lettuce, torn into bite-size pieces
½ cup thinly sliced red onion
3 ounces Stilton cheese, crumbled
¼ cup sage blossoms (optional)

1. In a small bowl, whisk together the mustard, vinegar, olive oil, salt and pepper. Set the vinaigrette aside.

2. In a large skillet, melt 1 tablespoon of the butter over moderate heat. Add the whole sage leaves. Increase the heat to moderately high and fry until the leaves are golden brown and crisp on the bottom, about 1 minute. Turn the leaves and cook until crisp on the second side, about 1 minute longer. Remove to a plate.

3. Melt the remaining 3 tablespoons butter in the same skillet over moderate heat. Add the bread cubes and sprinkle the chopped sage over them. Toss to coat the bread evenly with the butter and sage. Cook, turning occasionally, until the cubes are golden brown and crunchy, about 4 minutes. Season the croutons with an additional pinch of salt.

4. In a medium bowl, toss the pears with 2 tablespoons of the vinaigrette until evenly coated. In a large bowl, combine the lettuces and red onion and toss with the remaining vinaigrette.

5. Place an equal amount of lettuce on each of 4 large plates and arrange the pear slices on top. Sprinkle each salad with Stilton cheese and top with the toasted sage leaves, croutons and sage blossoms.

—Marcia Kiesel

SALAD OF FLAGEOLET BEANS WITH FENNEL AND WALNUTS

This salad is worth serving as a course by itself so that the flavors can be savored.

4 Servings

⅔ cup dried flageolet beans (4½ ounces), picked over
1 large shallot, finely chopped
4 teaspoons Champagne vinegar or white wine vinegar, or more to taste
½ teaspoon salt
2 tablespoons walnut oil
1 tablespoon cream or crème fraîche
1 small fennel bulb (½ pound)—quartered, cored and thinly sliced crosswise
2 teaspoons finely chopped fennel fronds
2 teaspoons minced parsley
6 walnut halves, broken into small pieces
¼ teaspoon freshly ground white pepper
1 teaspoon snipped fresh chives

1. In a medium saucepan, soak the beans in plenty of cold water for at least 6 hours or overnight. Pour off the water, re-cover the beans with fresh water and bring to a boil. Boil the beans vigorously for 5 minutes. Drain the beans in a colander and rinse well to remove any scum.

2. Return the beans to the saucepan, cover with cold water and bring to a boil over high heat. Reduce the heat to moderately low and simmer until the beans are tender, about 1¼ hours. Drain the beans and place in a medium bowl. Let cool to tepid. Reserve the broth for another use.

3. In a small bowl, combine the shallot, vinegar and salt and let stand for a few minutes. Whisk in the walnut oil and cream. Taste and add more vinegar if necessary.

4. Add the dressing to the beans along with the fennel, fennel fronds, parsley and walnuts. Toss gently. Add the pepper and season to taste with salt. Arrange the salad on 4 plates and garnish with the chives.

—Deborah Madison

BLACK-EYED PEA AND MINT MARIGOLD SALAD

If you cannot find fresh or frozen black-eyed peas, use dried black-eyed peas, first cooking them until just tender.
6 Servings

2 cups freshly shelled black-eyed peas or 1 package (10 ounces) frozen black-eyed peas, thawed
3 ounces sliced smoked bacon
1 large garlic clove, minced
1 small onion, finely diced
⅓ of a small red bell pepper, finely diced
2 serrano chiles, seeded and minced
1 tablespoon balsamic vinegar
2 teaspoons finely chopped chives
2 teaspoons finely chopped mint marigold or tarragon
1 tablespoon freshly ground black pepper
Salt

1. In a medium saucepan, bring 4 cups of water to a boil over moderately high heat. Add the fresh or thawed peas and boil until tender, 15 to 20 minutes. Drain, refresh under cold running water and set aside.

2. In a medium nonreactive skillet, cook the bacon over moderate heat, turning once, until crisp, about 5 minutes. Using tongs, transfer the bacon to paper towels to drain. Coarsely chop and set aside.

3. Heat the bacon fat over low heat. Add the garlic, onion, red bell pepper and serrano chiles and cook, stirring, until softened, about 5 minutes. Add the reserved peas and bacon. *(The recipe can be prepared up to 1 day ahead. Cover and refrigerate. Rewarm over moderate heat before proceeding.)* Add the vinegar, chives, mint marigold and black pepper. Season to taste with salt. Serve hot.
—*Kevin Rathbun*

INDONESIAN FRIED TOFU AND BEAN SPROUT SALAD (TAHU GORENG)

If you've never eaten fried tofu, the meaty, chewy texture may surprise you. Even people who are not particularly fond of bean curd often love it prepared this way. Hot, spicy peanut sauce provides the perfect counterpoint to the crisp, cool bean sprouts.
4 Servings

⅓ pound bean sprouts (about 2 cups)
4 firm Chinese tofu cakes*
¼ cup corn or peanut oil
3 tablespoons kecap manis* or homemade Indonesian Sweet Soy Sauce (recipe follows)
1 tablespoon fresh lime or lemon juice
1 tablespoon smooth peanut butter
1 small fresh hot red chile, seeded and very thinly sliced (about 2 teaspoons), or ½ teaspoon crushed red pepper
1 garlic clove, crushed through a press
1 teaspoon sugar
2 scallions, thinly sliced
*Available at Asian markets

1. Place the bean sprouts in a large heatproof bowl. Cover with boiling water and let stand for 2 minutes. Drain and rinse in cold water; drain well. Dry on paper towels. Refrigerate for 15 minutes or overnight.

2. Dry the tofu cakes on a paper towel. Cut into 12 (1-inch) cubes. Heat the oil in a wok or large skillet. Fry the cubes in 2 or 3 batches over moderate heat, tossing, until light brown all over, 5 to 7 minutes. Remove with a slotted spoon and drain on paper towels.

3. In a small bowl, combine the *kecap manis*, lime juice, peanut butter, chile, garlic and sugar. Stir to blend the dressing well.

4. To assemble the salad, place the bean sprouts on a serving platter. Cover with the fried tofu. Pour the sauce over all. Garnish with the scallions. Serve at room temperature.
—*Copeland Marks*

INDONESIAN SWEET SOY SAUCE (KECAP MANIS)

In a covered jar in the refrigerator, this sauce keeps almost forever. With a little oil, water, garlic and ginger, it makes a fabulous marinade for lamb chops and steaks.
Makes About 1½ Cups

1 cup sugar
1¼ cups light Chinese soy sauce*
2 garlic cloves, bruised
2 large star anise pods*
*Available at Asian markets

1. In a heavy medium saucepan, cook the sugar over moderately low heat, stirring occasionally, until it melts and is caramelized to a rich brown.

2. Add the soy sauce, garlic, star anise and ¼ cup of water. Cook stirring, until the caramel and soy are blended. Simmer until slightly thickened, about 15 minutes. Let cool, then pour into a wide-mouthed jar. (Include the garlic and anise.) Cover and refrigerate for up to 2 years.
—*Copeland Marks*

WARM ARTICHOKE SALAD WITH BACON AND MUSTARD VINAIGRETTE
4 Servings

2 teaspoons Dijon mustard
1 tablespoon red wine vinegar
¼ cup plus 2 tablespoons extra-virgin olive oil
1 large garlic clove, smashed
¾ teaspoon salt
¼ teaspoon freshly ground pepper
8 medium artichokes (about 4 pounds)
1 lemon, halved
¼ cup all-purpose flour
2 tablespoons fresh lemon juice
4 slices of bacon
2 tablespoons finely chopped parsley

1. In a bowl, whisk the mustard with the vinegar. Slowly whisk in the olive oil. Add the garlic, ¼ teaspoon of the salt and the pepper and 1 tablespoon of water; set aside. *(The vinaigrette can be prepared 1 day ahead; cover and refrigerate.)* Before serving, remove the garlic.

2. Trim the artichokes by first snapping off the tough outer leaves near the base. Using a sharp, stainless steel knife, cut off the stems; cut the crowns to within 1½ inches of the base. Scoop out the hairy chokes with a spoon and trim carefully to remove all the leaves. Immediately rub the surfaces with the cut lemon and set the artichoke hearts aside.

3. Place the flour in a large nonreactive pot and whisk in enough cold water to make a smooth paste. Then whisk in 4 more cups of cold water. When all the water has been added, whisk in the lemon juice and the remaining ½ teaspoon salt. Bring to a boil over high heat, and boil for 3 minutes. Add the artichoke hearts, and bring back to a boil. Reduce the heat to moderate and simmer until the artichoke hearts are just tender when pierced with a small sharp knife, 25 to 30 minutes. Remove from the heat and let the hearts cool to room temperature in the liquid.

4. In a medium skillet, cook the bacon over moderate heat, turning once, until cooked but not crisp, about 6 minutes. Drain on paper towels. Cut each slice crosswise into 6 pieces.

5. Drain the artichoke hearts well. Using a teaspoon, scoop out any remaining portions of the chokes. Slice the artichoke hearts ¼ inch thick. Whisk the vinaigrette to combine and add the sliced artichokes, bacon and parsley. Toss well. Serve at room temperature.
—David Rosengarten

SPICY CUCUMBER AND FRUIT SALAD
Here is an intriguing first course, guaranteed to whet the appetite. It has a light, but definite, cucumber taste.
4 Servings

4 Kirby cucumbers, peeled
½ of a cantaloupe
1½ cups strawberries, halved or quartered, depending on size
2 tablespoons fresh lime juice
1½ teaspoons finely chopped fresh coriander (cilantro)
1 teaspoon minced seeded jalapeño pepper
⅛ teaspoon salt
1 head of Bibb lettuce, leaves separated

1. Using a small melon-baller, cut the cucumbers into balls. Place in a large bowl.

2. Using the same melon-baller, scoop out balls from the cantaloupe; reserve the rind. Add the cantaloupe and strawberries to the cucumbers.

3. With your hands, squeeze the cantaloupe rind to collect 1 tablespoon of the juice, then discard the rind. Add the juice to the salad. Sprinkle on the lime juice, coriander, jalapeño pepper and salt; toss to coat well. *(The recipe can be prepared to this point up to 2 hours ahead. Cover and refrigerate.)*

4. To serve, arrange 4 small lettuce leaves in each of 4 salad bowls. Divide the salad evenly among the bowls. Spoon any juices on the bottom of the bowl over the salads.
—Diana Sturgis

GREEN PAPAYA SALAD
Som tam is as popular in Thailand as green mangoes and Thai boxing, and that's popular! This salad has its roots in the villages of northeast Thailand, yet it regularly appears on the menus of Bangkok's finest restaurants. If green papaya is unavailable, julienned carrots make a good substitute. To eat the salad, wrap a small amount of it in the cabbage and lettuce leaves.
6 Servings

¼ pound medium shrimp, shelled and deveined
1 garlic clove, chopped
2 serrano chiles, seeded and minced
1 tablespoon chopped roasted peanuts
1 tablespoon palm sugar* or light brown sugar
¼ teaspoon salt
½ pound long beans (asparagus beans) or green beans
1 firm-ripe tomato, sliced

1 green papaya (about 2 pounds)—
    peeled, seeded and cut into 2-by-
    ⅛-inch strips
¼ cup fresh lime juice
1 teaspoon nam pla*
1 head of leaf lettuce, leaves
    separated
¼ head of green cabbage, cored and
    cut into wedges
*Available at Asian markets

    1. In a small saucepan, bring 3 cups of water to a boil. Add the shrimp and cook until loosely curled and just opaque throughout, 2 to 3 minutes. Drain and rinse under cold water; drain. Cut the shrimp into ½-inch pieces.
    2. In a blender or food processor, combine the garlic, serrano chiles, peanuts, palm sugar and salt. Blend or process to a paste. Set aside.
    3. In a saucepan, blanch the long beans in boiling salted water until crisp-tender, about 5 minutes. Drain, rinse under cold water and drain again.
    4. In a large mixing bowl, combine the shrimp and tomato. Mash with a wooden spoon until blended. Alternatively, use a mortar and pestle to pound the ingredients. The shrimp will retain their shape for the most part, and the tomato should be almost pureed. Gradually add the papaya strips and mix until incorporated. Add the reserved garlic paste and blend well. Stir in the lime juice and *nam pla*.
    5. To serve, arrange some of the lettuce leaves on a platter. Mound the salad on top. Place the long beans, cabbage wedges and remaining lettuce leaves on separate plates and serve alongside.
—*Jeffrey Alford*

APRICOT-SHRIMP SALAD
6 Servings

1½ pounds medium shrimp—
    cooked, shelled and deveined
12 fresh apricots or 8 ounces dried
Small bunch of watercress, trimmed
5 tablespoons peanut oil
3 tablespoons cider vinegar
¾ teaspoon curry powder

    1. Divide the shrimp into two equal portions. Cut one portion crosswise into thirds; cut the remainder in half lengthwise. Place in a medium bowl.
    2. Pit the fresh apricots; quarter 6 of them and roughly chop the rest. (If using dried apricots, cover them with boiling water and let them soak for 20 minutes. Drain. Cut half of them in two; roughly chop the remainder.) Add the chopped and unchopped apricots to the shrimp. Coarsely chop half of the watercress and toss it with the shrimp and apricots.
    3. In a small bowl, whisk the oil with the vinegar and curry. Toss all but 1 tablespoon with the shrimp mixture. Mound in the center of a serving platter and surround with the remaining watercress. Drizzle the remaining dressing over the watercress. Serve the salad at room temperature.
—*Rosalee Harris*

WARM SHRIMP SALAD WITH CHAMPAGNE VINEGAR SAUCE
6 Servings

30 medium shrimp (about 1 pound),
    shelled
6 cups mixed salad greens such as
    watercress, radicchio, young
    chicory, arugula and mâche
6 large mushroom caps, peeled and
    cut into thin matchsticks
2 tomatoes—peeled, seeded and
    finely diced
3 shallots, minced
6 tablespoons Champagne vinegar
    or white wine vinegar
6 tablespoons heavy cream
1½ sticks (6 ounces) unsalted
    butter, at room temperature, cut
    into 12 pieces
¼ teaspoon salt
Pinch of cayenne pepper
Fresh chervil sprigs, for garnish

    1. Split the shrimp along the back, leaving a small portion attached at the top and bottom. Remove the veins. Flatten each shrimp to form a circle. Steam the shrimp until just opaque throughout, about 1½ minutes.
    2. Toss the salad greens and arrange on 6 salad plates. Scatter the mushrooms over the top. Place a small mound of diced tomatoes in the center of each salad and surround with 5 shrimp.
    3. In a small saucepan, combine the shallots and vinegar. Bring to a boil over moderately high heat and cook until the liquid is reduced to 2 tablespoons, about 2 minutes. Stir in the cream and boil until reduced to 3 tablespoons, 2 to 3 minutes longer.

4. Remove from the heat and whisk in the butter, a few pieces at a time, until thoroughly blended. Stir in the salt and cayenne. Spoon the sauce over the salads and garnish with a sprig of chervil. Serve warm.
—Jean-Georges Vongerichten

SHRIMP SALAD HILARY

In this delicately colored, creamy salad, the characteristic tartness of sour cream and plain yogurt is shaded with mustard and smoothed with a touch of Cognac. There is a lot of sauce, which doubles as a dressing for the bed of lettuce.

8 Servings

2½ tablespoons coarse (kosher) salt
2 pounds medium shrimp, in their shells
1 pound thinly sliced Canadian bacon
4 medium tomatoes—peeled, seeded and chopped
½ cup finely chopped scallions
2 cups sour cream
2 cups plain yogurt
¼ cup Cognac
¼ cup Dijon mustard
1 teaspoon freshly ground pepper
2 small heads of romaine lettuce—separated into leaves

1. Bring a large pot of water to a boil with 2 tablespoons of the salt. When the water has reached a rolling boil, add all the shrimp at once. Cook, uncovered, until the shrimp are pink and tender, about 3 minutes. Drain the shrimp into a colander and rinse immediately under cold running water to prevent further cooking; drain well. Shell the shrimp and devein, if desired. Split them in half lengthwise and place in a large bowl.

2. Cut the bacon into ½-inch pieces. Add to the shrimp. Add the tomatoes and scallions.

3. In a medium bowl, combine the sour cream, yogurt, Cognac, mustard, remaining ½ tablespoon salt and the pepper; stir until blended. Pour the dressing over the shrimp salad and toss gently until mixed. Refrigerate, covered, until chilled.

4. Before serving, arrange the lettuce in an attractive pattern on each salad plate. Divide the shrimp salad among the plates, adding as much sauce as desired. Let stand at room temperature for 15 minutes before serving.
—W. Peter Prestcott

CLASSIC SQUID SALAD

Cooking squid takes an extremely watchful eye. Some people prefer it almost raw in their salads; if you do, cook it for 15 seconds. Precision timing is all important here: You should taste while cooking, and to be really compulsive about it, use a stopwatch. Start testing for tenderness after 90 seconds. As soon as the rawness disappears, drain all the squid. It's okay if the squid does not seem completely tender. It's done. If the squid is cooked any longer, it will begin to toughen.

8 Servings

1½ pounds cleaned squid, bodies cut crosswise into ¼-inch rings and tentacles halved lengthwise if large
1 medium carrot, cut into fine 1½-inch-long strips
2 celery ribs, halved lengthwise and thinly sliced crosswise on the bias
½ of a medium red bell pepper, finely diced
½ of a small red onion, finely diced
3 tablespoons minced parsley
1 tablespoon white wine vinegar
1 tablespoon fresh lemon juice
1 teaspoon salt
½ teaspoon freshly ground black pepper
½ teaspoon minced garlic
¼ cup plus 1 tablespoon extra-virgin olive oil

1. Bring a large saucepan of water to a boil over high heat. Have ready a big bowl of ice water. Add the squid to the boiling water all at once, stir and cover the pot to help the water return to a boil. Cook, tasting for doneness, for up to 3 minutes (see headnote).

2. Drain the squid and immediately plunge it into the ice water. When cool, drain again and transfer to a large bowl. Add the carrot, celery, bell pepper, onion and parsley; toss to combine.

3. In a small bowl, whisk the vinegar with the lemon juice, salt, black pepper and garlic. Gradually whisk in the oil until incorporated. Pour over the squid and toss. Serve at room temperature.
—Mark Bittman

SCALLOP AND ORANGE SALAD

4 Servings

1 navel orange
About 1⅓ cups orange juice
1 pound bay or sea scallops
2 tablespoons thinly slivered ginger
1 garlic clove, minced
1 tablespoon honey
¼ cup sherry wine vinegar or white wine vinegar
½ cup olive oil
½ teaspoon salt
Freshly ground black pepper
1 tablespoon minced fresh coriander (cilantro) leaves
⅓ cup thinly sliced scallions
1 medium red bell pepper, cut into ¾-by-⅛-inch julienne strips

1. With a swivel-bladed peeler, remove the zest from the orange in pieces as large as possible and cut into thin strips; place in a bowl, cover with cold water and set aside. Working over a large bowl to collect the juices, section the orange, removing as much of the pith and membrane as possible. Reserve the orange sections; measure any juice and add as much as needed to measure 1⅓ cups.

2. If sea scallops are being used, cut them in half horizontally, across the grain, to make rounds; then cut each round into wedge-shaped quarters. Rinse the scallops and pat them dry. Place them in a nonreactive saucepan and add the orange juice. Cover and bring to a boil over moderate heat. Reduce the heat to low and poach for 30 to 60 seconds, or until just cooked through (do not overcook or they will toughen). Remove from the pan with a slotted spoon and set aside.

3. Add the ginger, garlic and honey to the orange juice in the pan. Bring the mixture to a boil over high heat; lower the heat to moderate and, stirring frequently, reduce the mixture to ½ cup. Cool to room temperature.

4. In a medium bowl, whisk together the vinegar, oil, salt, black pepper to taste and the coriander. Chop half the reserved orange zest and add it to the dressing, along with half the scallions and all the bell pepper.

5. Add the reserved scallops to the dressing and toss well. Cover and chill for at least 1 hour.

6. To assemble the salad, mound the scallop salad in the center of a chilled platter. Dip the reserved orange sections into the reduced orange juice mixture. Then, garnish the platter with "butterflies," using 2 glazed orange sections for the wings and 2 strips of orange zest for the antennae of each butterfly. Sprinkle the remaining scallions over the salad and top with the remaining strips of orange zest.
—F&W

ARUGULA AND
SCALLOP SALAD
2 Servings

1 small bunch of arugula
2 tablespoons olive oil, preferably extra-virgin
Salt and freshly ground black pepper
2 garlic cloves, minced
½ pound bay scallops, trimmed of connecting muscle
1 small or ½ of a large red or yellow bell pepper, cut into thin lengthwise strips
1 small jicama (about 3 ounces), peeled and cut lengthwise into thin strips
½ of a small fennel bulb—trimmed, cored and cut into thin lengthwise strips
1 tablespoon minced chives
2 teaspoons minced fresh tarragon or ½ teaspoon dried
2 teaspoons minced parsley

1. Toss the arugula with ½ tablespoon of the olive oil and divide between 2 plates. Season lightly with salt and black pepper.

2. In a large skillet, heat the remaining 1½ tablespoons olive oil. Add the garlic and sauté over moderately high heat for 30 seconds. Add the scallops and continue to cook, tossing, for 1 minute.

3. Add the bell pepper, jicama and fennel. Cook, tossing, until the scallops are opaque throughout and the vegetables are warmed through. Add the chives, tarragon, parsley, ½ teaspoon salt and ⅛ teaspoon black pepper. Toss well and arrange over the arugula.
—Anne Disrude

SCALLOP, MUSSEL AND
ASPARAGUS SALAD
WITH ORANGE-
SAFFRON DRESSING
❗ Dry-style Rieslings are a fine foil for this dish. For ripe flavor and depth, try a top Californian, such as Trefethen White Riesling or Firestone "Dry."
4 Servings

1 pound medium asparagus, tough ends snapped off, cut into 1-inch pieces
2 dozen small mussels, scrubbed and debearded
1 pound small sea scallops
8 saffron threads
2 tablespoons minced shallots
2 tablespoons fresh orange juice
½ teaspoon finely grated orange zest
½ teaspoon salt
¼ teaspoon freshly ground pepper
1 tablespoon rice vinegar or 2 teaspoons white wine vinegar diluted with 1 teaspoon water
2 tablespoons extra-virgin olive oil

1. Fill a medium saucepan with ½ inch of water and insert a steamer basket. Bring the water to a boil over high heat, add the asparagus pieces, cover and steam until bright green and just tender, about 4 minutes. Remove the basket and rinse the asparagus under cold running water until cool; set aside. Pour out the water.

2. Add another ½ inch of water to the saucepan and bring to a boil over high heat. Add the mussels to the steamer basket, cover and steam until they open, about 2 minutes. Remove the mussels from the basket and set aside; discard any that have not opened. Add the scallops to the basket, cover and steam until just opaque throughout, about 3 minutes. Remove the scallops and set aside. Measure 2 tablespoons of the steaming liquid into a small bowl. Crumble the

saffron threads into the liquid and set aside to steep.

3. Remove the mussels from their shells and briefly rinse the mussels if they are sandy. Cut the scallops to the same size as the mussels, if necessary. In a large bowl, combine the mussels, scallops and asparagus and set aside.

4. In a medium bowl, whisk together the shallots, orange juice, orange zest, salt, pepper, vinegar and olive oil. Whisk in the reserved saffron liquid. Set aside to blend the flavors for about 20 minutes.

5. Up to 1 hour before serving, pour the dressing over the seafood and asparagus and mix well; stir occasionally until ready to serve. Serve the salad at room temperature.
—Marcia Kiesel

LOBSTER SALAD WITH TARRAGON AND SWEET PEPPERS

This recipe was inspired by a delicious molded lobster salad at Park Bistro in New York City. For an extra-special presentation, serve the salad in an avocado half. For lunch or a light dinner, serve a generous mound atop tender lettuce with a side of blanched green beans and sliced steamed artichoke bottoms, or sandwich the salad between thick slices of brioche.
8 Servings

Four 1¼- to 1½-pound cooked
 lobsters
¼ cup mayonnaise
1 tablespoon sour cream or
 crème fraîche
1 tablespoon fresh lime juice
2 teaspoons minced fresh tarragon
¼ teaspoon freshly ground
 white pepper
1 small red bell pepper
1 small yellow bell pepper
2 tablespoons minced fresh chives

Freshly ground black pepper
Lime wedges, for serving

1. Giving a twist and a pull, break off the tails from the lobsters. Then break off the knuckles and claws in 1 piece. Place 1 of the tails belly-up on a work surface and use sharp kitchen shears to cut through the thin underside shell. Remove the tail meat and set aside. Repeat with the remaining tails.

2. Using a lobster cracker or a nutcracker, crack the claws and knuckles, extract the meat and add it to the tail meat. Extract any other meat and discard all the shells. Cut all the meat into ½-inch pieces and transfer to a large bowl.

3. In a small bowl, whisk the mayonnaise with the sour cream, lime juice, tarragon and white pepper. Fold the mixture into the lobster.

4. Roast the red and yellow bell peppers directly over a gas flame or under the broiler as close to the heat as possible, turning frequently, until charred all over. Transfer the peppers to a paper bag and set aside to steam for 10 minutes. Using a small sharp knife, scrape off the blackened skins and remove the cores, seeds and ribs. Rinse the peppers and pat dry. Cut the peppers into ¼-inch dice. *(The recipe can be prepared to this point up to 1 day ahead. Cover the lobster salad and peppers separately and refrigerate overnight. Let return to room temperature before proceeding.)*

5. Stir the chives and 2 tablespoons each of the diced red and yellow bell peppers into the lobster salad. Season with black pepper to taste. Garnish with the remaining diced peppers and lime wedges.
—Tracey Seaman

CITRUS'S CRAB COLESLAW

The crab and cabbage mixture is rolled up in blanched cabbage and then sliced into smaller sections that resemble sushi rolls.
6 Servings

1 small head of savoy cabbage
½ cup mayonnaise
2 tablespoons white wine vinegar
1 teaspoon minced fresh tarragon
 or ½ teaspoon dried
2 tablespoons ketchup
1 teaspoon grainy mustard
2 dashes of hot pepper sauce
¼ teaspoon curry powder
¼ teaspoon freshly ground black
 pepper
½ pound lump crabmeat, flaked
 and picked over to remove
 any cartilage
½ cup diced (¼-inch) bell peppers—
 red, yellow and/or green,
 for garnish

1. Bring a large pot of salted water to a boil over high heat. Remove 6 of the large dark green outer leaves of the cabbage. Plunge them into the boiling water and blanch until softened, about 3 minutes. Drain and rinse under cold water. Pat dry.

2. Quarter and core the remaining cabbage and shred enough to make 1 packed cup. Place in a medium bowl. Reserve the remaining cabbage for another use.

3. In a medium bowl, whisk together the mayonnaise and vinegar until smooth. Stir in the tarragon, ketchup, mustard, hot sauce, curry powder and black pepper until well blended. Add the shredded cabbage and toss until well coated. Add the crabmeat and toss again.

4. Remove the cores of the blanched cabbage leaves. Lay the leaves on a flat surface with the stem ends facing you. Fill each leaf with ¼ cup of the crab mixture. Roll up the leaves to enclose the filling. Using a serrated knife, cut each roll into 4 pieces. Stand 4 pieces on end in a row in the center of each plate. Garnish with the diced bell peppers.
—Michel Richard

SWEDISH WEST COAST SALAD

This can be served with a fresh herb sauce or with a vinaigrette.
12 Servings

1½ pounds small or medium shrimp, in their shells
2 lobsters (about 1½ pounds each)
1 cup dry white wine
1 large shallot, minced
Freshly ground pepper
1 pound small mussels, scrubbed and debearded
1 pound asparagus, trimmed and peeled
1 cup freshly shelled peas
10 Malpèque or Belon oysters
Lettuce leaves
1 pound lump crabmeat
2 tomatoes—peeled, seeded and cut into thin strips
3 hard-cooked eggs, cut into wedges
1 ounce salmon caviar
2 lemons, cut into wedges
Fresh dill and chervil sprigs, for garnish

1. Bring a large pot of salted water to a boil over high heat. Add the shrimp and cook until just opaque throughout, about 3 minutes. Remove the shrimp from the pot and drain well. Let cool completely, then cover and refrigerate.

2. Bring the water in the pot back to a boil. Add the lobsters and cook for 8 minutes. Drain and let cool completely, then cover and refrigerate until cold.

3. In a medium nonreactive saucepan, combine the wine, shallot and pepper and bring to a boil over high heat. Add the mussels, cover tightly and cook until they open, about 2 minutes. Discard any that do not open. Transfer the mussels to a bowl to cool completely. Cover and refrigerate. *(The recipe can be prepared to this point up to 1 day ahead.)*

4. Cook the asparagus in a pot of boiling salted water until just tender, about 4 minutes. Transfer them to a bowl of ice water to stop the cooking. In the same pot, boil the peas until tender, about 2 minutes. Drain and chill immediately in the ice water. Drain the asparagus and peas. Cover and refrigerate until cold.

5. Shortly before serving, crack the lobster shells. Remove the tail meat and slice ½ inch thick. Crack the claws and knuckles and remove the meat. Peel the shrimp. Remove the mussels from their shells. Shuck the oysters, leaving them on the half shell with their liquor.

6. Line a large platter with lettuce leaves. Arrange the lobster, shrimp, mussels, oysters, crabmeat, peas, asparagus and tomato strips in separate mounds on top. Garnish with the eggs, salmon caviar, lemon wedges and dill and chervil sprigs.
—Christer Larsson

WILTED SALAD WITH CHICKEN LIVERS, APPLES AND MUSTARD SEED

In this salad, affordable chicken livers stand in for the more expensive foie gras that might appear on a restaurant menu. The results are almost as luxurious, and the rich flavor of sherry vinegar and the pungent crunch of whole mustard seed make it a robust starter for a hearty but elegant winter meal.
6 Servings

1 bunch of watercress
1 head of red leaf lettuce
1 small head of romaine lettuce
2 teaspoons whole yellow mustard seed
¼ cup sherry wine vinegar
1 pound chicken livers
1 tart apple, preferably Granny Smith
¾ cup olive oil

1. Remove and discard the coarser watercress stems. Trim away the ends and any tough outer leaves from the red leaf and romaine lettuce. Separate into leaves, wash them well and dry. Tear the leaves into bite-size pieces, wrap and refrigerate. *(The greens can be prepared up to 1 day ahead.)*

2. In a small bowl, stir together the mustard seed and sherry vinegar and let stand for 1 hour.

3. Sort and trim the livers, discarding any that are not whole and firm. Separate the livers into lobes and thoroughly pat dry. Peel and core the apple and cut it into ½-inch chunks.

4. In a large skillet, warm 3 tablespoons of the oil over high heat until very hot. Add the apple and cook, tossing and stirring once or twice, until lightly browned, about 5 minutes. With a slotted spoon, transfer the apple chunks to a bowl.

5. Add 1 tablespoon oil to the skillet and set over moderate heat. Add the chicken livers and cook until stiffened and lightly browned on one side, about 2 minutes. Turn carefully and cook until browned on the other side but still pink and juicy inside, 2 to 3 minutes longer. With a slotted spoon, transfer the livers to the bowl with the apples and cover with foil to keep warm.

6. Put the greens in a large bowl. To the skillet, add the vinegar and mustard seed and the remaining ½ cup oil; bring just to the boil, stirring to scrape up any browned bits from the bottom of the pan. Pour the hot dressing over the greens and toss well.

7. Divide the salad among 6 plates. Spoon the apple chunks and livers over the greens. Pour any dressing remaining in the bowl over the salads and serve at once.
—Michael McLaughlin

WILTED ESCAROLE SALAD WITH PANCETTA AND GARLIC

Escarole is a wide-leafed cousin of the bitter green known as curly endive or chicory as well as of the currently fashionable radicchio. Offer the salad as a first course, if you wish, particularly if the menu is Italian. I like it almost as well, though, served alongside a plainly grilled or roasted meat entrée, rather like an elaborate hot vegetable. In any case, there is a delightful contrast between the coarse ribs, which remain crunchy, and the tender green leaves, which wilt considerably.

6 Servings

2 large heads of escarole
½ pound pancetta or slab bacon, cut into ¼-inch dice
½ cup olive oil
24 large garlic cloves, peeled
24 brine-cured black olives, preferably Calamata
2 tablespoons fresh lemon juice
2 lemons, cut into wedges

1. Trim away the ends and any wilted outer leaves of the escarole. Separate the leaves, wash them well and dry. Tear the leaves into bite-size pieces, wrap and refrigerate. *(The escarole can be prepared up to 1 day ahead.)*

2. In a medium saucepan of boiling water, blanch the pancetta for 3 minutes. Drain and pat dry.

3. Set a large deep nonreactive skillet or flameproof casserole over moderately low heat and add the pancetta. Cook uncovered, stirring occasionally, until crisp and golden brown, 20 to 30 minutes. With a slotted spoon, transfer the pancetta to a small bowl. Discard the fat and wipe the skillet clean.

4. Return the skillet to moderately low heat, add the olive oil and garlic cloves and cook, stirring frequently, until the garlic is lightly browned and tender, about 20 minutes. With a slotted spoon, transfer the garlic to the bowl with the pancetta; reserve the oil in the skillet. *(The salad can be prepared to this point up to 3 hours before serving. Leave the oil in the skillet and rewarm over moderate heat before proceeding.)*

5. Add the olives to the skillet and cook, stirring, for 1 minute. Add the escarole, pancetta and garlic cloves, toss to coat with oil, cover and cook over moderate heat for 1 minute. Sprinkle on the lemon juice, toss again, remove from the heat and let stand, covered, for 1 minute before serving.

6. Divide the salad among 6 plates and garnish with wedges of lemon to squeeze onto the salads to taste.
—Michael McLaughlin

WARM SALAD OF FETA CHEESE AND TOMATOES WITH GARLIC SHRIMP

With no shrimp, this salad makes an unusual warm vegetable to serve alongside roasted or charcoal-grilled lamb.

6 Servings

1 bunch of arugula
1 small head of curly endive
1 large head of red leaf lettuce
½ pound feta cheese
¼ cup plus 6 tablespoons olive oil
3 medium garlic cloves, crushed through a press
24 medium shrimp (about 1 pound), shelled but with tails left on
Salt
1 pint (about 20) cherry tomatoes
2 tablespoons minced fresh oregano (see Note)
¼ cup red wine vinegar
Freshly ground pepper

1. Trim away the ends and any wilted outer leaves of the arugula, endive and red leaf lettuce. Separate the leaves, wash them well and dry. Tear the leaves into bite-size pieces, wrap and refrigerate. *(The greens can be prepared up to 1 day ahead.)*

2. Rinse the feta of its salty brine, pat dry and crumble. Reserve at room temperature for up to 2 hours. When you are ready to complete the salad, toss the greens and feta together in a large bowl.

3. In a medium nonreactive skillet, combine ¼ cup of the olive oil and the garlic. Cook over moderately low heat, stirring frequently, until the garlic begins to sizzle gently, about 2 minutes. Add the shrimp, season with a pinch of salt and cook, stirring, until they are pink, curled and opaque, about 5 minutes. Do not let the garlic brown. With a slotted spoon, leaving as much garlic behind in the skillet as possible, transfer

the shrimp to a bowl and cover with foil to keep warm.

4. Add the remaining 6 tablespoons olive oil, the cherry tomatoes and oregano to the skillet. Cook over moderate heat, tossing and stirring gently, until the tomatoes are heated through and the oregano fragrant, about 5 minutes.

5. Add the vinegar to the pan, increase the heat to high and bring just to a boil. Immediately pour the contents of the skillet over the greens and feta in the bowl and toss well.

6. Divide the salad among 6 plates. Garnish each salad with 4 shrimp, season generously with pepper and serve at once.

NOTE: Fresh oregano is preferable here, but if it is unavailable, substitute 1½ teaspoons dried oregano soaked for 1 hour in 2 tablespoons of the olive oil. Add the soaked oregano, along with any oil it has not absorbed, to the skillet with the tomatoes in Step 4.
—Michael McLaughlin

BRAISED ARTICHOKE HEARTS
4 Servings

6 ounces slab bacon, sliced ¼ inch thick and cut crosswise into ¼-inch strips
1 tablespoon olive oil
2 small carrots, thinly sliced
2 small celery ribs, thinly sliced
1 small onion, thinly sliced
5 garlic cloves, minced
16 small uncooked artichoke bottoms, with trimmed stems attached
1½ cups dry white wine
½ teaspoon fresh thyme or ¼ teaspoon dried
1 bay leaf, crumbled
¼ teaspoon salt
¼ teaspoon freshly ground pepper

1. In a large, deep, nonreactive skillet or flameproof casserole, sauté the bacon in the olive oil over moderately high heat until browned, 3 to 4 minutes.

2. Add the carrots, celery, onion and garlic, reduce the heat to moderate and continue cooking until the celery and onion are softened but not browned, 2 to 3 minutes.

3. Place the artichokes stem-up in the skillet. Add the wine, thyme, bay leaf, salt, pepper and enough water to barely cover the bases of the artichokes. Bring to a simmer, cover and cook over moderately low heat until the artichokes are tender enough to be pierced easily with a knife, about 20 minutes.

4. Serve the artichokes in plates with the bacon, accompanying vegetables and some of the cooking liquid.
—*John Robert Massie*

ARTICHOKE, CORN AND OYSTER GRATIN
❢ Mâcon Blanc, such as Joseph Drouhin
2 Servings

1 cup heavy cream
8 oysters, with their liquor reserved
½ teaspoon Dijon mustard
¼ teaspoon salt
Pinch of white pepper
2 teaspoons fresh lemon juice
2 large cooked artichoke bottoms, cut into 10 wedges each
6 tablespoons cooked fresh or frozen corn kernels
2 tablespoons unsalted butter
⅔ cup coarse fresh white bread crumbs
½ teaspoon hot red pepper flakes, finely chopped

1. Preheat the oven to 425°. In a heavy, medium saucepan, bring the cream and oyster liquor to a boil over moderately high heat. Cook until reduced by half, about 15 minutes. Remove from the heat.

2. Season the sauce with the mustard, salt, white pepper and lemon juice. Add the artichokes and corn and toss to coat with sauce. Spoon one-quarter of this mixture into each of two ½-cup ramekins. Put 4 oysters in each and cover with the remainder of the artichoke-corn mixture.

3. In a small skillet, melt the butter over moderate heat until sizzling. Add the bread crumbs and red pepper flakes and sauté until the crumbs have absorbed the butter, about 1 minute.

4. Sprinkle the buttered crumbs over the tops of the ramekins, dividing evenly. Place on a cookie sheet and bake in the middle of the oven for about 5 minutes, until the sauce is bubbling and the crumbs are browned.
—*John Robert Massie*

ASPARAGUS BUNDLES
4 Servings

2 sheets of phyllo dough, cut lengthwise in half
About ¼ cup clarified butter
20 thin asparagus tips, 3 inches long
4 scallions, trimmed to 3 inches
Salt and freshly ground pepper

1. Preheat the oven to 375°. Lightly brush both sides of half a sheet of phyllo with butter. Arrange 5 of the asparagus tips and 1 scallion in a bundle along one short edge of the dough; there will be a 1-inch margin on either side.

2. Season the asparagus with salt and pepper. Fold in the edges of the phyllo and roll up. Place the bundle on a cookie sheet and brush with butter. Repeat with the remaining ingredients to make 3 more bundles.

3. Bake for 12 to 15 minutes, until golden brown. Serve hot.
—*F&W*

RED PEPPER AND EGGPLANT TERRINE

Serve slices of this terrine garnished with shreds of fresh mozzarella tossed with minced basil, cracked black pepper and any remaining pepper juices. The terrine needs to be refrigerated overnight (although it can be made up to three days ahead), so plan accordingly.

8 Servings

Salt
2 medium eggplants (about 1 pound each), peeled and sliced crosswise ⅛ inch thick
8 medium red bell peppers—halved lengthwise, cored and deribbed
⅓ cup plus 3 tablespoons extra-virgin olive oil

1. Lightly salt the eggplant and place in a colander to drain for 30 minutes.

2. Pat the eggplant dry. Layer the slices between paper towels. Cover with a cookie sheet and weigh down with a heavy pot for 1 hour. Preheat the oven to 350°.

3. Meanwhile, place the peppers in a roasting pan in a single layer, cut-side down. Drizzle on ⅓ cup of the oil. Roast the peppers until the skins are wrinkled and loose, 30 to 40 minutes. Remove from the oven and let stand until cool enough to handle. Peel off the skins. Tip the pan and pour the oil and juices into a bowl; reserve.

4. Pour the remaining 3 tablespoons oil into a small bowl. Dip a paper towel into the oil and very lightly grease a large skillet. Add as many of the eggplant slices as will fit in a single layer and fry, over moderately high heat, turning once, until tender and lightly browned, 2 to 3 minutes on each side. Drain on paper towels. Keep the skillet lightly oiled by rubbing it with the oil-dampened towel between batches. It is important not to allow the eggplant to absorb too much oil or the terrine will not hold together. *(The recipe can be prepared to this point up to 2 days ahead. Refrigerate the peppers and eggplant separately.)*

5. To assemble the terrine, preheat the oven to 350°. Lightly coat a nonreactive 6-by-3-by-2-inch loaf pan with the reserved oil and pepper juices. Line the bottom and sides with slightly overlapping eggplant slices, pressing into the corners to form a rectangular shape. Cover with an even layer of red peppers, pressing into corners and cutting pieces to fit as needed. Continue layering eggplant and peppers until the mold is filled, ending with a layer of eggplant.

6. Bake the terrine for 30 minutes. Use a paper towel to blot any liquid that has accumulated on top of the terrine during baking. Cover the terrine with plastic wrap and weight the top with a loaf pan of the same size filled with pie weights or beans. Let cool to room temperature. If additional oil rises to the surface, blot it with a paper towel. Refrigerate the weighted terrine overnight or for up to 3 days.

7. Unmold the terrine. With a sharp thin knife, cut into 8 equal slices. (If slices become distorted, reform by pressing on the sides with a knife.) To serve, place a slice of the terrine on each of 8 plates.

—Anne Disrude

EGGPLANT AND MUSHROOMS WITH BALSAMIC GLAZE

▼ The sweet and sour elements in this dish would be united by a young Cabernet Sauvignon, such as Beringer Knights Valley, Silverado Vineyards or a vigorous one from the Médoc, like Château Léoville Barton.

4 Servings

2½ tablespoons extra-virgin olive oil
4 small Japanese eggplants (about 1½ pounds total), halved lengthwise
3½ tablespoons balsamic vinegar
Salt and freshly ground pepper
½ pound fresh shiitake mushrooms, stems discarded, caps quartered
1 tablespoon coarsely chopped flat-leaf parsley

1. In a large nonreactive skillet, heat 1 tablespoon of the oil over moderate heat until almost smoking. Reduce the heat to low. Add half of the eggplants, cut-sides down, cover and cook until browned and soft, about 4 minutes. Sprinkle in 1 tablespoon of the balsamic vinegar and boil to glaze the eggplants and evaporate the vinegar, about 10 seconds. Using a spatula, transfer the eggplant halves to a large serving platter, cut-sides up, and set aside. Repeat with another tablespoon of oil, the remaining eggplants and another tablespoon of vinegar. Season the eggplants with salt and pepper to taste.

2. Add the remaining ½ tablespoon olive oil to the skillet and when hot, add the mushrooms in an even layer. Cover and cook until wilted and browned, about 3 minutes. Uncover, sprinkle in the remaining 1½ tablespoons balsamic vinegar and cook to reduce slightly, about 10 seconds. There should be some

liquid remaining. Season the mushrooms with salt and pepper to taste and pour the mushrooms over the eggplants, sprinkle the parsley on top and serve.
—Marcia Kiesel

CARAMELIZED ONION WITH SALSA
6 Servings

1 pound plum tomatoes—peeled, seeded and coarsely chopped
4 medium scallions, trimmed to 6 inches and thinly sliced
1 jalapeño pepper, seeded and minced
2 tablespoons minced flat-leaf parsley
1½ teaspoons minced fresh coriander (cilantro)
3 tablespoons extra-virgin olive oil
4 teaspoons sherry wine vinegar
½ teaspoon salt
⅛ teaspoon cayenne pepper
6 large slices of Spanish onion, about ¾ inch thick and 3½ inches in diameter
Freshly ground black pepper
Fresh coriander (cilantro) leaves, for garnish

1. In a large bowl, combine the tomatoes, scallions, jalapeño pepper, parsley, minced coriander, 2 tablespoons of the olive oil, 2 teaspoons of the vinegar, the salt and cayenne. Let marinate at room temperature for 1 hour.
2. Meanwhile, in a large heavy skillet, heat the remaining 1 tablespoon oil over moderately high heat. Reduce the heat to moderate, add the whole onion slices in a single layer and cook, turning once and pressing down on them occasionally with a spatula, until they are deep golden brown and shiny and tender throughout, about 45 minutes.
3. Add the remaining 2 teaspoons vinegar and 1 tablespoon of water. Flip the onion slices twice, being careful to keep them whole, to coat well. Season with salt and black pepper to taste.
4. To serve, place one onion slice on each plate and top with ¼ cup salsa. Garnish with fresh coriander leaves.
—Anne Disrude

BRAISED PEAR, CELERY AND ENDIVE WITH PARMESAN AND BASIL
4 Servings

1 Bosc pear
Juice of ½ lemon
1 head of celery
4 small Belgian endives
1 tablespoon unsalted butter
1 cup rich chicken or veal stock, preferably homemade
½ teaspoon salt
¼ teaspoon freshly ground pepper
4 teaspoons freshly grated Parmesan cheese
2 teaspoons minced fresh basil

1. Preheat the oven to 350°. Peel, quarter and core the pear. Sprinkle with the lemon juice. Remove and discard the tough outer ribs of celery until you reach the light-green inner ribs, or heart. Trim the celery heart to 5 inches, core and cut lengthwise into quarters.
2. Place the celery heart and endive in a single layer in a large flameproof baking dish. Dot with the butter and pour in the stock. Bring to a boil over high heat. Loosely cover with parchment or wax paper and bake in the oven for 15 minutes. Turn the vegetables over and bake for another 10 to 15 minutes, until almost tender. Add the pear quarters and braise until tender, about 15 minutes longer.
3. Preheat the broiler. There should be only 3 to 4 tablespoons of liquid in the bottom of the baking dish. If there's more, boil over high heat to reduce.
4. Tip the baking dish and baste the vegetables with the reduced stock. Sprinkle with the salt, pepper and Parmesan cheese and broil 3 to 4 inches from the heat until the cheese begins to brown, 1 to 2 minutes.
5. Place a piece of celery heart, pear and endive on each of 4 warmed plates. Drizzle over any remaining juices. Sprinkle with the basil and serve.
—Anne Disrude

PAPILLOTE OF LEEKS WITH PARMESAN CHEESE
Leeks are a grand enough vegetable to be served as a first course.
1 Serving

½ tablespoon vegetable oil
2 teaspoons unsalted butter, softened
3 small or 2 medium leeks, trimmed to 6 inches of white and tender green
¼ cup heavy cream
1½ tablespoons freshly grated Parmesan cheese
2 sprigs of fresh thyme or ⅛ teaspoon dried
Salt and freshly ground pepper

1. Preheat the oven to 400°. Fold a 15-by-20-inch sheet of butcher's paper, parchment or aluminum foil in half crosswise to make a 15-by-10-inch rectangle. Using scissors, cut the rectangle into a heart shape with the fold running vertically down the center. Open up the heart and brush with the oil. Spread the softened butter over the middle of half the papillote.

FIRST COURSES

2. Cut the small leeks in half lengthwise and rinse well. (If using medium leeks, cut lengthwise into quarters.) Place on top of the butter, cut-sides up.

3. Pour the cream over the leeks and sprinkle with the Parmesan cheese. Remove some of the leaves from the thyme sprigs and sprinkle on top. Add the sprigs with the remaining leaves. Season lightly with salt and pepper.

4. Fold the paper over the leeks and beginning at the top of the heart, make a series of tight overlapping folds to seal the papillote.

5. Place the papillote on a cookie sheet and bake 20 minutes. Serve hot.
—Anne Disrude

ZUCCHINI RIBBONS WITH ARUGULA AND CREAMY GOAT CHEESE SAUCE
4 Servings

1¼ cups heavy cream
½ cup finely chopped onion
3 sprigs of fresh thyme plus ½ teaspoon minced fresh thyme (or a total of 1½ teaspoons dried thyme)
3 parsley stems
3 black peppercorns
2 medium garlic cloves, unpeeled and lightly crushed
3 ounces mild goat cheese
4 teaspoons balsamic vinegar
Salt and coarsely cracked pepper
16 lengthwise slices of zucchini cut ⅛ inch thick (from about 3 medium)
3 tablespoons extra-virgin olive oil
1 bunch of arugula, large stems removed

1. In a heavy medium saucepan, combine the cream, onion, thyme sprigs (or 1 teaspoon dried), parsley stems, peppercorns and garlic. Bring just to a simmer, cover and cook over low heat until the garlic is very soft, about 20 minutes. Strain into a small saucepan; discard the solids.

2. Place the saucepan over low heat and whisk in the goat cheese, 1 teaspoon of the vinegar and the minced thyme (or ½ teaspoon dried). Stir until the sauce is smooth. Season with salt and pepper to taste.

3. Meanwhile, heat a large heavy skillet over high heat. Brush the zucchini slices with the olive oil and place in the skillet. Cook until lightly browned, about 1 minute on each side. Remove and keep warm.

4. To assemble, toss the remaining 3 teaspoons vinegar with the arugula. Gather together 3 arugula leaves and place crosswise on one of the zucchini slices; fold the zucchini slice over the arugula. Repeat with the remaining zucchini and arugula. Arrange four zucchini bundles decoratively on each serving plate. Spoon one-fourth of the sauce over each serving and top with more cracked pepper.
—Anne Disrude

STUFFED WHITE MUSHROOMS IN PHYLLO
A garnish of fried carrot, leek and zucchini matchsticks adds a sweet crunch to Joachim Splichal's mushroom-stuffed mushrooms.
4 Servings

1 stick (4 ounces) plus 2 tablespoons unsalted butter
20 large white mushrooms, stemmed
6 ounces medium white mushrooms, finely chopped
4 ounces shiitake mushrooms, stemmed and finely chopped
1 small shallot, minced
1 small garlic clove, minced
1 tablespoon heavy cream
2 teaspoons finely chopped parsley
Salt and freshly ground pepper
10 sheets of phyllo dough
2 cups vegetable oil, for frying
1 carrot, cut into 2-by-¼-inch matchsticks
1 leek (white and tender green), cut into 2-by-¼-inch matchsticks
1 zucchini, cut into 2-by-¼-inch matchsticks

1. In a large heavy skillet, melt 1 tablespoon of the butter over high heat. Add the large mushroom caps, rounded-side down, and cook until browned, about 2 minutes. Add ½ cup of water, cover and cook until tender, about 4 minutes longer. Using a slotted spoon, remove the mushroom caps and set aside. Pour the cooking liquid into a small saucepan and set aside. You should have about ½ cup of liquid.

2. In a large heavy skillet, melt 1 tablespoon of the butter over high heat. When hot, add the chopped white and shiitake mushrooms, the shallot and garlic and cook, stirring constantly, until softened, about 3 minutes. Stir in the cream and 1 teaspoon of the parsley. Season to taste with salt and pepper. Spoon this mixture into the mushroom caps and set aside to cool completely.

3. In a small saucepan, melt 6 tablespoons of the butter over low heat and set aside. On a work surface, cut each sheet of phyllo dough into 4 rectangles of equal size. Cover all but 2 of the rectangles with a damp towel. Using a pastry brush, lightly brush the 2 rectangles with some of the melted butter and stack them to make one layer. Center a stuffed mushroom cap, stuffed-side up, on the phyllo rectangles and bring up the sides to enclose the mushroom. Pinch to seal and separate the top layers of dough like flower petals. Place on a baking sheet. Repeat with the remaining stuffed mushrooms, butter and phyllo dough.

4. Preheat the oven to 300°. Bake the mushrooms until the pastry is golden brown and the packages are cooked through, about 45 minutes.

5. Meanwhile, in a heavy skillet, heat the oil over moderate heat to 350°. Add the carrot matchsticks and fry until brown and crisp, about 2 minutes. Using a slotted spoon, transfer to paper towels to drain thoroughly. Fry the leek matchsticks until crisp, about 1 minute; transfer to paper towels. Next, fry the zucchini matchsticks until brown and crisp, 2 to 3 minutes. *(The vegetables can be fried up to 1 hour ahead; keep at room temperature.)*

6. Boil the reserved mushroom cooking liquid over high heat until reduced to 3 tablespoons, about 6 minutes. Remove from the heat and whisk in the remaining 2 tablespoons butter and 1 teaspoon parsley.

7. Place 5 stuffed mushrooms in the center of each large plate and spoon a generous tablespoon of the sauce around them. Sprinkle the mushrooms with the fried vegetables and serve immediately.
—Baba S. Khalsa

BRAISED WILD MUSHROOMS

At Joseph Phelps Vineyards, this dish is made with the same wine as the one to be served with the meal.
8 Servings

1 long, narrow loaf of French bread, cut on the diagonal into ¼-inch slices
1 stick (4 ounces) unsalted butter, softened to room temperature
3 large garlic cloves, minced
⅛ teaspoon crushed red pepper
1 tablespoon chopped fresh thyme or ¾ teaspoon dried
1 tablespoon chopped fresh marjoram or ¾ teaspoon dried
1 pound shiitake mushrooms, stems removed
½ pound chanterelles, stems removed
½ pound white tree mushrooms, stems removed
¼ teaspoon salt
¼ teaspoon freshly ground black pepper
1 cup dry white wine, preferably Chardonnay or Gewürztraminer
8 paper-thin slices of red onion, for garnish

1. Preheat the oven to 350°. Butter the bread with a total of 2 tablespoons of the butter. Set the slices on a baking sheet and bake until lightly browned, 15 to 20 minutes.

2. Meanwhile, in a large heavy skillet, melt the remaining 6 tablespoons butter over moderately high heat. Add the garlic, red pepper, thyme and marjoram. Sauté until the garlic is fragrant and softened, about 2 minutes.

3. Add all of the mushrooms and toss quickly to coat with the herbed butter. Increase the heat to high and sauté until the mushrooms begin to brown slightly, about 3 minutes. Season with the salt and black pepper and add the wine. Bring to a boil and cook until the liquid is reduced by half and the sauce is slightly thickened, about 6 minutes.

4. Divide the slices of toasted French bread among 8 plates and top with the mushrooms and sauce. Garnish with the red onion slices.
—Joseph Phelps Vineyards, St. Helena, California

WILD MUSHROOMS EN PAPILLOTE

Wild mushrooms are perfect for this type of cooking because their heady perfume is released in a burst of aroma at the table. The large papillote makes a dramatic presentation. Depending on the season, you can use cèpes, shiitakes or morels. Chanterelles give off too much liquid for this type of enclosed cooking.
❧ St-Emilion such as Château Pavie
4 Servings

1 pound fresh cèpes, shiitakes or morels
¼ cup olive oil
½ cup crème fraîche
¼ cup chopped fresh tarragon or 1 teaspoon dried
2 garlic cloves, minced
1 teaspoon salt
¼ teaspoon freshly ground pepper
1 tablespoon unsalted butter, melted

1. Preheat the oven to 400°. Rinse the mushrooms thoroughly and dry them on paper towels. If they are very large, quarter them; otherwise leave them whole.

2. In a large bowl, combine the oil, crème fraîche, tarragon, garlic, salt and pepper. Add the mushrooms and toss to coat.

3. Cut a 12-inch heart out of parchment paper or heavy-duty aluminum foil. Brush melted butter over the paper. Put the mushrooms on one half of the heart. Fold the other half over the morels and with a series of overlapping crimp folds, seal the papillote. Place on a baking sheet.

4. Bake the papillote in the center of the oven for 20 minutes. Serve at once, cutting the paper open at the table.
—Lydie Marshall

POTATO CASES WITH SHIITAKE AND MOREL FILLING
4 Servings

8 baking potatoes (10 ounces each), about 5 inches long and 2 inches wide
½ ounce dried morels
2 tablespoons unsalted butter, melted (preferably clarified)
3 tablespoons extra-virgin olive oil
3 garlic cloves, sliced
¾ pound fresh shiitake mushrooms—stemmed, caps cut crosswise into slices and then halved
¼ cup dry white wine
¼ cup chicken stock or canned broth
⅓ cup heavy cream
1 teaspoon fresh lemon juice
½ teaspoon salt
⅛ teaspoon freshly ground pepper
1 tablespoon minced parsley

1. With a paring knife, trim the potatoes to make straight-sided rectangular boxes (3 by 1½ by 1½ inches). Holding a paring knife vertically, cut completely around the inside of the potato box, leaving a shell ⅛ to ¼ inch thick. Insert the paring knife horizontally ⅛ to ¼ inch from the bottom of the box. Without enlarging the cut, work the knife back and forth in a swiveling motion to loosen the inside piece. If necessary, insert the paring knife in several different spots. Keep the potato cases submerged in cool water while you make the filling.

2. Preheat the oven to 450°. Soak the morels in 2 cups of hot water until softened, about 10 minutes. Remove the morels, squeezing gently. Strain the soaking liquid through a double layer of dampened cheesecloth into a small saucepan. Rinse the morels and trim. Chop coarsely and add to the saucepan. Boil gently until the liquid is completely absorbed, about 20 minutes. Set aside.

3. Meanwhile, pat the potato cases dry and brush completely with the butter. Place on a baking sheet and bake for 20 minutes, or until golden brown, turning the cases on a different side every 5 minutes.

4. In a large skillet, heat the oil. Add the garlic and cook over moderately high heat until light golden, about 1 minute. Add the morels and the shiitakes. Cook, stirring, until the shiitakes are softened, about 2 minutes.

5. Add the wine and stock and cook over moderately high heat until the liquid is slightly reduced, about 1 minute. Stir in the cream and boil until thickened, about 2 minutes. Add the lemon juice, salt and pepper. Keep warm.

6. To serve, season the potato cases lightly with salt. Stir the parsley into the filling. (If the filling is too thick, stir in 1 to 2 tablespoons additional cream, stock or water.) Place 2 potato cases on each serving plate. Spoon filling into each potato and serve.
—*Anne Disrude*

PEAS AND MUSHROOMS IN CROUSTADES WITH CHERVIL
These vegetable-filled croustades have a lovely chervil flavor. They make a good first course, but can also be served as an elegant light lunch.
4 Servings

5 tablespoons unsalted butter
8 thin slices of firm-textured white bread, crusts removed
¼ pound mushrooms, diced
3 tablespoons minced fresh chervil
½ pound fresh shelled or frozen young peas
¾ cup heavy cream
¼ cup crème fraîche or sour cream
1 teaspoon fresh lemon juice
¼ teaspoon salt
¼ teaspoon freshly ground pepper

1. Preheat the oven to 400°. Melt 4 tablespoons of the butter. Brush over both sides of the bread slices. Gently press the bread into eight 2½-inch muffin tins. Bake until golden brown and crisp, about 20 minutes. Carefully lift the croustades from the tin and set aside. (The croustades can be made up to 5 hours ahead. Store at room temperature in an airtight tin.)

2. In a medium saucepan, melt the remaining 1 tablespoon butter over high heat. Add the mushrooms and sauté without stirring, until brown, about 2 minutes. Reduce the heat to moderate and cook until tender, about 1 minute longer.

3. Remove the mushrooms to a small bowl and toss with 1 tablespoon of the chervil. Set aside for at least 10 minutes and up to 1 hour.

4. In a medium saucepan with a steamer basket, steam the peas until tender, 4 to 8 minutes depending on their size.

5. Remove the basket and discard the cooking water. Return the peas to the saucepan. Add the heavy cream and boil over high heat, stirring, until the cream is thick and reduced by half, about 3 minutes. Add the mushrooms and crème fraîche and stir to combine.

6. Remove from the heat, season with the lemon juice, salt, pepper and remaining 2 tablespoons chervil. Divide the warm mixture evenly among the croustades and serve at once.
—*Marcia Kiesel*

White Lima Beans with Sorrel and Parsley (p. 184).

Salad of Flageolet Beans with Fennel and Walnuts (p. 38).

Chickpea Stew with Greens and Spices (p. 102).

Chilled Minted Pea Soup (p. 61).

ROOT VEGETABLE CAKES

At Patina in Los Angeles, Joachim Splichal serves these "cakes" of finely diced root vegetables with an unusual thyme-scented sauce made from reduced potato cooking liquid, potato puree and olive oil.

4 Servings

¼ *of a small celery root, peeled and cut into ¼-inch dice (1 cup)*
1 medium parsnip, peeled and cut into ¼-inch dice (1 cup)
¼ *of a small rutabaga, peeled and cut into ¼-inch dice (1 cup)*
1 medium kohlrabi, peeled and cut into ¼-inch dice (1 cup)
2 large red cabbage leaves
2 large savoy cabbage leaves
2 tablespoons unsalted butter
1 large beet, peeled and cut into ¼-inch dice (1 cup)
Salt and freshly ground pepper
1 pound Idaho potatoes
2 tablespoons olive oil
1 teaspoon chopped fresh thyme

1. Bring a large saucepan of salted water to a boil over high heat. Add the celery root, parsnip, rutabaga and kohlrabi; blanch until almost tender, 2 to 3 minutes. Using a slotted spoon, transfer the vegetables to a colander and refresh under cold running water. Drain well.

2. Bring the water back to a boil and add the red and savoy cabbage leaves. Cook until tender, about 4 minutes. Refresh under cold water and drain thoroughly. Transfer to a clean kitchen towel and pat dry. Using a 2-inch round cookie cutter, cut out 4 neat circles from the red cabbage leaves; repeat with the savoy cabbage leaves. Cut the remaining cabbage into thin julienne strips.

3. In a large skillet, melt the butter over moderately high heat. Add the reserved blanched root vegetables, the beet and julienned cabbage and cook, stirring, until just tender, about 5 minutes. Season with salt and pepper to taste.

4. Line the bottom of 4 straight-sided ½-cup ramekins with a savoy cabbage circle. Fill the ramekins with the diced root vegetables, packing them tightly. Top each ramekin with a red cabbage circle.

5. Peel and quarter the potatoes. In a medium saucepan, cover the potatoes with 4 cups of water; bring to a boil over moderately high heat. Cook until the potatoes are tender, about 20 minutes. Using a slotted spoon, transfer the potatoes to a plate. Reserve the cooking water; you should have about 1½ cups.

6. Mash one-fourth of the cooked potatoes in a medium bowl; discard the remaining potatoes. Boil the potato cooking liquid over moderately high heat until reduced to 1 cup, about 5 minutes. Whisk in the mashed potatoes and the olive oil. The sauce should be the consistency of pancake batter. Set aside.

7. Preheat the oven to 350°. Place the ramekins in a small roasting pan. Pour enough hot water into the pan to reach halfway up the sides of the ramekins. Cover with foil and bake until the vegetables are heated through and the beets are tender, 25 to 30 minutes.

8. To serve, reheat the potato sauce over moderate heat. Season with salt and pepper to taste and add the thyme. Invert the vegetable cakes onto serving plates and ladle the warm potato sauce around them.
—*Baba S. Khalsa*

NOODLE GALETTES WITH BASIL WATER AND TOMATOES

Charlie Trotter's rectangles of sautéed capellini are served floating in a basil-infused vegetable broth. The pasta is cooked the day before and refrigerated overnight before being cut into compact noodle galettes.

6 Servings

¼ *pound capellini*
2 tablespoons plus 1½ teaspoons pure olive oil
Salt and freshly ground pepper
½ *pound celery root, peeled and coarsely chopped*
½ *pound leeks, white and tender green portions, coarsely chopped*
1 large onion, chopped
4 celery ribs, chopped
2 parsnips, chopped
2 large bunches of basil (about ½ pound) plus 1 tablespoon finely shredded basil leaves
1 medium red tomato—peeled, seeded and diced
1 medium yellow tomato—peeled, seeded and diced
2 tablespoons extra-virgin olive oil

1. Bring a large pot of salted water to a boil. Add the pasta, bring back to a boil and cook, stirring, until al dente, about 4 minutes. Drain in a colander and toss with 2 tablespoons of the pure olive oil. Season with salt and pepper. In a baking pan, shape the capellini into a 10-by-9-inch rectangle about ⅓-inch thick. Cover and refrigerate overnight.

2. In a large saucepan, combine the celery root, leeks, onion, celery and parsnips and 3 quarts of water. Simmer over low heat for 2 hours. Strain the broth into a clean saucepan and discard the solids. Boil the broth over high heat, skimming occasionally, until reduced to 2 cups, about 20 minutes. Add the basil bunches and remove from the heat.

Cover and let steep for 2 minutes. Strain and set aside.

3. In a medium bowl, combine the red and yellow tomato with the extra-virgin olive oil and shredded basil. Set aside.

4. Cut the chilled capellini into six 5-by-3-inch rectangles. In a large nonstick skillet, heat the remaining 1½ teaspoons pure olive oil over moderately high heat. Add the noodle galettes and fry until lightly browned and crisp, about 5 minutes per side. Transfer to paper towels to drain. Meanwhile, reheat the basil water, if necessary.

5. To serve, place a noodle galette in each of 6 large, shallow soup bowls. Place a spoonful of the tomato mixture on top of each galette and pour the hot basil water into each bowl. Serve hot.
—*Baba S. Khalsa*

JALAPENO RISOTTO WITH SONOMA DRY JACK CHEESE

This risotto, made with fresh hot chiles, Italian arborio rice and California Dry Jack cheese, isn't from the menu of any trendy hot spot, but it might as well be. It is an amalgam of two cuisines that says everything there is to say about the assimilation of southwestern ingredients. If you cannot get the cheese called for, a mixture of equal parts of grated Monterey Jack and Parmesan can be substituted—not the same, but savory nonetheless.
6 Servings

6 to 7 cups unsalted chicken stock or 5 cups canned broth diluted with 2 cups water
1 stick (4 ounces) unsalted butter
1 cup minced onion
6 medium jalapeño peppers, seeded and minced
1 garlic clove, minced
1½ cups arborio rice
1 cup grated Vella Sonoma Dry Jack cheese (about 4 ounces)

1. In a heavy medium saucepan, bring the stock to a boil over high heat. Remove from the heat and keep warm.

2. In a large heavy saucepan, melt the butter over moderately low heat. Add the onion, jalapeños and garlic and cook, stirring occasionally, until softened, 6 to 8 minutes. Add the rice and stir to coat well with butter. Stir in 1 cup of the hot stock and cook, stirring, until the liquid is absorbed, 10 to 12 minutes.

3. Continue to cook the risotto, adding the hot stock, ½ cup at a time, and stirring until it is absorbed and the grains are just tender but still firm to the bite, 30 to 40 minutes.

4. Stir ⅓ cup of the cheese into the risotto, cover and let it stand for 3 minutes. Serve on plates and pass the remaining cheese and a pepper mill separately.
—*Michael McLaughlin*

AMARONE RISOTTO WITH PANCETTA AND PRUNES

The prunes in this risotto underscore the wine used in cooking it, as Amarone can taste somewhat pruny.
❢ In this case, the wine that's in the recipe doesn't go well with the completed dish. The prunes make this sweet-and-sour risotto a little too sweet for a glass of Amarone. Try a California Zinfandel bursting with sweet fruit, such as a Ridge Howell Mountain or a Sutter Home California.
6 Servings

2 cups plus 2 tablespoons rich, young Amarone
1½ cups chicken stock or low-sodium canned broth
2 tablespoons unsalted butter
1 tablespoon olive oil
2 ounces thinly sliced pancetta, shredded
2 garlic cloves, minced
¼ pound (scant 1 cup) arborio rice
½ cup pitted prunes (about 10), thickly sliced
¼ cup freshly grated Parmesan cheese
Salt and freshly ground pepper

1. In a medium nonreactive saucepan, combine 2 cups of the Amarone with the chicken stock and ⅔ cup of water. Bring just to a simmer over moderate heat; do not boil. Reduce the heat to low to keep at a bare simmer.

2. Meanwhile, in a large, heavy nonreactive saucepan or casserole, melt 1 tablespoon of the butter in the olive oil over moderately high heat. Add the pancetta and cook until it begins to brown, 2 to 3 minutes. Reduce the heat to moderate, add the garlic and cook until soft but not brown, 1 minute.

3. Add the rice and stir well to coat with the fat. Add ½ cup of the simmering liquid and cook, stirring, until completely absorbed by the rice, about 2 minutes. Continue adding more liquid, ½ cup at a time, as it is absorbed by the rice. When you have used about half of the liquid, add the prunes. Add the remaining liquid, in smaller amounts, stirring constantly, until the mixture is creamy but not soupy and the rice is tender but still firm, 25 to 30 minutes (there may be some liquid left over).

4. Stir in the Parmesan cheese and the remaining 1 tablespoon butter. Stir in the remaining 2 tablespoons Amarone. Cook for 1 minute. Season to taste with salt and pepper. Pass extra grated Parmesan cheese on the side if desired.
—*David Rosengarten*

3 SOUPS & CHOWDERS

GERMAN SUMMER SALAD SOUP

This tasty mélange of fresh vegetables—half soup, half salad—is a sort of Teutonic cousin of Spanish gazpacho. Best served the same day it's made, it can be ladled into shallow, individual bowls to be eaten with a fork and spoon, or into chilled mugs to be eaten with a spoon alone. The guiding rule here is the colder the better.

8 Servings

½ cup thinly sliced scallions
4 medium tomatoes, diced
1 large green bell pepper, diced
1 cup thinly sliced radishes, cut into half-rounds
1 cup diced celery heart, with some of the leaves
3 cups finely shredded romaine or iceberg lettuce, trimmed into 2-inch lengths
1½ teaspoons salt
1 teaspoon freshly ground black pepper
4 teaspoons sugar
¼ cup cider vinegar
1 medium cucumber, seeded and diced

1. Combine the scallions, tomatoes, bell pepper, radishes, and celery heart in a large bowl. Stir in the lettuce shreds.

2. In a small bowl, combine the salt, black pepper, sugar, vinegar and 1 cup of water. Stir to dissolve the salt.

3. Pour the seasoned liquid over the vegetables. Stir gently until the vegetables are well coated. Cover the soup and refrigerate it until very cold, at least 4 hours.

4. To serve: Add the diced cucumber just before serving. Serve cold in shallow soup bowls.

—F&W

SPICY POTATO-CUCUMBER SOUP WITH DILL

This refreshing, cool soup tastes even better when chilled overnight and served the next day with a simple tossed salad and a loaf of good bread. If the soup is too thick, thin it with a little chicken stock.

4 to 6 Servings

2½ tablespoons olive oil
1 medium onion, chopped
2 garlic cloves, minced
¼ teaspoon cayenne pepper, or more to taste
1½ pounds baking potatoes, peeled and diced
1 medium cucumber—peeled, seeded and diced
2 cups chicken stock or canned broth
1 cup dry white wine
1½ teaspoons cumin seeds
¼ teaspoon salt
½ teaspoon freshly ground black pepper
½ cup plain yogurt
1 cup milk
3 tablespoons chopped fresh dill
1 large tomato—peeled, seeded and chopped

1. In a large saucepan, heat the olive oil over moderate heat. Add the onion, garlic and cayenne and reduce the heat to low. Cook until the onion is soft and golden, 10 to 15 minutes.

2. Add the potatoes, cucumber, chicken stock, wine and 1 cup of water. Bring to a boil over high heat. Reduce the heat to low and simmer until the potatoes are tender, about 20 minutes.

3. Meanwhile, put the cumin seeds in a small skillet. Toast over moderately high heat, shaking the pan, until fragrant and dark, about 1 minute. Finely chop the seeds.

4. Puree the soup in a food processor, in batches if necessary, just until smooth. Transfer to a large bowl, add the salt and black pepper and let cool to room temperature. Whisk in the yogurt, milk, dill, toasted cumin seeds and tomato. Cover the soup with plastic wrap and store in the refrigerator; serve cold.

—Marcia Kiesel

SWEET POTATO VICHYSSOISE

I think you will like this takeoff on a traditional favorite. Although I think this soup is best served chilled, some of my friends also like it warm.

4 to 6 Servings

1¼ pounds sweet potatoes
1 cup (lightly packed) sliced scallions (white and 2 inches of the green)
2½ cups chicken stock or canned broth
Salt and freshly ground white pepper
¼ cup heavy cream
1 tablespoon chopped chives

1. Preheat the oven to 400°. Prick the sweet potatoes several times with a fork and bake for about 1 hour, until soft. Let cool; then scoop out the potato pulp from the skins.

2. Meanwhile, in a medium saucepan, combine the scallions with 1 cup of the chicken stock. Simmer over moderate heat, stirring occasionally, until the scallions are tender, about 10 minutes.

3. Pour the scallion-stock mixture into a food processor. Add the sweet potato pulp and ½ cup more of the chicken stock. Puree until smooth, about 30 seconds.

4. Return to the saucepan and stir in the remaining 1 cup stock. Bring to a boil over moderate heat, reduce the heat to low and simmer for 5 minutes, to al-

low the flavors to blend. Season with salt and white pepper to taste. Let the soup cool to room temperature, then cover with plastic wrap. Refrigerate until chilled, about 4 hours.

5. Before serving, beat the cream in a medium bowl with a whisk until slightly thickened. Pour the soup into chilled bowls and swirl into each a tablespoon of the thickened cream. Garnish with the chives.
—*Lee Bailey*

CORN SOUP WITH
SPICY PUMPKIN
SEEDS
Based on traditional corn chowder, this soup is thickened with pureed corn and potatoes instead of the usual flour or egg yolks. The spicy pumpkin seeds are more than a garnish; they are an intense final flavoring. The light cool cream added just before serving helps to mellow the fiery pumpkin seeds.
♥ California white Zinfandel, such as Beringer
4 to 6 Servings

1 tablespoon olive oil
1 medium onion, chopped
1 garlic clove, minced
1 small jalapeño pepper, seeded and minced, or 1 small dried hot red pepper, seeded and crushed
¼ teaspoon cumin
1 medium baking potato, peeled and cut into ¼-inch dice
3 cups corn kernels, fresh or frozen (see Note)
1 cup light cream or half-and-half
½ teaspoon salt
½ teaspoon freshly ground black pepper
Spicy Pumpkin Seeds (p. 16)

1. In a large saucepan or flameproof casserole, heat the olive oil over moderate heat. Add the onion, garlic, jalapeño pepper and cumin. Reduce the heat to low and cook until the onion is softened but not browned, about 5 minutes.

2. Add the potato and 4 cups of water. Bring to a boil over high heat. Add the corn, reduce the heat to low and simmer until the potato is tender, about 20 minutes.

3. Measure out 3 cups of the soup and puree in a food processor. Combine the pureed soup with the rest of the soup in a large serving bowl and let cool to room temperature.

4. Just before serving, stir in the light cream and season with the salt and pepper. Ladle the cooled soup into bowls and sprinkle each serving with some of the Spicy Pumpkin Seeds. Pass the remaining pumpkin seeds on the side.

NOTE: If you use fresh corn for this, reserve the corn cobs and throw them into the soup in Step 2 for added flavor. Remove and discard them at the end of Step 2.
—*Marcia Kiesel*

CHILLED MINTED
PEA SOUP
On the hottest day of the year, I can't resist this cool, beautiful soup.
4 Servings

2 tablespoons unsalted butter
1 small onion, finely chopped
1 head of Bibb lettuce, coarsely shredded
3 cups tiny tender fresh peas (about 3 pounds in the pod) or 2 packages (10 ounces each) frozen, thawed
1½ cups chicken stock or canned broth
½ teaspoon sugar
¼ teaspoon salt
½ cup heavy cream
1 tablespoon finely chopped fresh mint
⅛ teaspoon freshly ground pepper

1. In a large heavy saucepan or flameproof casserole, melt the butter over moderate heat. Add the onion and cook, stirring, until softened but not browned, 8 to 10 minutes. Add the lettuce; cook, stirring, until wilted, about 1 minute.

2. Add the peas, chicken stock, sugar, salt and 1½ cups of water. Bring to a boil over high heat, reduce the heat to moderate, cover and simmer until the peas are very tender, 8 to 10 minutes for fresh peas, 3 to 5 minutes for frozen.

3. Strain the soup. In a food processor, puree the solids with 1 cup of the cooking liquid. Pass this puree through a sieve into a large bowl. Gradually stir the rest of the liquid into the puree, mixing well.

4. Add the cream, mint, pepper and additional salt to taste. Let cool to room temperature. Cover and refrigerate until well chilled, at least 6 hours, or overnight. Stir and adjust the seasoning, if necessary, before serving.
—*Leslie Newman*

COOL ASPARAGUS
SOUP
This creamy chilled soup makes a wonderful starter for a lazy Sunday lunch.
6 Servings

1½ pounds asparagus
4 tablespoons unsalted butter
1 small onion, chopped
2 medium leeks, chopped
1 medium celery rib, chopped
1 large baking potato, peeled and cut into ½-inch dice
3½ cups chicken stock or canned broth
1 teaspoon fresh lemon juice
Salt
¾ teaspoon freshly ground white pepper
½ cup half-and-half or light cream

¼ cup plus 2 tablespoons crème
 fraîche or sour cream
Paprika, for garnish

1. Cut off the asparagus tips and set aside. Coarsely chop the stems.

2. In a large skillet, melt the butter over moderate heat. Add the asparagus tips, onion, leeks, celery and potato. Cover, reduce the heat to low and cook until soft, about 25 minutes.

3. Meanwhile, in a medium saucepan, bring the stock to a boil over high heat. Add the asparagus stems, cover and cook over low heat until soft, about 25 minutes. Strain into a large bowl pressing on the stalks to extract as much liquid as possible. Discard the stems.

4. Put the sautéed vegetables in a food processor and puree until smooth, about 30 seconds. Whisk the vegetable puree into the strained stock. Season with the lemon juice, salt to taste and the pepper. Cover with plastic wrap and refrigerate until chilled, about 1 hour.

5. When ready to serve, stir in the half-and-half; mix well. Top each soup bowl with 1 tablespoon of the crème fraîche and garnish each with a dash of paprika.
—Lee Bailey

AVOCADO SOUP WITH PAPAYA-PEPPER RELISH
8 to 10 Servings

2 teaspoons unsalted butter
1 small onion, minced
1 carrot, minced
2 small celery ribs, minced
1 garlic clove, minced
4 cups chicken stock or
 canned broth
1 cup heavy cream, chilled
1 small red bell pepper
1 poblano pepper
2 large avocados, preferably Hass,
 cut into ¼-inch dice
2 tablespoons fresh lime juice
5 tablespoons minced fresh
 coriander (cilantro)
¼ teaspoon salt
¼ teaspoon freshly ground
 black pepper
1 small papaya—peeled, seeded and
 cut into ¼-inch dice
1 teaspoon walnut oil
3 jalapeño peppers, seeded and
 minced

1. In a large saucepan or flameproof casserole, melt the butter over moderate heat. Add the onion, carrot, celery and garlic and cook, stirring, until softened but not browned, about 6 minutes. Add the chicken stock and bring to a boil. Remove from the heat and let cool to room temperature. Stir in the cream and refrigerate until well chilled, about 2 hours or overnight.

2. Roast the red bell pepper and poblano over a gas flame or under the broiler, turning, until charred all over. Put the peppers in a paper bag and let steam for 10 minutes. Peel and seed the peppers and discard the cores. Cut the peppers into ¼-inch dice.

3. Whisk the diced avocados into the chilled soup until they begin to break up and thicken it slightly. Stir in 1 tablespoon of the lime juice, 2½ tablespoons of the coriander, ⅛ teaspoon of the salt and ⅛ teaspoon of the black pepper. Refrigerate until the flavors are well blended.

4. In a small bowl, combine the papaya and the roasted red pepper and poblano with the remaining 1 tablespoon lime juice and 2½ tablespoons coriander. Stir in the walnut oil and the remaining ⅛ teaspoon each of salt and black pepper. Serve the soup chilled, with a dollop of the papaya relish and a sprinkling of the minced jalapeños.
—Robert Del Grande

CHILLED CREAM OF SUMMER SQUASH SOUP
6 to 8 Servings

3 medium zucchini
2 medium yellow summer squash
3 tablespoons plus 1 teaspoon
 coarse (kosher) salt
2 tablespoons olive oil
2 tablespoons unsalted butter
2 garlic cloves, minced
2 medium onions, coarsely chopped
¼ teaspoon freshly ground
 black pepper
Pinch of cayenne pepper
2 tablespoons chopped parsley
½ cup sliced mushrooms
4 cups chicken stock or low-sodium
 canned broth
⅓ cup plus ½ cup plain yogurt
1 tablespoon plus 1 teaspoon fresh
 lime juice
4 plum tomatoes—peeled, halved
 and seeded
1 teaspoon sugar
2 tablespoons coarsely chopped
 basil

1. Using a coarse grater, grate the zucchini and yellow squash. Alternate layers of grated squash and the 3 tablespoons coarse salt in a colander or large strainer set over a bowl. Allow the squash to drain for 15 to 30 minutes (this removes excess moisture).

2. Meanwhile, in a large flameproof casserole, heat the oil and butter. Add the garlic and onions and sauté over high heat, stirring, until the onions are softened, about 5 minutes.

3. One handful at a time, squeeze the moisture from the grated squash and add each handful in turn to the casserole. Add the black pepper, cayenne and parsley. Toss the mixture over high heat for a couple of minutes to evaporate excess juices.

4. Add the mushrooms and toss over high heat for a minute. Add the chicken stock and bring to a boil. Cover the pot, lower the heat, and simmer until the vegetables are tender, about 10 minutes.

5. In a food processor, puree the soup in batches. Add ⅓ cup of the yogurt to the last batch. (For a smoother soup, strain the puree through a fine-mesh sieve.) Stir in 1 teaspoon of the lime juice and chill thoroughly.

6. Sprinkle the seeded plum tomato halves with the remaining 1 teaspoon salt, the sugar and remaining 1 tablespoon lime juice. Place on a rack or in a strainer and leave for about 10 minutes. Squeeze gently to remove as much liquid as possible and then coarsely chop.

7. To serve, top each bowl of soup with a dollop of the remaining ½ cup yogurt. Top with a large spoonful of the chopped marinated tomato and sprinkle with the chopped fresh basil.
—F&W

SPICED TOMATO SOUP

This soup is made with a food mill, which allows you to cook the tomatoes with their skins, as the food mill will remove them later. If, however, you would like to puree the soup in a food processor, the tomatoes should be peeled first. To do this, plunge them into boiling water for 1 minute, then rinse under cold water until cool enough to handle.
8 to 10 Servings

4 tablespoons unsalted butter
2 tablespoons olive oil
4 medium onions, sliced
3 medium leeks (white and tender green), sliced
3 garlic cloves, minced
2 carrots, sliced
2 celery ribs, with leaves, sliced
2 teaspoons coarse (kosher) salt
Freshly ground black pepper
⅛ teaspoon cayenne pepper
½ teaspoon chopped fresh tarragon or basil, or ¼ teaspoon dried tarragon or basil
3 sprigs of parsley, coarsely chopped
4 pounds tomatoes
½ teaspoon sugar
Small pinch of freshly grated nutmeg
½ teaspoon cumin
¼ teaspoon ground coriander (optional)
1 tablespoon tomato paste
5 cups chicken stock or canned broth
Plain yogurt and chopped fresh coriander (cilantro), for serving

1. In a large saucepan, heat the butter and oil until sizzling. Add the onions and toss to coat. Cover the saucepan and "sweat" the onions over medium heat, tossing occasionally, until slightly softened, 5 to 6 minutes.

2. Add the leeks and garlic and toss to coat. Cover and cook for about 3 minutes. Add the carrots, celery, salt, black pepper, cayenne, tarragon and parsley. Cover and cook until the vegetables are slightly softened, about 5 minutes.

3. Meanwhile, prepare the tomatoes. Core the tomatoes, cut them in half crosswise and squeeze out as many seeds as you can. Cut each tomato half in two.

4. Add the tomato quarters, sugar, nutmeg, cumin, ground coriander and tomato paste to the saucepan. Bring the mixture, uncovered, to a boil; then lower the heat to a simmer and cook, stirring to break up the tomatoes, until the vegetables are beginning to cook down and thicken, 10 to 12 minutes.

5. Add the chicken stock and bring to a boil over moderately high heat. Reduce the heat to low and simmer, partially covered, for 15 minutes.

6. Pass the soup through the medium disk of a food mill (if using a food processor, process to a coarse puree). Return the soup to the pan, taste and add more salt, black pepper and cayenne to taste.

7. Serve the soup hot with a dollop of yogurt and a sprinkling of coriander.
—F&W

BUTTERNUT SQUASH CONSOMMÉ WITH LEEK RAVIOLI

Tom Colicchio of Mondrian in Manhattan makes a stock with squash and onion, which he refrigerates overnight and skims of fat before clarifying with egg white and leek. The clear consommé is served garnished with creamy leek ravioli.
4 Servings

2 medium butternut squash (about 3½ pounds total)
2 tablespoons unsalted butter
1 medium onion, halved and thinly sliced
4 medium leeks
2 egg whites
2 teaspoons vegetable oil
⅓ cup heavy cream
Pinch of cinnamon
Pinch of ground cardamom
Salt and freshly ground pepper
Twelve 3-inch round gyoza or wonton skins*
*Available at Asian markets

1. Slice off the narrow neck portion of both squash, about the top 5 inches. Halve the neck portions crosswise, peel and cut into 1-inch cubes. Set all but 2 of the cubes aside. Cut the 2 reserved cubes into ¼-inch dice for garnish and set aside separately.

2. Trim, peel and coarsely chop the remaining bulbous portions of the squash; set aside.

3. In a medium saucepan, heat 1 tablespoon of the butter over low heat. Add the onion and cook, stirring occa-

sionally, until translucent, about 10 minutes; set aside.

4. In a large saucepan, heat the remaining 1 tablespoon butter over moderate heat. Add the large squash cubes and cook, stirring occasionally, until lightly browned, about 5 minutes. Stir in the onion and add 8 cups of water. Bring just to a simmer, carefully ladling the scum from the surface of the liquid. Reduce the heat to low and cook for 1 hour. Strain the stock into a large bowl and discard the solids; you should have about 4 cups. Let cool, cover and refrigerate overnight.

5. Using a slotted spoon, remove the solidified fat from the surface of the stock. Transfer the stock to a medium saucepan and set aside.

6. In a food processor, finely chop 1 of the leeks. Add the reserved coarsely chopped squash and pulse until finely chopped. Add the egg whites and pulse to mix. Whisk this mixture into the cold stock in the saucepan and bring to a simmer over low heat, whisking occasionally. Once the egg whites begin to coagulate on the surface of the stock, stop whisking. Simmer the stock very gently for 20 minutes from the time it starts to boil lightly. Carefully strain the consommé through fine-mesh cheesecloth or a clean piece of muslin into another large saucepan. You should have about 2¼ cups.

7. While the stock is being clarified, cut the 3 remaining leeks into ¼-inch dice. Reserve 2 tablespoons for garnish. In a large skillet, heat the oil over low heat. Add the diced leeks, cover and cook, stirring occasionally, until tender, about 10 minutes. Add the cream and cook until thickened, about 4 minutes. Set aside to cool. Add the cinnamon and cardamom; season to taste with salt and pepper.

8. Lay out the gyoza skins on a work surface. Moisten the edges with water and put 1 teaspoon of the leek filling in the center of each. Fold the skins over the filling to make half circles; press the edges to seal.

9. In a large saucepan of boiling water, blanch the reserved diced squash and leek garnishes over high heat until tender, about 3 minutes. Using a slotted spoon, transfer to a small bowl. Reduce the heat to moderately high and add the ravioli. Cook, stirring gently, until tender, about 5 minutes. Drain well.

10. Meanwhile, gently reheat the consommé over moderate heat; do not let it boil. Season to taste with salt. To serve, place 3 ravioli in each of 4 large, shallow soup bowls. Ladle the hot consommé into the bowls and garnish each serving with the blanched diced leek and squash.
—Baba S. Khalsa

ASPARAGUS SOUP
Here's a good use for the tough ends and peelings removed from asparagus spears. Save them in the freezer until you have enough.
4 Servings

6 cups tough asparagus ends (from about 4 pounds of asparagus), cut into 1-inch pieces
1 teaspoon salt
6 cups asparagus peelings
¼ teaspoon freshly ground white pepper
2 tablespoons heavy cream (optional)
2 tablespoons unsalted butter (optional)
8 cooked asparagus spears, cut into small pieces, for garnish

1. In a large saucepan or stockpot, place the asparagus ends and enough water to half cover them. Add ½ teaspoon of the salt and bring to a boil over high heat. Reduce the heat and simmer, uncovered, until the asparagus is very soft, 15 to 20 minutes.

2. Add the peelings and cook, stirring, until the peelings are limp and have colored the liquid bright green, 3 to 5 minutes.

3. In a blender or food processor, puree the stems and peelings with their cooking liquid, in batches if necessary. Press through a fine sieve to remove stringy or coarse pieces. Return the soup to the saucepan. If the soup is too thick, thin it with a little water.

4. Season with the remaining ½ teaspoon salt and the pepper. For a richer soup, stir in the cream and/or the butter. Garnish each serving with the asparagus spears.
—F&W

RED CABBAGE SOUP
The shredding, slicing, dicing and mincing of various ingredients provides texture and contributes flavor to this slightly pungent, savory soup.
6 Servings

4 tablespoons unsalted butter
1 large red onion, very thinly sliced
1 large garlic clove, minced
1 teaspoon sugar
1 tablespoon all-purpose flour
6 cups shredded red cabbage (1½- to 2-pound head)
¼ teaspoon thyme
6 cups beef broth
¼ cup dry red wine
2 tablespoons red wine vinegar
¾ teaspoon salt
¼ teaspoon freshly ground white pepper
½ pound kielbasa (Polish sausage), sliced ¼ inch thick
½ cup sour cream
1 tart apple—peeled, cored and coarsely shredded

1. In a large saucepan, melt the butter over low heat. Add the onion and sauté for 5 minutes. Stir in the garlic, cover and cook gently for 3 minutes.

2. Stir in the sugar and flour and cook for 1 minute. Add the cabbage, thyme, broth, wine, 1 tablespoon of the vinegar, the salt and pepper; cook over moderate heat, stirring occasionally, for 2 minutes.

3. Increase the heat slightly and bring the soup to a boil. Reduce the heat so that the soup is simmering, and cook for 20 minutes.

4. Meanwhile, sauté the kielbasa in a heavy skillet over medium heat for about 5 minutes, or until cooked through and crisp.

5. Stir the remaining 1 tablespoon vinegar into the soup and remove from the heat. Ladle into six shallow bowls and garnish each with five or six slices of kielbasa, a dollop of sour cream and some of the shredded apple.
—F&W

POTATO SOUP WITH GREENS AND CRISP POTATO SKIN CROUTONS
8 to 10 Servings

3 Russet or other baking potatoes
¼ pound smoked slab bacon with rind
1 medium onion, coarsely chopped
2 garlic cloves, coarsely chopped
½ pound kale, large stems removed, coarsely chopped (about 8 cups)
2 cups chicken stock or canned broth
3 tablespoons olive oil
1 teaspoon salt
¼ teaspoon freshly ground pepper
1 bunch of watercress, coarsely chopped
2 tablespoons unsalted butter (optional)

1. Preheat the oven to 400°. Pierce the potatoes several times with a fork. Bake for about 50 minutes, or until tender when pierced with a fork. Leave the oven on.

2. Meanwhile, remove the rind from the bacon in one piece and reserve. Slice the bacon into ¼-inch-thick slices and then slice crosswise into ¼-inch-wide matchsticks. In a stockpot, fry the bacon over moderate heat until browned. Remove with a slotted spoon and drain on paper towels; reserve the bacon for garnish.

3. Pour off all but 2 tablespoons of the bacon fat. Add the onion and garlic to the pan. Cover and cook over low heat until softened but not browned, about 5 minutes.

4. Add the kale, chicken stock, reserved bacon rind and 4 cups of water. Bring to a simmer and cook, partially covered, for 1 hour. Remove and discard the bacon rind.

5. When the potatoes are done, cut them in half and scoop out the insides; set aside. Halve each potato skin lengthwise. Brush liberally on both sides with the oil. Place on a baking sheet and bake at 400°, turning once, for 15 minutes, until browned and crisp. Remove from the oven and sprinkle with ½ teaspoon salt. Chop into bite-size croutons.

6. Mash half of the reserved potato pulp and stir into the soup. Break the remainder into small pieces and drop into the soup. Simmer until warmed through, about 10 minutes. Season with the remaining ½ teaspoon salt and the pepper.

7. Stir the watercress into the soup. Swirl in the butter. Ladle the soup into individual soup bowls and sprinkle the potato skin croutons and bacon on top.
—Anne Disrude

COLLARD GREENS AND BLACK-EYED-PEA SOUP WITH CORNMEAL CROUSTADES

In the South, greens are traditionally simmered with smoked pork and spices. In this recipe the side dish is turned into a soup simply by using more liquid. Cooking time for the greens is about one hour, two to three hours less than down-home cooked greens. The shorter cooking time preserves the texture, flavor and color of the greens; cooked longer, they tend to become a rather soft, gray-green mass that only a southerner can love.

The soup improves if made a day in advance and can be frozen and reheated in a microwave oven.
6 to 8 Servings

¾ pound smoked ham hocks, hog jowls or pork knuckles
4 carrots, halved crosswise
4 celery ribs, halved crosswise
15 to 20 sprigs of parsley, tied in a bundle with kitchen string
1 teaspoon salt
4 whole cloves
4 medium onions
10 garlic cloves, unpeeled
8 black peppercorns
6 small dried hot red peppers
2½ teaspoons thyme
2 imported bay leaves
2 cups dry white wine
½ pound dried black-eyed peas, rinsed and picked over
1½ pounds collard greens
12 to 16 thin slices of Cornmeal Yeast Bread (recipe follows)
½ cup finely grated Cheddar cheese
Cider vinegar and hot pepper sauce, for serving

1. In a large stockpot, place the ham hocks, carrots, celery, parsley and salt. Stick a clove into each onion and add to the pot. Thread the garlic cloves onto toothpicks or short wooden skewers (for easy removal when the soup is done) and add to the pot. Tie the peppercorns, hot peppers, thyme and bay leaves in a double thickness of cheesecloth and add to the pot. Add the wine and 5 quarts of hot water. Bring to a boil over high heat. Reduce the heat to a simmer and cook, partially covered, for 1 hour, skimming the surface occasionally.

2. Meanwhile, place the black-eyed peas in a large saucepan and add cold water to cover by at least 2 inches. Bring to a boil over moderately high heat and boil for 2 minutes. Remove from the heat, cover and let stand for 1 hour; drain.

3. Wash the collard greens and discard any large, coarse stems. Stack the greens and cut into ½-inch-wide strips. Cut across the strips to make ½-inch squares.

4. When the broth has simmered for 1 hour, add the greens and simmer, partially covered, until wilted, about 15 minutes. Add the black-eyed peas and simmer, uncovered, until the peas are tender but not mushy, about 45 minutes.

5. With a slotted spoon, remove the ham hocks, carrots, celery, onions and garlic from the soup and set aside. Remove the herb bouquet and the parsley bundle and discard. Remove the cloves from the onions and discard the cloves.

6. Place the carrots, celery and onions in a food processor. Remove the garlic cloves from the toothpicks and squeeze the garlic into the food processor, discarding the skins. Puree the vegetables with ½ cup of the broth until smooth, about 30 seconds. Return the puree to the soup.

7. Preheat the broiler. Pick over the ham hocks to remove any meat. Crumble the lean pieces with your fingers and return them to the soup. Season with additional salt to taste.

8. Place the slices of Cornmeal Yeast Bread on a baking sheet and broil 4 inches from the heat for about 1 minute on each side, until lightly browned and crisp. Sprinkle each slice with the cheese and broil for about 20 seconds, until melted and bubbly.

9. To serve, ladle the soup into bowls and float 2 croustades on top of each. Pass a cruet of vinegar and a bottle of hot sauce on the side.
—Sarah Belk

CORNMEAL YEAST BREAD
"Why not," I asked myself one day, "create a bread with a light cornmeal flavor and the crackling, crisp crust of a French baguette?" So I did. These loaves have a lighter crumb than standard southern corn breads so that they are somewhat more refined. You can slice the bread on the diagonal as you would French bread and serve it with lunch or dinner or with butter and damson plum preserves for breakfast or afternoon tea. Or slice it very thin, toast and serve with pork rillettes, foie gras or chicken liver mousse to accompany aperitifs.
Makes 4 Loaves

2 envelopes (¼ ounce each) active dry yeast
Pinch of sugar
1½ cups lukewarm water (105° to 115°)
3 to 3¾ cups all-purpose flour
1 cup yellow cornmeal
1 tablespoon salt
1 egg yolk
1 tablespoon milk

1. In a small bowl, mix the yeast and sugar with 1 cup of the warm water and let stand until bubbly, about 5 minutes.

2. In a large bowl, mix 3 cups of the flour with the cornmeal and salt. Add the remaining ½ cup warm water and the dissolved yeast mixture and stir until well combined.

3. Turn the dough out onto a lightly floured work surface and knead, adding additional flour as needed to avoid stickiness, until smooth and elastic, about 10 minutes. Place the dough in a lightly oiled bowl and turn to coat the dough with oil. Cover with plastic wrap and let rise in a warm place until the dough is doubled in bulk, about 2 hours.

4. Punch down the dough, cover and let rise again in a warm place until doubled, about 1½ hours.

5. Preheat the oven to 425°. Divide the dough into 4 equal portions and roll each portion with the palms of your hands into a narrow, 12-inch-long loaf. Place the loaves on a baking sheet lightly sprinkled with cornmeal.

6. In a small bowl, beat the egg yolk and milk together. Brush the top of each loaf with the egg glaze (see Note). Using a sharp knife, cut diagonal slashes 2 inches apart along the length of the loaves. Let the dough rest 5 minutes. Brush again with the egg glaze. Bake in the center of the oven for 20 to 25 minutes, or until the loaves sound hollow when tapped on the bottom.

NOTE: The egg glaze results in a shiny golden crust. For a crackly-crisp crust, omit the glaze and sprinkle the uncooked loaves with water just before baking.
—Sarah Belk

ACORN SQUASH AND TURNIP SOUP

Whenever I prepare this soup, with its very light and gentle flavor, I almost always make a few extra servings for later. With good crusty bread and a crisp salad, it will make a perfect little lunch.

6 Servings

- 1½-pound acorn squash, halved and seeded
- 4½ tablespoons unsalted butter
- 1 pound leeks (white part only), coarsely chopped
- 1 medium onion, coarsely chopped
- 2 small carrots, grated
- 1 garlic clove, chopped
- 1 teaspoon sugar
- ¾ pound turnips, peeled and cut into eighths
- 5½ cups chicken stock or canned broth
- ½ teaspoon salt
- ¼ teaspoon freshly ground white pepper
- ⅛ teaspoon ground coriander
- Crème fraîche or sour cream, for serving

1. Preheat the oven to 375°. Place the squash in a foil-lined baking pan. Rub the exposed areas with ½ tablespoon of the butter and bake until fork tender, about 1 hour.

2. Meanwhile, in a large flameproof casserole, melt the remaining 4 tablespoons butter over moderate heat. Add the leeks, onion, carrots and garlic. Sprinkle the sugar over all. Reduce the heat to low, cover and cook until the vegetables are soft, about 20 minutes.

3. Add the turnips, 2 cups of the stock, the salt, white pepper and coriander. Simmer uncovered over moderate heat until the turnips are tender, about 25 minutes.

4. When the squash is cooked, scoop out the flesh and add it to the casserole. Puree the soup, in batches if necessary, in a blender or food processor. Return to the pot and stir in the remaining 3½ cups stock. Heat through. Serve with a dollop of crème fraîche or sour cream.
—Lee Bailey

RUTABAGA SOUP WITH CURRIED COUSCOUS DUMPLINGS

Because of their heft and hardness, rutabagas are the very devil to cut. Choose your heaviest, sharpest knife and cut a small slice off one side of the rutabaga so that it won't wobble on the work surface. Using a gentle seesaw motion, halve the root lengthwise. Cut each half into ½-inch-thick slices and then peel them and cut into dice.

8 Servings

- 6 slices of bacon (about ¼ pound), cut crosswise into thin strips
- ½ of a medium red bell pepper, cut into thin julienne strips
- 1 small rutabaga (about 1 pound), peeled and cut into ½-inch dice
- 1 small carrot, thinly sliced
- 3 medium leeks (white and tender green), thinly sliced
- 1 medium onion, chopped
- ½ of a medium celery root, peeled and cut into thin julienne strips
- 1 teaspoon minced fresh ginger
- ¾ teaspoon crumbled thyme
- ½ teaspoon crumbled marjoram
- ¼ teaspoon crumbled rosemary
- ¼ teaspoon freshly ground black pepper
- 10 cups rich beef stock, preferably homemade
- 3 tablespoons unsalted butter
- 2 shallots, minced
- 1 teaspoon curry powder
- ¼ teaspoon freshly grated nutmeg
- 1 tablespoon all-purpose flour
- ½ teaspoon salt
- 1 egg
- ½ cup couscous
- 2 tablespoons chopped flat-leaf parsley

1. In a large heavy saucepan, cook the bacon over moderately low heat until crisp, 10 to 12 minutes. With a slotted spoon, transfer the bacon to paper towels. Pour all but 1 tablespoon of the drippings into a heatproof bowl and reserve.

2. Add the bell pepper to the saucepan and stir-fry over moderate heat until crisp-tender, 2 to 3 minutes; transfer to paper towels to drain.

3. Add 2 more tablespoons of the drippings to the saucepan and add the rutabaga, carrot, leeks, onion, celery root, ginger, ½ teaspoon of the thyme, the marjoram, rosemary and black pepper. Cook over moderate heat, stirring occasionally, until the leeks and onion are translucent but not browned, about 10 minutes.

4. Add the beef stock and bring to a simmer. Cover and cook over moderately low heat until the rutabaga is tender, about 45 minutes.

5. Meanwhile, prepare the dumplings. In a small skillet, melt 1 tablespoon of the butter. Add the shallots and cook over moderate heat until soft, 2 to 3 minutes. Stir in the curry powder, nutmeg, flour, salt and remaining ¼ teaspoon thyme; stir over low heat for 3 minutes.

6. In a food processor, cream the remaining 2 tablespoons butter with the egg. Add the couscous and process for 10 seconds. Scrape down the sides of the bowl. Process for 10 seconds longer. Add the curry mixture and process for 15 seconds. (The couscous is still very granular at this point, but will soften when cooked.) Scrape the dumpling mixture onto a piece of foil and flatten slightly. Wrap and refrigerate until firm enough to shape, about 45 minutes. (The

recipe can be prepared 1 day ahead to this point. Return the soup to a simmer before proceeding.)

7. Pinch off small bits of the dumpling dough and roll, with lightly buttered hands, into ½-inch balls. Increase the heat to moderate so that the soup bubbles gently. Add all of the dumplings, cover and simmer gently—without peeking—for 30 minutes, or until the dumplings are tender and fluffy.

8. Stir the chopped parsley and the reserved bacon and bell pepper into the soup. Season with salt. Ladle into large shallow soup plates, making sure no one is slighted on dumplings.
—Jean Anderson

CAULIFLOWER CRESS SOUP WITH CORN WAFERS

In early summer, there are a number of varieties of cress that you might like to experiment with. A favorite of mine is referred to locally as "upland cress" and is cultivated in an ordinary garden instead of in running water. If you can, in this recipe, substitute a local type for the ubiquitous watercress.

4 Servings

1½ cups cauliflower florets
1½ cups low-fat milk
6 tablespoons unsalted butter
¾ cup chopped scallions
1½ cups baking potatoes, peeled and finely diced
2½ cups chicken stock or canned broth
10 cups (lightly packed) watercress sprigs (about 3 bunches)
Salt and freshly ground white pepper
¼ cup crème fraîche
Corn Wafers (recipe follows)

1. In a medium saucepan, combine the cauliflower florets and milk. Bring to a boil over moderate heat. Reduce the heat to moderately low and simmer until the cauliflower is tender, about 5 minutes. Drain the cauliflower mixture over a bowl, reserving the cauliflower and the milk separately.

2. In a large saucepan, melt the butter. Add the scallions and sauté over moderately low heat until softened but not browned, 3 to 4 minutes. Add the potatoes and chicken stock. Increase the heat to moderately high and bring to a boil. Reduce the heat to moderately low, cover and simmer until the potatoes are tender, about 10 minutes.

3. Reserve 12 sprigs of watercress for garnish. Stir the remaining watercress into the stock and cook until the cress is tender, about 5 minutes. Add the cauliflower to the soup and cook until heated through, about 3 minutes.

4. Working in batches, if necessary, puree the soup in a blender or food processor. Return to the saucepan and add the reserved milk. Reheat over moderately low heat, about 5 minutes. Season with salt and white pepper to taste. (If the soup is too thick, thin with a little more milk or water.)

5. Serve warm and garnish each bowl with 1 tablespoon of crème fraîche and a few sprigs of watercress. Accompany with Corn Wafers.
—Lee Bailey

CORN WAFERS

These crisp cornmeal wafers have a dignified shape but the down-home taste of corn bread.
Makes About 16 Wafers

¾ cup white cornmeal
¼ teaspoon salt
1 cup boiling water
2 tablespoons margarine or unsalted butter, cut into small pieces

1. Preheat the oven to 425°. Coat 2 cookie sheets with nonstick cooking spray. In a medium bowl, mix the cornmeal and salt. Stir in the boiling water. Add the margarine and mix thoroughly, until smooth. Let stand for 5 minutes.

2. Restir the batter to blend well and then spoon 1 tablespoon onto the prepared cookie sheet. It should spread into a 3-inch circle. If it's too watery and thin to hold its shape, stir in 1 more tablespoon of the cornmeal. If it's too thick, add more water by teaspoons. Using a tablespoon, spoon the remaining batter for each wafer onto the cookie sheets.

3. Bake the wafers for 20 minutes, or until they are crisp and golden brown around the edges.
—Lee Bailey

AU PIED DE COCHON'S ONION SOUP

Is there any place in Paris that's better known for its onion soup than the 24-hour bistro Au Pied de Cochon? Almost all of us who love Paris have at one time or another made the ritualistic late-night trip to this restaurant in Les Halles, the former French food market, to indulge in this Parisian gastronomic pastime.

▼ Serve a fruity, direct, refreshing red, such as Jaffelin Beaujolais-Villages, or an off-dry white, such as Kenwood Chenin Blanc, as a contrast to the mélange of savory flavors in this classic soup.

6 Servings

1 pound white or other sweet onions (about 2 large), thinly sliced
2 cups dry white wine, such as Muscadet or Mâcon-Villages
2 tablespoons unsalted butter
6 cups unsalted beef stock, or 3 cans (13¾ ounces each) beef broth diluted with 1 cup water

6 slices of French or Italian bread, cut ½ inch thick, preferably stale
2 cups freshly grated imported French Gruyère cheese (about ½ pound)

1. Preheat the oven to 425°. Combine the onions, white wine and butter in a medium flameproof gratin dish. Bake, uncovered, stirring once or twice, until the onions are very soft and most of the liquid is evaporated, about 1 hour. Remove from the oven and set the onions aside.

2. Preheat the broiler with the rack set about 6 inches from the heat. In a large saucepan, bring the stock to a simmer over high heat.

3. Arrange 6 deep ovenproof soup bowls on a baking sheet. Evenly distribute the cooked onions among the bowls and ladle the simmering stock over the onions. Place a round of bread on top of each. Sprinkle the grated cheese over all.

4. Place the baking sheet under the broiler and cook for 2 to 3 minutes, until the cheese is just melted and lightly browned. Serve hot.
—Patricia Wells

BROCCOLI, ONION
AND CHEESE SOUP
Serve this soup with homemade croutons or lots of crusty bread.
▼ The sweetly aromatic onion and cheese flavors of this soup match with the fruity-pungent taste of Pinot Noir, such as Joseph Drouhin Laforet Bourgogne Rouge, served slightly cool.
8 to 10 Servings

5 tablespoons unsalted butter
2 large Bermuda onions, thinly sliced
¼ teaspoon freshly ground black pepper
3 sprigs of fresh thyme or ½ teaspoon dried
1 tablespoon sugar
¾ cup Riesling or other sweet white wine
3 cups canned reduced-sodium chicken broth
3 tablespoons all-purpose flour
1½ cups milk
½ cup heavy cream
¼ teaspoon dry mustard
⅛ teaspoon freshly grated nutmeg
Dash of cayenne pepper
½ pound Jarlsberg cheese, grated
½ cup grated Gruyère cheese
1 bunch of broccoli, separated into 1½-inch florets (about 4 cups)
1 red bell pepper, cut into ¼-inch dice

1. In a large flameproof casserole, melt 2 tablespoons of the butter over moderately high heat. Add the onions, black pepper, thyme and sugar. Reduce the heat to moderately low, place a circle of wax paper directly over the top of the onions and cover tightly with a lid. Simmer, stirring occasionally, until the onions are very soft and golden, about 1 hour and 15 minutes.

2. Remove the cover and the wax paper and increase the heat to moderate. Cook until the onions are golden brown, about 25 minutes.

3. Add the wine and boil until reduced by half, about 5 minutes. Add the chicken stock and 2 cups of water. Bring to a boil and simmer for 15 minutes.

4. In a large saucepan, melt the remaining 3 tablespoons butter over moderate heat. Whisk in the flour and cook, stirring, for 2 minutes without browning. Whisk in the milk, cream, dry mustard, nutmeg and cayenne and bring to a boil. Boil, stirring, for 2 minutes. Remove from the heat and stir in the Jarlsberg and Gruyère until smooth.

5. Scrape the cheese mixture into the hot soup, add the broccoli and cook until crisp-tender, about 8 minutes. Stir in the bell pepper and cook until heated through, about 2 minutes.
—Mimi Ruth Brodeur

THREE-MUSHROOM
SOUP WITH PORT
AND TARRAGON
6 Servings

1 cup dried porcini mushrooms (1 ounce), rinsed in cold water
About 8 cups chicken stock or canned low-sodium broth
4 tablespoons unsalted butter
1 medium shallot, minced
½ pound fresh shiitake mushrooms, stems discarded, caps sliced into ¼-inch strips
½ pound white button mushrooms, sliced lengthwise ¼ inch thick
¼ cup all-purpose flour
¼ cup vintage or tawny port
½ teaspoon tarragon
Salt and freshly ground pepper
6 tablespoons crème fraîche, for serving
3 tablespoons snipped chives, for garnish

1. In a large nonreactive saucepan, combine the dried porcini mushrooms and 8 cups of chicken stock and bring to a boil over high heat. Reduce the heat to moderate and simmer, partially covered, for 40 minutes. Pour the broth through a cheesecloth-lined strainer. Rinse the mushrooms in the strainer to remove any grit. Cut off any tough stems. Coarsely chop the mushrooms; set the mushrooms and broth aside.

2. Wipe out the saucepan. Add the butter and melt over moderate heat. Add the shallot and cook, stirring, until softened, about 1 minute.

3. Add the sliced shiitake and white button mushrooms and increase the heat

to moderately high. Cook, stirring frequently, until the mushrooms have exuded most of their liquid, 5 to 7 minutes.

4. Reduce the heat to moderately low and sprinkle the mushrooms with the flour. Cook for 1 minute, stirring constantly and scraping the bottom of the saucepan with a wooden spoon. Gradually whisk in the reserved mushroom broth and the port. Add the reserved porcini and the tarragon and bring to a boil over high heat. Reduce the heat to moderate and simmer, partially covered, for 20 minutes. Season with salt and pepper to taste. *(The recipe can be prepared to this point up to 2 days ahead. Reheat over moderately low heat before serving.)*

5. Serve the soup piping hot, garnished with the crème fraîche and a sprinkling of chives.
—Rick Rodgers

CREAM OF ASPARAGUS AND MOREL SOUP

Asparagus and morels complement each other, and they represent springtime at its best. I made this soup with morels from Oregon during May and June. Substitute various mushrooms in subsequent months, since asparagus now is widely available throughout the year.

8 Servings

2¼ pounds asparagus
5 tablespoons unsalted butter
3 tablespoons all-purpose flour
6 cups chicken stock, preferably homemade
½ pound fresh morels or other wild mushrooms, well washed
1¼ cups crème fraîche or heavy cream
Salt and freshly ground pepper

1. Snap off the tough end of each asparagus spear and peel the stalk. Bring a large pot or 2 skillets of salted water to a boil. Add the asparagus and cook until just tender, 3 to 5 minutes. Drain and rinse under cold running water to refresh. Cut off 16 asparagus tips and reserve. Coarsely chop the remaining asparagus.

2. In a large heavy saucepan or flameproof casserole, melt 3 tablespoons of the butter over moderate heat. Add the flour and cook, stirring, for 1 minute. Whisk in the chicken stock and bring to a boil, stirring occasionally. Add the chopped asparagus and morels, reserving the 8 smallest morels for garnish. Reduce the heat, cover and simmer for 30 minutes.

3. With a slotted spoon, transfer the asparagus and morels to a food processor. Add 1 cup of the stock and puree until smooth. Blend the puree and the remaining stock and pass the soup through a fine-mesh sieve to remove any tough fibers.

4. In a small saucepan, heat the remaining 2 tablespoons butter and ¼ cup of the crème fraîche until simmering. Add the 8 reserved morels and cook over moderately low heat for 10 minutes. Add the reserved asparagus tips and cook for 2 minutes longer.

5. To serve, reheat the soup if necessary. Whisk in the remaining 1 cup crème fraîche. Season with salt and pepper to taste. Add the morels and asparagus tips with their liquid. Serve very hot, garnishing each bowl with 1 morel and a couple of asparagus tips.
—Lydie Marshall

MUSHROOM-BARLEY SOUP

The inspiration for this soup came from my sister-in-law Linda. Mine is a meatless version.

8 to 10 Servings

2 tablespoons vegetable oil
2 medium onions, coarsely chopped
2 large celery ribs, cut into small dice
¾ cup pearl barley
½ ounce imported dried mushrooms, rinsed and coarsely chopped
4 cups (packed) coarsely chopped escarole
1 can (16 ounces) Italian peeled tomatoes, drained and coarsely chopped, juice reserved
2 large carrots, sliced ¼ inch thick
½ pound fresh white mushrooms, sliced ¼ inch thick
Salt and freshly ground pepper

1. In a large nonreactive saucepan, heat the oil over moderate heat. Add the onions and cook, stirring occasionally, until softened and lightly browned, about 10 minutes. Add the celery and cook until slightly softened, about 3 minutes.

2. Stir in 2 quarts of water. Add the barley, dried mushrooms, escarole and the tomatoes with their juice. Increase the heat to high and bring to a boil. Reduce the heat to moderately low, cover and simmer gently until the barley is tender and the broth is flavorful, about 1 hour and 15 minutes.

3. Add the carrots and simmer, covered, for 15 minutes. Add the fresh mushrooms and continue to simmer, covered, for 30 minutes. Season with the salt and pepper to taste. Serve immediately. *(The soup can be made up to 2 days ahead. Let cool, cover and refrigerate. Reheat slowly before serving.)*
—Susan Shapiro Jaslove

MUSHROOM HAZELNUT SOUP
4 to 6 Servings

⅓ cup hazelnuts (about 2 ounces)
4 tablespoons unsalted butter
1 large onion, chopped
1 pound mushrooms, coarsely chopped
About 4 cups chicken stock, preferably homemade
Salt and freshly ground pepper
2 tablespoons minced flat-leaf parsley

1. Preheat the oven to 350°. Place the hazelnuts on a baking sheet and toast until golden brown, 8 to 10 minutes. Place the hot nuts in a sieve and rub with a dish towel to remove as much skin as possible. Let the nuts cool to room temperature. Grind the nuts in a nut grater or food processor.

2. In a large saucepan, melt the butter over moderately high heat. Add the onion and sauté until softened and translucent, about 5 minutes. Add the mushrooms and cook, stirring frequently, for 5 minutes. Add 4 cups of stock and bring to a boil; reduce the heat and simmer for 5 minutes.

3. Working in batches, transfer the solids and a small amount of the liquid from the soup to a blender or food processor. Add some of the hazelnuts to each batch and puree until smooth. Pour into a large saucepan. Cook over moderate heat until warmed through, about 5 minutes. Season with salt and pepper to taste. Thin with more stock if desired. Pour into a warm tureen or individual soup bowls and garnish with a sprinkling of parsley.
—*Joyce Goldstein, Square One Restaurant, San Francisco*

SOUTHWESTERN BLACK BEAN SOUP

Inspired by several soups I tasted throughout New Mexico and Arizona, this main-course soup is rich in spices typical of southwestern fare and full of the zesty heat of their local green chiles. (Reduce the amount of jalapeños if you wish, but the soup is meant to be quite hot.) Serve with hot corn bread and a green salad with avocados and oranges.
6 Servings

¼ cup rendered bacon fat or olive oil
4 medium onions, chopped
8 garlic cloves, minced
2 cans (13¾ ounces each) beef broth
2 cans (13¾ ounces each) chicken broth
1 pound dried black beans
1 meaty ham hock or ham bone
1 pig's foot, split (optional)
3 tablespoons cumin
2 tablespoons oregano, preferably Mexican
1 tablespoon thyme
¼ teaspoon ground cloves
5 pickled jalapeños, minced (about ¼ cup)
Sour cream, diced tomatoes and chopped scallions, for serving

1. In a large saucepan or flameproof casserole, heat the bacon fat. Add the onions and garlic, cover and cook over moderately low heat, stirring occasionally, until tender, about 20 minutes.

2. Add the beef and chicken broths, the black beans, ham hock, pig's foot, cumin, oregano, thyme, cloves, jalapeños and 2 cups of water. Bring to a boil, reduce the heat to moderately low and simmer, partially covered, stirring and skimming occasionally, for about 2½ hours, or until the beans are tender.

3. Remove and discard the pig's foot. Remove and reserve the ham hock. In a food processor, puree half of the soup; return the puree to the pan with the remaining soup.

4. Remove the meat from the ham hock, shred and add to the soup. Simmer for 10 minutes.

5. To serve, ladle the soup into bowls. Top each with a dollop of sour cream and a generous sprinkling of diced tomato and scallion.
—*Michael McLaughlin*

FLAGEOLET AND LEEK SOUP

Pale green, gray-green and shale white, flageolets are the Champagne, the ne plus ultra, of dried beans. Their appearance is as delicate as their taste. If you like, serve this soup with croutons fried in butter or light olive oil.
4 Servings

1½ cups dried flageolet beans (10½ ounces), picked over
2 medium leeks—greens coarsely chopped, whites quartered lengthwise and thinly sliced crosswise
1 large carrot, coarsely chopped
1 small onion, chopped
1 celery rib, chopped
5 parsley sprigs
10 black peppercorns
2 bay leaves
1½ teaspoons salt
2 tablespoons unsalted butter
3 tablespoons minced parsley or chervil
½ cup milk or cream
¼ teaspoon freshly ground pepper

1. In a large saucepan, soak the beans in plenty of cold water for at least 6 hours or overnight. Pour off the water, re-cover the beans with fresh water and bring to a boil. Boil the beans vigorously for 5 minutes. Drain the beans in a colander and rinse well.

2. In a stockpot, combine the leek greens, carrot, onion, celery, parsley sprigs, peppercorns, 1 of the bay leaves and 1 teaspoon of the salt. Add 10 cups of water and bring to a boil over high heat. Reduce the heat to moderately low and simmer for 25 minutes. Strain the broth and reserve. Discard the solids.

3. In a large saucepan, melt the butter over moderately high heat. Add the leek whites, the remaining bay leaf, 2 tablespoons of the minced parsley and ½ cup of water. Simmer for 5 minutes. Add the beans and the reserved vegetable broth and bring to a boil over high heat. Reduce the heat to moderately low, cover partially and simmer very gently until the beans are almost tender, about 1¼ hours.

4. Add the remaining ½ teaspoon salt and continue cooking until the beans are very tender, about 10 minutes longer. If necessary, add enough boiling water to keep the beans amply covered.

5. Remove 1 cup of beans and broth from the pot and puree in a blender or food processor until smooth. Gently stir the puree back into the beans in the pot. Stir in the milk. Reheat the soup and season with salt to taste. Stir in the pepper and the remaining 1 tablespoon minced parsley just before serving.
—Deborah Madison

BACON, POTATO, WHITE BEAN AND RED PEPPER SOUP
6 Servings

½ pound slab of hickory-cured bacon, cut into ⅜-inch dice
1 tablespoon unsalted butter
1 small onion, minced
3 large leeks (white and tender green), cut into ½-inch-thick slices
½ of a large carrot, coarsely chopped
1 small celery rib, coarsely chopped
1 medium red bell pepper, cut into ¼-inch-thick slices
4 cups chicken stock or canned low-sodium broth
¾ teaspoon salt
¼ teaspoon freshly ground white pepper
2 large all-purpose potatoes (about 1 pound), peeled and cut into ½-inch cubes
1 can (19 ounces) white kidney beans, drained and rinsed

1. Place the bacon in a large saucepan and cover with water. Simmer over moderately high heat for 5 minutes. Drain and pat the bacon dry with paper towels. Wipe out the saucepan and return the bacon to the pan. Fry over moderately high heat until golden and crisp, about 4 minutes. Drain the bacon on paper towels.

2. Discard all the fat in the pan and wipe out with paper towels. Add the butter to the pan and melt over moderate heat. Add the onion, leeks, carrot, celery and red bell pepper. Cook until the vegetables are lightly browned and softened, about 5 minutes.

3. Add the bacon, chicken stock, salt and white pepper. Simmer, skimming occasionally, for 5 minutes.

4. Add the potatoes and simmer, skimming, until the potatoes are just tender when pierced with a knife, about 15 minutes.

5. Add the white beans and simmer 1 to 2 minutes more. Serve hot. (*This recipe can be made up to 1 day ahead. Cover and refrigerate. Remove any accumulated fat before reheating.*)
—Lee Bailey

CORN-AND-LEEK SOUP
8 to 10 Servings

6 tablespoons unsalted butter
2 teaspoons finely chopped garlic
4 medium leeks (white and tender green), chopped
3 medium boiling potatoes, peeled and diced
2 bay leaves
4 cups chicken stock
4 cups fresh corn kernels (from 8 to 10 large ears)
About 1 cup heavy cream
¼ cup fresh lemon juice
Salt and freshly ground pepper

GARNISH
1½ tablespoons unsalted butter
1 cup fresh corn kernels (from 2 to 3 large ears)

1. In a large soup pot, melt the butter over moderate heat. Add the garlic and cook for 1 minute. Stir in the leeks, cover and cook over low heat until the leeks are softened but not browned, about 10 minutes.

2. Add the potatoes, bay leaves and chicken stock and bring to a boil over moderately high heat. Lower the heat and gently simmer until the potatoes are tender, 5 to 10 minutes. Add the 4 cups of corn and cook another 5 minutes.

3. Remove the bay leaves and, working in batches, puree the soup in a blender or food processor. Return the soup to the pot and blend in 1 cup of the cream, the lemon juice, and salt and pepper to taste; reheat thoroughly.

4. Meanwhile, prepare the garnish: In a small skillet, melt the butter over moderate heat. Add the 1 cup corn and cook, stirring, until heated through, 1 to 2 minutes. Ladle the soup into individual bowls and garnish each with a heaping tablespoon of the sautéed corn.
—F&W

Richly Glazed Pearl Onions (p. 165).

Clockwise from lower right: Grapefruit and Avocado Salad with Tomato-Cumin Dressing (p. 221), fruit salad, California Rancho-Style Hominy (p. 159), baked ham, Flagstaff Green Rice (p. 197).

Garlic-Braised Eggplant and Chickpea Casserole (p. 101).

GRILLED CORN SOUP WITH CILANTRO AND ANCHO CHILE CREAMS

This colorful soup is a fine-tuned blend of southwestern flavors. The Cilantro Cream and Ancho Chile Cream are swirled in at the end, not only for flavor but for a striking presentation.

8 Servings

8 ears of corn, shucked but with the last, inner layer of husk left on
4 cups chicken stock or canned low-sodium broth
2 medium carrots, chopped
1 large celery rib, chopped
2 small onions, chopped
4 garlic cloves, peeled
2 serrano chiles, seeded and minced
2 cups heavy cream
Salt
Cilantro Cream and Ancho Chile Cream (recipes follow)

1. Prepare a slow charcoal fire or preheat the oven to 375°. Arrange half of the corn on the grill and cook on one side for 5 minutes; turn the corn and grill until browned and flecked with black spots, about 5 minutes more. Alternatively, arrange all the corn on a large heavy baking sheet and roast for 7 minutes. Turn and continue roasting until the leaves are dry and falling off, about 7 minutes more. Set aside to cool, then remove and discard the remaining leaves and silk from the corn.

2. In a large saucepan, combine the chicken stock, carrots, celery, onions, garlic and serrano chiles and bring to a boil over moderate heat. Reduce the heat to low and simmer until the vegetables soften slightly, about 5 minutes.

3. Meanwhile, using a small knife, cut the kernels from the ears of corn. Add the kernels to the stock and simmer for 10 minutes.

4. In a blender, puree the contents of the saucepan in two batches until smooth. Strain the soup back into the saucepan. *(The soup can be prepared to this point and refrigerated, covered, overnight.)*

5. Bring the soup to a simmer over moderate heat. Stir in the heavy cream and season to taste with salt. Reduce the heat to low and simmer for 5 minutes. Ladle the soup into warmed, shallow soup bowls. For each serving swirl in about 1 tablespoon each of the Cilantro and Ancho Chile Creams. Serve hot.
—Stephan Pyles

ANCHO CHILE CREAM

Ancho chiles are dried, ripened poblano chiles and are available at Spanish markets or specialty food stores.

Makes About ⅔ Cup

1 small ancho chile—halved, stemmed and seeded
3 tablespoons half-and-half or milk
2 tablespoons sour cream or crème fraîche

1. Heat a small skillet over high heat until quite hot, then add the ancho chile and toast until fragrant, about 30 seconds per side. Transfer the chile to a small bowl and add hot water to cover. Set aside for 10 minutes to soften.

2. Pat the chile dry and place in a blender with the half-and-half; puree until smooth. Strain the mixture through a fine sieve into a bowl; whisk in the sour cream.
—Stephan Pyles

CILANTRO CREAM

Makes About ⅔ Cup

5 large spinach leaves, stemmed
1 cup (loosely packed) fresh coriander (cilantro) leaves
3 tablespoons half-and-half or milk
2 tablespoons sour cream or crème fraîche

1. In a small saucepan, bring 2 cups of water to a boil over high heat. Add the spinach and cook, stirring, until soft, about 1 minute. Drain and plunge into cold water. Squeeze the spinach dry by wringing it in a clean kitchen towel.

2. In a blender, puree the fresh coriander, spinach and half-and-half. Blend until smooth. Strain through a fine sieve into a bowl. Add the sour cream.
—Stephan Pyles

CREAM OF CELERY SOUP

4 Servings

5 tablespoons unsalted butter
4 cups chopped celery, including some of the inner leaves
1½ cups chopped onions
1 large garlic clove, peeled
1 bay leaf
¾ cup dry white wine
4 cups chicken stock
½ teaspoon salt
¼ teaspoon freshly ground white pepper
½ teaspoon basil, crumbled
1 cup half-and-half or light cream
2 tablespoons all-purpose flour

1. In a medium saucepan, melt 3 tablespoons of the butter over moderate heat. Add the celery, onions and garlic and sauté, stirring occasionally, until the onion is translucent, about 10 minutes. Add the bay leaf, ½ cup of the wine and the chicken stock. Bring the mix-

ture to a boil, reduce the heat slightly and simmer for 20 minutes. Remove and discard the bay leaf.

2. Force the mixture through a food mill fitted with a medium disk and set over a medium saucepan, or puree in a food processor and then force through a medium sieve set over the saucepan. Stir in the remaining ¼ cup white wine, the salt, pepper, basil and half-and-half. Place the soup over low heat.

3. In a small skillet, melt the remaining 2 tablespoons butter over moderate heat. Stir in the flour and cook, stirring constantly, for about 1 minute. Remove from the heat. Stirring constantly, blend about 1 cup of the soup into the skillet. Transfer the mixture to the pan of soup, increase the heat slightly and cook, stirring, until the soup simmers and thickens slightly. Serve hot.
—Jim Fobel

FRESH PEA SOUP GARNIE

The rich texture and sweet flavor of fresh pea soup goes well with a number of accompaniments, such as the variety of garnishes included here.
4 to 6 Servings

1 teaspoon fresh lemon juice
1 teaspoon dried mint
5 tablespoons unsalted butter, softened
½ pound Weisswurst (veal sausage)
2 teaspoons capers, rinsed and patted dry
4 tablespoons sour cream
2 scallions (white portion only), cut into ½-inch pieces
¼ cup grated carrots
¼ cup finely chopped walnuts
4 cups shelled fresh peas (from about 3 pounds in the pod)
2 medium onions, sliced ¼ inch thick
2 garlic cloves, peeled
2 cups chicken stock, preferably homemade
Salt

1. In a small bowl, sprinkle the lemon juice over the mint and let stand to allow the mint to absorb the juice. Blend 3 tablespoons of the butter into the soaked mint. Let the mint butter sit at room temperature for 1 hour before using.

2. In a saucepan of simmering water, cook the Weisswurst for 15 minutes. When cool enough to handle, peel the casing off and cut the sausage into very thin slices on the diagonal. Set aside, covered, until ready to serve.

3. In a small bowl, combine the capers and sour cream; refrigerate until serving time. In another small bowl, combine the scallions, carrots and walnuts. Set aside until serving time.

4. In a medium saucepan, melt the remaining 2 tablespoons butter over moderate heat. Add the peas, onions and garlic and toss them well until coated with the butter. Tightly cover the pan, lower the heat slightly and cook, shaking the pan occasionally to prevent the vegetables from sticking, until the peas are crisp-tender and the onion translucent, 4 to 5 minutes.

5. Add the chicken stock and simmer the vegetables gently, uncovered, until the peas are completely cooked, 4 to 5 minutes.

6. In a food processor or blender, puree the mixture to as fine a texture as possible, working in batches if necessary. Press the puree through a fine strainer into a clean saucepan. Rewarm the soup over low heat, stirring occasionally. Taste for seasoning, adding salt if necessary.

7. Serve hot in warmed soup bowls. Pass the mint butter, capers, Weisswurst and vegetable-nut garnish on the side.
—F&W

GREEN PEA SOUP

Buttermilk adds both richness and zest to this smooth pea soup.
6 Servings

4 medium leeks (white part only)— trimmed, washed and cut into thin julienne strips
4 tablespoons unsalted butter
1 large head of lettuce, outer leaves and core removed, shredded
6 fresh whole mint leaves
4 cups fresh shelled green peas or 2 packages (10 ounces each) frozen peas
¾ teaspoon salt
½ teaspoon freshly ground white pepper
2 teaspoons fresh lemon juice
1½ cups chicken stock or canned broth
½ cup buttermilk

1. In a steamer, cook the leeks until almost tender, about 7 minutes.

2. In a large heavy saucepan, melt the butter over moderate heat. Add the lettuce, mint, peas, salt, pepper and the leeks and stir to combine. Increase the heat to moderately high and cook, stirring, for 4 minutes. Reduce the heat to low, cover and cook until the peas are soft, 10 to 13 minutes.

3. Transfer the cooked vegetables to a food processor. Add the lemon juice and puree until smooth.

4. Return the puree to the pot. Add the stock and bring to a simmer over moderate heat. Stir in the buttermilk and heat through but do not boil. Serve hot.
—Lee Bailey

RICH PEA SOUP
6 to 8 Servings

6 packages (10 ounces each) frozen peas, thawed
8 medium scallions
½ bunch of parsley (or stems from 1 bunch)
6 medium garlic cloves, unpeeled
10 black peppercorns
3 cups canned chicken broth
⅔ cup heavy cream
¼ teaspoon salt
⅛ teaspoon freshly ground white pepper
3 tablespoons all-purpose flour
3 tablespoons unsalted butter, softened

1. In a large saucepan or flameproof casserole, combine 3 packages of the peas, the white parts of the scallions (reserve the green tops), the parsley, garlic, peppercorns, chicken broth and 3 cups of water. Bring to a boil, reduce the heat to a simmer and cook uncovered over low heat for 30 minutes.

2. While the soup base is simmering, puree the remaining peas with ⅓ cup of the heavy cream. Press through a fine sieve to remove all the bits of skin.

3. Strain the soup base and discard the solids. Return the broth to the saucepan. Add the pureed peas to the broth, whisk to blend and heat through but do not boil. Season with the salt and pepper. *(The recipe can be prepared to this point up to 1 day ahead.)*

4. Beat the remaining ⅓ cup heavy cream until stiff. Cut the scallion greens into thin strips.

5. Knead the flour and butter until smooth to make a *beurre manié*. Whisk into the soup. Cook over moderately high heat, whisking, until the soup thickens and comes just to a boil. Remove from the heat and ladle into soup bowls. Garnish each serving with a small dollop of whipped cream and a pinch of scallion strips.
—Anne Disrude

GREENPORT SOUP
This soup is named after the little village where I spend my weekends and vacations, but I love to serve it to friends in my city apartment as well.
6 Servings

4 cans (13¾ ounces each) chicken broth
1 package (10 ounces) frozen tiny peas
1 head of leafy lettuce (about ½ pound), shredded
½ cup chopped scallions
½ cup (packed) flat-leaf parsley
4 tablespoons unsalted butter
3 tablespoons all-purpose flour
¼ cup dry sherry
½ cup sour cream
Salt and freshly ground pepper
Parmesan Croutons (recipe follows)

1. In a large heavy saucepan, combine the chicken broth, peas, lettuce, scallions and parsley. Bring to a boil over moderate heat. Reduce the heat to low and simmer for 15 minutes.

2. In a blender or food processor, puree the soup in batches until very smooth.

3. In a large saucepan, melt the butter over low heat. Add the flour and cook, stirring, for 2 to 3 minutes without letting the roux color. Whisk in the pureed soup and the sherry. Bring to a simmer over moderate heat. Cook for 5 minutes to blend the flavors. Remove from the heat. *(The recipe can be prepared to this point up to 1 day ahead. Let cool, cover and refrigerate. Reheat before proceeding.)*

4. Stir the sour cream into the hot soup. Season with salt and pepper to taste. Serve topped with the Parmesan Croutons.
—W. Peter Prestcott

PARMESAN CROUTONS
These croutons may be made in a variety of shapes and sizes. Cubed, they are perfect for soup, but when they are cut into larger triangles, these cheese-sprinkled toasts become a wonderful crisp base for canapés or a savory accompaniment to salads.
Makes About 2 Cups

4 tablespoons unsalted butter
1 tablespoon vegetable oil
4 thin slices of firm-textured white bread, crusts removed
1 tablespoon plus 1 teaspoon freshly grated Parmesan cheese

1. In a large heavy skillet, preferably nonstick, melt the butter in the oil over moderately high heat. When the foam subsides, add the bread slices and fry until lightly browned, about 2 minutes. Turn the bread and fry the other side for 2 minutes.

2. Sprinkle on the cheese, turn the bread one more time and cook until the cheese is lightly browned, about 1 minute. Drain on paper towels. Let cool for 5 minutes, then cut into ½-inch cubes.
—W. Peter Prestcott

CHARRED CARROT SOUP

Charring shredded carrots in a cast-iron skillet caramelizes the vegetable's natural sugars and adds a deep, rich, nutty flavor to the soup. This slight sweetness is brought back into balance by a small amount of red wine vinegar added at the end.

4 Servings

½ tablespoon vegetable oil
6 medium carrots, peeled and shredded
2 shallots, coarsely chopped
2 garlic cloves, coarsely chopped
½ teaspoon thyme
1 small Idaho potato, peeled and coarsely chopped
3½ cups rich chicken stock
½ cup heavy cream
1 tablespoon red wine vinegar
¼ teaspoon salt
¼ teaspoon coarsely cracked pepper
1 tablespoon unsalted butter, softened
1 tablespoon chopped parsley or chives, for garnish

1. Preheat a 12-inch cast-iron skillet over high heat for 5 minutes. Add the oil, then the carrots. Stir to coat the carrots with oil and cook, stirring frequently, until the carrots are partially charred, about 15 minutes.
2. Reduce the heat to moderate and add the shallots, garlic and thyme. Cook until the shallots are softened, 2 to 3 minutes.
3. Add the potato and stock. Simmer over low heat until the carrots and potato are very soft, about 15 minutes.
4. In a food processor or blender, puree the soup until smooth. (For an extra-silky texture, press the soup through a fine sieve after pureeing.)
5. Transfer the soup to a saucepan and add the cream (use extra chicken stock or water for a thinner consistency if desired). Add the red wine vinegar, salt and pepper. Stir in the butter and serve hot, sprinkled with the chopped parsley.

—*Anne Disrude*

CREAMY CARROT SOUP

This hot soup is especially good when thickened with heavy cream, but it can also be served plain.

6 to 8 Servings

2 tablespoons unsalted butter
2 cups coarsely chopped onions
4 garlic cloves, minced
½ teaspoon salt
½ cup dry white wine
3 pounds carrots, peeled and coarsely chopped
About 4 cups Vegetable Stock (recipe follows)
½ teaspoon freshly ground pepper
½ cup heavy cream (optional)

1. In a large saucepan or flameproof casserole, melt the butter over moderately low heat. Add the onions, garlic and salt. Cook until the onions are soft and translucent but not browned, about 10 minutes.
2. Add the wine and continue to cook until the liquid has evaporated, about 5 minutes.
3. Add the carrots and enough Vegetable Stock to cover. Bring to a boil; reduce the heat to moderate and cook, uncovered, stirring frequently, until the carrots are soft, about 30 minutes.
4. Puree the soup in a blender or food processor. Return the puree to the saucepan and add enough additional stock to bring the soup to a desirable consistency.
5. Season with the pepper and additional salt to taste. Add the cream, if desired, heat through and divide the soup among individual bowls.

—*Annie Somerville*

VEGETABLE STOCK

This flavorful stock can be used to make all varieties of vegetable soup. Store it in the freezer for up to 1 week.

Makes About 1½ Quarts

2 medium carrots, peeled and coarsely chopped
4 celery ribs with leaves, coarsely chopped
1 large potato, coarsely chopped
½ pound mushrooms, coarsely chopped
6 garlic cloves
1 bunch of parsley
2 bay leaves
1 teaspoon black peppercorns
1 teaspoon salt

In a large heavy saucepan, combine the carrots, celery, potato, mushrooms, garlic, parsley, bay leaves, peppercorns and salt. Add 8 cups of cold water and bring to a boil over high heat. Reduce the heat to moderately low and simmer, uncovered, for 1 hour. Strain the stock. Cover and refrigerate for up to 3 days, or freeze for longer storage.

—*Annie Somerville*

CREAM OF CARROT SOUP WITH BRANDY AND CHERVIL

For this simple recipe, chicken wings and carrots are cooked together to form a stock base for this delicious soup.

4 Servings

1 pound chicken wings, sectioned
1 tablespoon all-purpose flour
2 tablespoons unsalted butter
3 large shallots, chopped
1 pound carrots, peeled and chopped
½ cup brandy
½ cup dry white wine
1 cup plus 2 tablespoons chopped fresh chervil
½ cup heavy cream

1 teaspoon salt
½ teaspoon freshly ground pepper

1. Toss the chicken wings with the flour. In a large saucepan or flameproof casserole, melt the butter over high heat. Add the chicken wings and cook, stirring occasionally, until the chicken is browned, about 6 minutes.

2. Add the shallots and cook until softened and fragrant, about 2 minutes. Add the carrots, brandy and wine and cook over high heat for 1 minute. Add 2½ cups of water, return to a boil and reduce the heat to low. Cover and simmer until the carrots are very tender, about 45 minutes. Remove from the heat.

3. With a slotted spoon, remove and discard the chicken pieces. Stir in 1 cup of the chervil. Puree the soup in batches in a blender or food processor until very smooth. Return to the pan, stir in the cream and season with the salt and pepper. Divide the soup among 4 bowls, garnish each with the remaining chopped chervil and serve at once.
—Marcia Kiesel

ROSEMARY-SCENTED SWEET RED PEPPER BISQUE

Serve this soup hot or cold at the start of an elegant meal or for a light supper.
🍷 Crisp white, such as Dry Creek Fumé Blanc
6 Servings

4 medium sprigs of rosemary plus 6 tiny sprigs for garnish
2 medium sprigs of sweet marjoram
3 tablespoons fruity olive oil
6 large red bell peppers, cut into ½-inch strips
6 medium leeks (white part only), thinly sliced
2 large garlic cloves, crushed through a press
¼ teaspoon freshly ground black pepper
2½ cups beef stock or canned broth
¾ cup plus 6 tablespoons crème fraîche or sour cream, at room temperature
Salt

1. Tie the medium sprigs of rosemary and the marjoram in a double thickness of cheesecloth. Wring lightly to release the volatile oils.

2. In a large heavy saucepan, heat the oil over high heat until almost smoking, about 1 minute. Add the peppers and cook, tossing, until slightly softened, about 2 minutes. Add the leeks and garlic and cook, tossing, for 2 minutes longer.

3. Add the cheesecloth bag of herbs and the black pepper. Reduce the heat to very low, cover tightly and cook until considerable juices have accumulated and the peppers are very soft, about 1 hour.

4. Remove and discard the cheesecloth bag. Scrape the vegetables into a food processor or blender and puree until smooth.

5. Strain the puree into a medium saucepan and add the stock. Cook over moderate heat until just hot, 2 to 3 minutes. Stir in ¾ cup of the crème fraîche and season to taste with salt.

6. Ladle the soup into heated soup plates and garnish each serving with 1 tablespoon of the crème fraîche and a tiny rosemary sprig.
—Jean Anderson

GREAT LAKES CORN CHOWDER
6 Servings

6 medium ears of corn
6 strips of bacon, cut into ½-inch pieces
1 small onion, finely chopped
1 small green bell pepper, finely chopped
1 jalapeño pepper—seeded, deveined and chopped
1 small celery rib, finely chopped
3 medium tomatoes—peeled, seeded and finely chopped
2 medium boiling potatoes (about 1 pound), peeled and cubed
1 teaspoon salt
⅛ teaspoon ground allspice
Pinch of sugar
1 small bay leaf
2 cups half-and-half or light cream, at room temperature
1 cup milk
Freshly ground black pepper
Chopped parsley, for garnish

1. Working over a bowl, cut the corn kernels from the cobs at about half their depth. Then, using the back of the knife, scrape the cobs over the bowl to release all the "milk"; set aside.

2. In a large nonreactive saucepan, fry the bacon over moderately high heat, stirring occasionally, until crisp, about 10 minutes. Transfer the bacon to paper towels to drain. Crumble and set aside.

3. Discard all but 3 tablespoons of the bacon drippings from the pan. Add the onion and cook over moderate heat until golden, 4 to 5 minutes. Add the green bell pepper, jalapeño and celery and cook until slightly softened, about 2 minutes. Add the tomatoes, potatoes, salt, allspice, sugar, bay leaf and the reserved corn kernels with their "milk" and stir well. Cook over moderate heat until the mixture begins to sizzle.

4. Reduce the heat to low. Cover and cook, stirring occasionally, until the potatoes are tender, 35 to 45 minutes. Stir in the cream and milk and bring just to a boil. Remove from the heat and season with black pepper to taste. Ladle the chowder into bowls and garnish with the reserved bacon and the parsley.
—Phillip Stephen Schulz

NEW JERSEY TOMATO AND SWEET POTATO CHOWDER

Named for the home of the world's tastiest tomatoes, this chowder is one for which ripe, fresh tomatoes are a must. So save this recipe for the appropriate season.

4 Servings

2 tablespoons unsalted butter
1 medium onion, chopped
1 large celery rib, minced
2 medium sweet potatoes (1 pound), peeled and diced
1⅔ cups beef stock or canned broth
4 large tomatoes (2 pounds)— peeled, seeded and chopped
Pinch of sugar
½ teaspoon salt
¼ teaspoon freshly ground pepper
⅛ teaspoon cinnamon
Pinch of freshly grated nutmeg
3 tablespoons chopped parsley, for garnish

1. In a large nonreactive saucepan, melt the butter over moderately low heat. Add the onion and cook until slightly softened, about 5 minutes. Stir in the celery and sweet potatoes and cook for 5 minutes longer.

2. Add the beef stock, tomatoes and sugar and bring to a boil over moderately high heat. Reduce the heat to moderately low, cover and cook until the potatoes are tender, about 30 minutes.

3. Transfer 1 cup of the chowder to a blender or food processor and puree until smooth. Return the puree to the chowder and add the salt, pepper, cinnamon and nutmeg. Cook for 5 minutes to blend the flavors. Season with additional salt and pepper to taste. Ladle the chowder into bowls, garnish with the parsley and serve immediately.
—Phillip Stephen Schulz

ROASTED GARLIC AND LEEK CHOWDER

The garlic cloves in this chowder can be roasted unpeeled in a 350° oven for 25 minutes, then peeled and mashed. Or cook as specified below. Either way, the garlic loses much of its sting as the chowder cooks, but it's still strong enough to produce a redolent soup.

❦ This pungent soup would find its complement in a fruity but textured red, such as a light Bordeaux. Try Barton & Guestier Merlot or Château St-Georges St-Emilion.

3 to 4 Servings

1 tablespoon unsalted butter
6 large garlic cloves, peeled
⅓ cup diced salt pork (1 ounce)
3 medium leeks, white and tender green, chopped
2 medium carrots, diced
1 celery rib with leaves, finely chopped
4 cups chicken stock or canned broth
2 medium baking potatoes (about 1 pound), peeled and cut into ½-inch dice
1 teaspoon chopped fresh sage
Salt and freshly ground pepper
1 tablespoon chopped parsley

1. In a small skillet, melt the butter over very low heat. Add the garlic, cover and cook, turning once, until tender, about 25 minutes; the garlic will be golden, but should not burn. Using a fork, mash the garlic to a paste.

2. Meanwhile, blanch the salt pork in a small saucepan of boiling water for 3 minutes. Drain well.

3. In a large saucepan, fry the salt pork over moderate heat until golden, about 4 minutes. Discard all but 1 tablespoon of the fat, but leave the salt pork in the pan.

4. Add the leeks and cook, stirring constantly, for 1 minute. Stir in the garlic paste and cook for 1 minute longer. Stir in the carrots and celery. Reduce the heat to moderate, cover and cook, stirring occasionally, until the vegetables have softened, about 10 minutes.

5. Stir in the chicken stock and bring to a boil over moderately high heat. Add the potatoes and return to a boil. Reduce the heat to moderate and simmer until the potatoes are tender, about 15 minutes.

6. Transfer 1 cup of the chowder to a blender or food processor and puree until smooth. Return the puree to the chowder and stir in the sage. Season with salt and pepper to taste. Ladle the chowder into bowls and garnish with the parsley.
—Phillip Stephen Schulz

4 MAIN COURSES VEGETABLES & GRAINS

MOREL AND FONTINA PIZZA
1 Serving

1 ounce dried morel mushrooms
2 tablespoons olive oil
⅛ teaspoon salt
⅛ teaspoon coarsely cracked pepper
4 ounces (1 ball) Basic Pizza Dough (recipe follows)
Cornmeal
3 ounces Italian Fontina cheese, shredded
Freshly minced chives, for garnish

1. Preheat the oven and a pizza stone or tiles (see Note) to 500° 1 hour before use. Soak the morels in 2 cups of warm water until soft, about 15 minutes. Scoop them out, reserving the soaking liquid, and let drain on paper towels. Trim off the sandy bottoms. Taste a morel to see if the sand is all gone; if not, soak again, discarding the liquid from the second soaking.

2. Strain the soaking liquid through several layers of dampened cheesecloth into a saucepan. Boil over high heat until reduced to ¼ cup, about 20 minutes.

3. In a medium skillet, heat 1 tablespoon of the oil. Add the morels and sauté over moderate heat, stirring, until softened, about 3 minutes. Add the reduced soaking liquid and cook, uncovered, until the liquid is absorbed. Season with the salt and pepper.

4. On a lightly floured surface, roll out the pizza dough into a round 7 inches in diameter and ¼ inch thick. Dust a pizza peel or a baking sheet with some cornmeal and lay the dough on top. Lightly brush the pizza crust with about 1 teaspoon of the oil, then prick all over with a fork.

5. Slide the pizza onto the hot pizza stone. Bake for 5 minutes until a light crust has formed. Remove from the oven and sprinkle with the cheese. Place the morels on top and drizzle with the remaining 2 teaspoons oil. Return to the oven and bake for 5 more minutes, or until the cheese is melted and the bottom of the crust is browned. Garnish with the chives before serving.

NOTE: For a crisp crust, we recommend using a pizza stone (available in specialty cookware stores) or unglazed stone tiles (available from floor covering stores). Of course, it's perfectly possible to make a pizza on a baking sheet, preferably a heavy one.
—Anne Disrude

BASIC PIZZA DOUGH
Makes About 1¼ Pounds

1 envelope (¼ ounce) active dry yeast
1 tablespoon sugar
1½ cups warm water (110° to 115°)
3¼ cups unbleached flour, preferably bread flour
½ teaspoon salt
¼ cup olive oil, preferably extra-virgin

1. In a small bowl, combine the yeast and sugar. Add the water and stir to mix. If the yeast is not active and bubbling within 5 minutes, discard it and repeat the procedure with a new envelope of yeast.

2. Measure 3 cups of flour by spooning into a 1-cup measure and leveling off with a knife, and place in a large bowl. Stir in the salt, then form a well in the center of the flour.

3. Pour the yeast mixture into the well and add the oil. Stir in the flour, beginning in the center and working toward the sides of the bowl. When all the flour is incorporated and the dough is still soft but begins to mass together, turn it out onto a lightly floured work surface.

4. Using a dough scraper to lift any fragments that cling to the work surface, knead the dough, adding just enough of the remaining ¼ cup flour until the dough is no longer sticky. (It is better that the dough be too soft than too stiff.) Continue to knead until smooth, shiny and elastic, 10 to 15 minutes.

5. Shape the dough into a ball and place it in a large, oiled bowl; turn the dough over to coat with the oil. Cover with plastic wrap, set in a warm draft-free place and let rise until doubled in bulk, 1 to 1½ hours. Punch the dough down and reshape into a ball. Cover and refrigerate until doubled in bulk, 20 minutes to 1 hour.

6. Divide the dough into 7 balls of equal weight (4 ounces) and use one for each flat pizza. To freeze, wrap each ball well in plastic; let the dough thaw before proceeding.
—Anne Disrude

THREE MUSHROOM AND CHEESE PIZZA
Any pizza lover will become addicted to these thin crisp rounds of dough, which are topped with wild mushrooms, various cheeses and lots of seasonings.
Makes Three 8-Inch Pizzas

1 envelope (¼ ounce) active dry yeast
⅓ cup warm milk (105° to 115°)
⅔ cup warm water (105° to 115°)
1½ tablespoons plus 1 teaspoon olive oil
Pinch of sugar
2 cups unbleached all-purpose flour
3 tablespoons whole wheat flour
1½ teaspoons salt
1½ ounces dried porcini mushrooms
1½ ounces dried shiitake mushrooms
4 cups boiling water
1 tablespoon unsalted butter
3 pounds fresh mushrooms, sliced
½ teaspoon freshly ground pepper

1½ cups dry red wine
2 to 3 teaspoons minced garlic
1 teaspoon dried thyme
1 teaspoon dried oregano
1 teaspoon dried marjoram
1 teaspoon dried sage
6 ounces Provolone cheese, shredded
6 ounces Italian Fontina cheese, shredded
2 tablespoons freshly grated Parmesan cheese
3 tablespoons mixed fresh herbs, preferably a combination of chopped parsley, chives, marjoram and oregano

1. Dissolve the yeast in the warm milk and stir in the warm water, olive oil and sugar. Let stand for 5 minutes.

2. In a large bowl, mix together the all-purpose and whole wheat flours and ½ teaspoon of the salt. Add the yeast mixture and stir well to form a soft workable dough. (If the dough is too sticky, add small amounts of all-purpose flour.)

3. Turn the dough out onto a lightly floured surface and knead for 10 minutes by hand (or 5 minutes in a mixer with a dough hook), until the dough is smooth and elastic.

4. Place the dough in a large bowl, cover with a damp cloth and let rise in a warm place until doubled, about 1 hour.

5. Meanwhile, soak the porcini and shiitake mushrooms in the boiling water until they are very soft and the water has cooled to room temperature, 30 to 45 minutes. Remove the mushrooms with a slotted spoon and set aside. Strain the soaking liquid through a sieve lined with several layers of dampened cheesecloth and reserve. Trim the tough stems from the shiitake mushrooms and the sandy root areas from the porcini. Thinly slice the mushrooms.

6. In a large skillet, melt the butter over moderately high heat. Add the fresh mushrooms and sauté, tossing, until slightly wilted, 3 to 5 minutes. Season with the remaining 1 teaspoon salt and the pepper.

7. In a small heavy saucepan, combine the reserved mushroom soaking liquid with the wine. Boil over moderate heat until the liquid is reduced to ⅓ cup, 15 to 20 minutes.

8. In a bowl, toss together the sautéed and dried mushrooms, the garlic, dried thyme, oregano, marjoram and sage and the reduced wine-mushroom liquid.

9. Preheat the oven to 500°. Punch down the dough and divide it into 3 equal parts, about 6 ounces each. Form each piece of dough into a ball and roll it out on a floured surface to form a pizza round 8 to 9 inches in diameter. Place each pizza round on a 10-by-14-inch baking sheet and set aside to rise for 10 minutes.

10. Top each round of dough with about ⅔ cup of the mushroom mixture, spreading to cover evenly, leaving a ½- to ¾-inch border. Combine the Provolone and Fontina cheeses and sprinkle over the mushrooms.

11. Bake the pizzas for 7 minutes; switch the pizzas on the bottom rack to the top and continue to bake until the crust is crisp and golden brown, 3 to 5 minutes longer. Remove from the oven and sprinkle with the Parmesan cheese and fresh herbs. Serve hot.
—Annie Somerville

EGGPLANT AND RED PEPPER PIZZA WITH ROSEMARY

❣ The far-from-subtle peppery, cheesy notes of this dish are most easily matched with a gutsy, direct-flavored red wine, such as Louis M. Martini Barbera from California or Jaboulet Gigondas from the Rhône.
Makes Eight 9-Inch Pizzas

1 envelope (¼ ounce) active dry yeast
½ cup plus 2 tablespoons olive oil
5 cups all-purpose flour
2¾ teaspoons salt
2 large eggplants, thinly sliced lengthwise
4 large red bell peppers
2 large baking potatoes, thinly sliced
1 tablespoon plus 1 teaspoon cornmeal
1 tablespoon plus 1 teaspoon finely chopped rosemary
¾ pound Bucheron goat cheese, crumbled
2 teaspoons freshly ground black pepper

1. Pour 1½ cups of warm water into a large bowl. Sprinkle the yeast over the water and let stand until it dissolves, about 3 minutes. Stir in 2 tablespoons of the olive oil and 2 cups of the flour. Add 1 teaspoon of the salt and the remaining 3 cups flour. Turn the dough out onto a lightly floured work surface and knead until the dough is firm and pliable, about 5 minutes. Put the dough into a lightly oiled bowl and cover with plastic wrap. Set aside in a warm place until doubled in size, about 1½ hours. Punch the dough down and cut it into 8 equal pieces. Flatten each piece into a small disk and wrap individually in plastic wrap. Refrigerate until ready to use. (*The recipe can be prepared to this point and frozen for up to 1 month.*)

2. In a large nonreactive colander, layer one-third of the sliced eggplant and sprinkle with ¼ teaspoon of the salt. Repeat this procedure with two more layers of eggplant and ½ teaspoon of the salt. Place a plate over the eggplant and weigh it down with cans or a heavy pot; let drain for 1 hour.

3. Roast the red peppers directly over a gas flame or under the broiler as close to the heat as possible, turning, until charred all over. Place the peppers in a paper bag to steam for 5 minutes. Peel the peppers and remove the cores, seeds and membranes. Cut the peppers lengthwise into ¼-inch-wide strips. Place in a bowl; toss with 1 teaspoon of the oil.

4. Pat the eggplant dry with paper towels. In a large skillet, heat 1 tablespoon of the olive oil. Add one-fourth of the eggplant slices and cook over high heat, turning once, until browned, about 2 minutes on each side. Remove the cooked eggplant to a bowl. Repeat with the remaining 3 batches of eggplant and 3 tablespoons of olive oil.

5. Cook the potato slices in lightly salted boiling water until just tender, about 5 minutes; drain well.

6. Preheat the oven to 500°. Sprinkle a baking sheet with 1 teaspoon of the cornmeal.

7. On a lightly floured surface, roll out 1 piece of the pizza dough to form a 9-inch round ⅛ inch thick. Place the dough on the prepared baking sheet and brush lightly with some of the remaining olive oil.

8. Place a layer of potato in the center of the pizza, leaving about a ½-inch border. Sprinkle with ¼ teaspoon of the rosemary and 3 tablespoons of the goat cheese. Place 3 or 4 slices of eggplant over the cheese and brush the top with some of the remaining olive oil. Scatter one-eighth of the pepper strips over the pizza and season with an additional ¼ teaspoon rosemary, ⅛ teaspoon salt and ¼ teaspoon black pepper. Repeat with the remaining dough and toppings and bake 2 pizzas at a time. For each new batch of pizzas, sprinkle the baking sheet with 1 teaspoon of cornmeal.

9. Bake the pizzas on the lowest rack of the oven until crisp and golden, about 8 minutes. Cut into wedges and serve hot or at room temperature.
—Marcia Kiesel

CARAMELIZED ONION AND BACON PIZZA
2 Servings

¼ cup plus 2 tablespoons olive oil
3 Spanish onions (about 2 pounds), thinly sliced
¾ teaspoon thyme
2 imported bay leaves
½ cup dry red wine
2 tablespoons red wine vinegar
½ teaspoon salt
¼ teaspoon coarsely cracked pepper
¼ pound thickly sliced slab bacon, cut crosswise into ¼-inch strips
8 ounces (2 balls) Basic Pizza Dough (p. 84)
Cornmeal
Chopped parsley, for garnish

1. Preheat the oven and a pizza stone or tiles (see Note) to 500° 1 hour before use. In a large heavy nonreactive skillet, heat ¼ cup of the olive oil until shimmering. Add the onions, thyme and bay leaves and stir to coat well. Cover and cook over low heat until the onions are soft and translucent, about 10 minutes. Uncover and cook, stirring occasionally, until the onions are golden, 30 to 40 minutes.

2. Stir in the wine and vinegar. Cook until the liquid evaporates, about 10 minutes. Season with the salt and pepper. Remove from the heat and set aside.

3. In a medium skillet, fry the bacon until lightly browned. Drain on paper towels.

4. On a lightly floured surface, roll out each ball of pizza dough into a round 7 inches in diameter and ¼ inch thick. Dust a pizza peel or a baking sheet with some cornmeal and place the dough on top. Lightly brush the pizza crusts with about 2 teaspoons of olive oil. Spread the onions on top and sprinkle with the bacon. Drizzle the remaining oil over all.

5. Slide the pizza onto the hot pizza stone. Bake for 8 to 10 minutes, or until the bottom of the crust is browned. Sprinkle with parsley before serving.

NOTE: For a crisp crust, we recommend using a pizza stone (available at specialty cookware stores) or unglazed stone tiles (available from floor covering stores). Of course, it's perfectly possible to make a pizza on a baking sheet, preferably a heavy one.
—Anne Disrude

CORNMEAL PIZZA WITH GREENS AND FONTINA
Bitter greens such as broccoli rabe or arugula are common pizza toppings in Italy. Collards and other southern greens provide the same pleasant bitterness as their Italian counterparts.
▼ This savory dish calls for an equally piquant red, such as Chianti, to point up its rich, cheesy flavor. Try the elegant Antinori Pèppoli or the straightforward Frescobaldi.
4 Servings

2 envelopes (¼ ounce each) active dry yeast
Pinch of sugar
1¼ cups lukewarm water (105° to 115°)

2¾ cups all-purpose flour
2 teaspoons salt
¾ cup yellow cornmeal
¼ cup plus 2 tablespoons extra-virgin olive oil
1 medium onion, finely chopped
2 large garlic cloves, minced
2 pounds watercress or arugula, tough stems removed, or kale (see Note)
Pinch of crushed red pepper
2 ounces very thinly sliced Smithfield ham or prosciutto
½ pound Italian Fontina cheese, shredded
¼ pound mozzarella cheese, shredded or cut into thin slices

1. In a medium bowl, combine the yeast, sugar and warm water. Stir gently to mix, then let stand until bubbly, about 5 minutes.

2. Add 1 cup of the flour and the salt and mix thoroughly. Add the cornmeal and another 1 cup of flour and mix well.

3. Sprinkle the remaining ¾ cup flour on a work surface. Turn out the dough and knead, incorporating the flour a little at a time, until smooth and elastic, 5 to 10 minutes. Place the dough in a lightly oiled bowl and turn to coat. Cover the bowl with plastic wrap and let rise in a warm place until doubled in bulk, about 1 hour.

4. Punch the dough down, cover and let rise until doubled in bulk, about 1 hour longer.

5. Meanwhile, in a large skillet, heat 2 tablespoons of the olive oil over moderately low heat. Add the onion and cook until it begins to soften, about 5 minutes. Add the garlic and cook until fragrant, about 30 seconds.

6. Increase the heat to moderate. Add the watercress and cook, partially covered, tossing frequently, until the watercress is wilted and the liquid it exudes has evaporated, about 5 minutes.

7. Stir in the red pepper. Drain the greens in a colander and press with a wooden spoon to force out any excess liquid. When cool enough to handle, chop coarsely and set aside.

8. Preheat the oven to 500°. Working quickly, divide the dough into 4 equal pieces. Form each piece of dough into a ball and roll it out on a floured surface to form a pizza round ¼ inch thick and 7 to 8 inches in diameter.

9. Place a large heavy ungreased baking sheet in the oven for 5 minutes. Place two dough rounds on the hot baking sheet, brush each lightly with 1½ teaspoons of the oil and return to the oven. Bake until pale golden brown, 8 to 12 minutes. Repeat with the other 2 rounds.

10. Spread the ham on the pizza crusts, leaving a ½- to 1-inch border. Divide the greens equally among the pizzas, covering the ham. Top with the cheeses, distributing them evenly and covering the greens completely. Drizzle on the remaining 2 tablespoons oil. Bake the pizzas, two at a time, until the cheese is melted and bubbly and the crust is golden brown, 5 to 8 minutes.

NOTE: To substitute kale for the watercress, wash and trim 1 pound of kale. Stack the leaves and cut into ¼-inch-wide strips. Blanch for 1 minute in boiling salted water. Drain thoroughly, then squeeze out any excess water. Cook the onion and garlic as the recipe directs. Add the kale and the crushed red pepper and toss over moderate heat for 1 minute. Proceed as directed in Step 7.
—Sarah Belk

TOMATO AND
MUSHROOM PIE
The base for this delicious pie is a thick, but light and tasty, pizza-like dough that is flavored with freshly grated Parmesan cheese.
Makes Two 12-Inch Pies

DOUGH
1½ tablespoons active dry yeast
¼ teaspoon sugar
1 cup warm water (105° to 115°)
About 3 cups all-purpose flour
½ cup (2 ounces) freshly grated Parmesan cheese
4 tablespoons olive oil
½ teaspoon salt

TOPPING
2 pounds (about 18) plum tomatoes—peeled, seeded and cut into ½-inch dice
2 teaspoons salt
8 medium mushrooms, thinly sliced
1 teaspoon oregano
1 teaspoon basil
½ teaspoon freshly ground pepper
8 ounces mozzarella cheese, coarsely shredded
6 tablespoons freshly grated Parmesan cheese
4 tablespoons olive oil

1. Make the dough: In a small bowl, combine the yeast and sugar. Add ¼ cup of the warm water and stir briefly to dissolve the yeast; let rest to proof for about 5 minutes, until bubbly.

2. In a large bowl, combine 3 cups of flour, the cheese, 3 tablespoons of the oil and the salt. Mix well with a fork, then stir in the remaining ¾ cup warm water and the yeast mixture. Stir until well blended. Turn the dough out onto a lightly floured surface and knead until

smooth and elastic, about 10 minutes; add a small amount of additional flour if necessary to prevent sticking. The dough should be soft and moist.

3. Grease a large bowl with the remaining 1 tablespoon oil. Place the dough in the bowl, turn it over once to grease the top, cover with a towel and let rise in a warm, draft-free place until doubled in bulk, about 1½ hours.

4. Meanwhile, toss the tomatoes with the salt and let drain on several thicknesses of paper towel for 15 minutes. Transfer to a sieve placed over a bowl and let drain until you are ready to assemble the pie.

5. Lightly coat two 12-inch round pizza pans or small baking sheets with olive oil. Punch the dough down and divide it in half. Place each piece on one of the pizza pans and pat out into an 11- or 12-inch circle, pinching the outside edge all around to raise the edge slightly. Cover loosely with kitchen towels or wax paper and let rise in a warm place until doubled in bulk, about 45 minutes. After about 30 minutes, preheat the oven to 450°.

6. If both pies will fit on the floor of your oven at the same time, then assemble both before baking. If only one will fit at a time, assemble one and, while it is baking, assemble the remaining pie. For each pie, arrange half of the tomato pieces evenly over one of the dough rounds. Arrange half of the mushroom slices over the tomatoes. Sprinkle evenly with ½ teaspoon each of the oregano and basil, and ¼ teaspoon of the pepper. Top with half the mozzarella and half the Parmesan; then drizzle 2 tablespoons of the olive oil over the pie. Repeat with the remaining ingredients and dough round for the second pie.

7. Bake in the center of the oven for 12 to 15 minutes, or until the crust has puffed, the bottom is lightly browned and the cheese is melted. Transfer the pie, still on the pan, to the floor of the oven if using a gas stove, or to the lowest shelf setting if using an electric stove, and bake for 2 to 3 minutes, until the bottom is golden brown and crisp. Serve hot, cut into wedges.
—*Jim Fobel*

DEEP-DISH
SPINACH PIE
When cut into small squares, this wonderful pie makes a great hors d'oeuvre. Serve warm or chilled.
8 Servings

PASTRY
1⅓ cups all-purpose flour
½ teaspoon salt
4 tablespoons unsalted butter, chilled and cut into thin slices
4 tablespoons lard or vegetable shortening
3 tablespoons ice water
1 egg yolk

FILLING
2 pounds spinach
½ teaspoon tarragon
¼ cup fresh lemon juice
1 cup (about 4 ounces) crumbled feta cheese or coarsely shredded sharp Cheddar cheese
4 whole eggs, at room temperature
2 cups heavy cream
½ teaspoon salt
¼ teaspoon freshly grated nutmeg

1. Make the pastry: In a large mixing bowl, combine the flour with the salt. Cut in the butter and lard until the mixture resembles coarse meal. Tossing with a fork, sprinkle on the ice water. Gather the dough into a ball. If any of the flour doesn't adhere, add a few more drops of ice water. Shape into a 6-inch disk. Wrap and refrigerate for at least 1 hour.

2. Preheat the oven to 400°. On a lightly floured surface, roll out the pastry to a 13-inch square. Loosely drape into an 8-inch square baking pan; press the dough to conform to the pan. Roll the overhanging pastry all around to make a raised edge and crimp decoratively. Place a sheet of aluminum foil slightly larger than the pan over the dough and press to conform to the pastry. Fill with pie weights. Bake until the crimped edge is set, about 10 minutes. Carefully remove the foil and pie weights, return the shell to the oven and bake, tapping the bottom with a wooden spoon if it bubbles up, for about 8 minutes, until the bottom is dry and firm.

3. Beat the egg yolk in a small bowl with a fork. Generously paint enough of the yolk over the bottom of the shell to coat. Return the shell to the oven until set, about 2 minutes. Let cool on a wire rack. Reduce the oven temperature to 350°.

4. Meanwhile, make the filling: Stem the spinach and rinse the leaves well. Place the leaves, with just the water clinging to them, in a large pot. Cook, covered, over moderately high heat, stirring once or twice, until wilted and greatly reduced in bulk, about 3 minutes. Drain the spinach into a colander; press with the back of a spoon to extract the excess liquid. Coarsely chop the spinach; there will be 1½ to 2 cups.

5. In a large bowl, combine the spinach with the tarragon, lemon juice and cheese. Toss the mixture to blend.

6. In another bowl, whisk together the eggs, cream, salt and nutmeg.

7. Arrange the spinach mixture in the partially baked pastry shell and pour the egg mixture over the top. Place the pie on a cookie sheet and bake until puffed all over and lightly speckled with golden brown, about 50 minutes. Cool on a rack for 1 hour, or refrigerate until chilled, before cutting.
—*Jim Fobel*

ASPARAGUS TART A LA BICYCLETTE

Asparagus, nature's deluxe harbinger of spring, is glorious in all its forms. This rich custard tart is delicious warm or at room temperature.

6 Servings

1½ pounds thin asparagus
Caraway Pie Dough (recipe follows)
4 ounces cream cheese, cut into small pieces, at room temperature
1 egg yolk
1 cup heavy cream
3 whole eggs
¾ teaspoon salt
½ teaspoon freshly ground white pepper
¼ pound thinly sliced boiled ham, cut into 2-by-⅛-inch strips
⅓ cup freshly grated Parmesan cheese

1. Peel the asparagus stems. Cut the tips to 2½ inches; thinly slice 2½ inches of the stalks. Discard the remainder.

2. In a medium saucepan of boiling water, blanch the asparagus until just tender, about 3 minutes. Drain and rinse under cold water; drain well.

3. Roll out the Caraway Pie Dough to a 12-inch round. Fit into a flour-dusted 10-inch tart pan with a removable bottom. Trim off the excess dough. Prick with a fork and freeze for 10 minutes.

4. Meanwhile, preheat the oven to 425°. In a bowl, blend the cream cheese with the egg yolk. Gradually mix in the cream; the texture will be slightly lumpy. Beat in the whole eggs, one at a time. Blend in the salt and white pepper.

5. Evenly distribute the sliced asparagus stalks and ham over the pastry. Pour in half of the custard mixture. Bake on the lowest shelf of the oven for 15 minutes.

6. Remove the tart from the oven and pour in the remaining custard. Arrange the asparagus tips around the pie like the spokes of a wheel. Sprinkle the Parmesan cheese over the top.

7. Reduce the oven temperature to 375°. Bake the tart on the bottom shelf of the oven to 40 minutes, or until a skewer inserted in the center comes out clean. Let the tart rest for 15 minutes before serving.
—W. Peter Prestcott

CARAWAY PIE DOUGH
Makes a Single 10-Inch Tart Shell

1½ cups all-purpose flour
⅛ teaspoon salt
½ teaspoon caraway seeds
5 tablespoons cold unsalted butter, cut into tablespoons
2 tablespoons chilled vegetable shortening, cut into tablespoons
2 to 3 tablespoons ice water

1. In a food processor, combine the flour, salt and caraway seeds. Process for 3 seconds. Add the butter and shortening and process, turning the machine quickly on and off, until the mixture resembles oatmeal.

2. With the machine on, pour in 2 tablespoons of the ice water and process just until the dough begins to mass together; do not overprocess. Add more water if the pastry does not mass together. Gather the dough into a ball and pat into a 6-inch disk. Cover with waxed paper and refrigerate until firm.
—W. Peter Prestcott

THREE-ONION TART
6 Servings

PASTRY
1½ cups all-purpose flour
½ teaspoon salt
1 stick (4 ounces) cold unsalted butter, cut into thin slices
4 to 5 tablespoons ice water
1½ tablespoons Dijon mustard

FILLING
5 tablespoons unsalted butter
4 medium yellow onions, chopped
2 medium red onions, sliced ⅛ inch thick
¼ pound Gruyère cheese, grated
12 white onions, each about 1 inch in diameter
1 tablespoon sugar
2 whole eggs
1 egg yolk
1 cup heavy cream
¼ teaspoon salt
¼ teaspoon freshly ground white pepper

1. Make the pastry: Place the flour, salt and butter in a large bowl. Using a pastry blender or two knives, cut the butter into the flour until the mixture resembles coarse meal. Quickly stir in 4 tablespoons of the ice water with a fork. If the dough does not hold together easily when squeezed, blend in as much as 1 tablespoon more of ice water. Do not overmix. Pat the dough into a flat round about 6 inches in diameter; wrap in waxed paper and refrigerate for at least 30 minutes.

2. Preheat the oven to 425°. Place the pastry dough between two sheets of wax paper and roll it into a 12-inch round. Line an 11-inch quiche pan with the pastry, fitting it carefully into the pan without stretching. Prick the dough

all over with a fork and fit a piece of aluminum foil into the pastry shell. Fill the foil with dried beans or rice and bake for 15 minutes.

3. Remove the pastry shell from the oven; remove the foil and beans or rice. Brush the interior of the pastry shell with the mustard and return it to the oven for 5 minutes to crisp the shell. Remove and reduce the oven temperature to 350°.

4. Make the filling: In a large skillet, melt 3 tablespoons of the butter over low heat and cook the yellow onions, stirring, until they are soft but not brown, about 5 minutes. Transfer the onions to a plate.

5. Melt 1 more tablespoon of the butter in the skillet. Add a layer of red onion slices and cook over moderate heat until they are slightly colored on the underside. (Do not disturb the slices or they may separate into rings.) Remove with a slotted spoon and set aside. Repeat with the remaining onion slices and butter.

6. Layer the yellow onions over the pastry shell and sprinkle the cheese evenly over them. Arrange the red onion slices on top, cooked-sides up, leaving a 1-inch border around the edge.

7. Add the remaining 1 tablespoon of butter to the skillet and cook the white onions for 2 minutes, tossing them so that they cook on all sides. Sprinkle the sugar over them and, tossing again, cook them until the sugar caramelizes, about 2 minutes. Reserving the glaze in the pan, remove the onions and cut them in half lengthwise. Arrange them, cut-sides down, around the edge of the tart. Place the tart on a baking sheet.

8. In a bowl, beat together the whole eggs, egg yolk, cream, salt and pepper until blended. Pour the mixture into the tart shell over the onions.

9. Bake the tart for 20 minutes. Reduce the temperature to 325° and bake for an additional 10 minutes, or until the filling is set.

10. Meanwhile, add ½ cup of water to the skillet and, stirring to blend in the caramelized sugar, reduce the liquid to about 2 tablespoons. When the tart is done, brush the surface with the glaze reserved in the skillet. Serve warm or at room temperature.
—F&W

VIDALIA ONION TART

Like a pizza, this onion tart, which owes its gentle character to sweet Vidalia onions, can bring happiness to a sizable crew. It's rich with eggs, cream and milk, highlighted by the fragrance of nutmeg.

❗ Beaujolais, such as a Morgon from Duboeuf, has the fruitiness and acidity to enhance the onion and bacon flavors.

4 to 6 Servings

1½ cups all-purpose flour
Salt
1½ sticks (6 ounces) cold unsalted butter
¼ cup ice water
3 large Vidalia onions (about 2½ pounds), thinly sliced
¼ teaspoon paprika
⅛ teaspoon cayenne pepper
½ pound sliced bacon, cut crosswise into ¼-inch strips
2 whole eggs
1 egg yolk
½ cup heavy cream
¼ cup milk
⅛ teaspoon freshly grated nutmeg

1. In a large bowl, combine the flour with a pinch of salt. Cut 1 stick of the butter into ½-inch cubes. Cut the butter cubes into the flour until the mixture resembles coarse meal. Add the ice water in a thin stream, tossing until the dough begins to mass together. Add a few more drops of water if necessary. Gather the dough into a ball and press into a 6-inch disk. Wrap tightly; refrigerate until cold but still malleable, 30 to 60 minutes.

2. Meanwhile, in a large heavy saucepan or flameproof casserole, melt the remaining 4 tablespoons butter over moderate heat. Add the onions and reduce the heat to low. Cook, stirring occasionally, until the onions are very soft and golden, about 40 minutes. Pour the onions into a colander and drain off any excess liquid. Place in a large bowl and let cool. Season with ½ teaspoon salt, the paprika and the cayenne.

3. On a lightly floured surface, roll out the dough ⅛ inch thick. Fit into an 11-by-1-inch tart pan with a removable bottom, pressing the dough against the sides of the pan without stretching it. Trim the crust even with the rim of the pan. Refrigerate or freeze for at least 30 minutes.

4. Preheat the oven to 350°. Put the chilled tart shell on a baking sheet. Line the crust with aluminum foil and fill with pie weights or dried beans. Bake until almost cooked through, about 40 minutes. Remove the foil and weights and cook until golden brown and crisp, 15 minutes. Let cool slightly.

5. Meanwhile, in a large skillet, fry the bacon strips over moderately high heat, stirring occasionally, until lightly browned, about 5 minutes. Drain on paper towels.

6. In a medium bowl, blend the whole eggs and egg yolk together. Mix in the cream and milk. Strain the custard through a fine sieve. Season with the nutmeg. Arrange the onions in the tart shell. Scatter the bacon over the onions and pour the custard mixture over all. Return to the oven and bake for 25 minutes, or until the custard sets.
—*Brill Williams*

RATATOUILLE PIE
WITH BASIL CRUST

This is one for the vegetarian crowd to be sure, but it's not to be passed up even if you aren't a card-carrying member. ❡ Highly seasoned and topped with tangy Parmesan cheese, this pot pie calls for the contrast of a clean, striking white, such as the Ruffino Libaio or Antinori Galestro.

6 Servings

BASIL CRUST
1½ cups plus 1 teaspoon all-purpose flour
1 teaspoon sugar
½ teaspoon salt
6 tablespoons vegetable shortening
1 egg yolk
1 teaspoon red wine vinegar
2 to 3 tablespoons cold water
2 tablespoons chopped fresh basil

RATATOUILLE
1 large eggplant (about 1½ pounds)
¾ teaspoon salt
½ cup extra-virgin olive oil
2 small zucchini (about 1 pound), sliced
1 tablespoon unsalted butter
1 large onion, thinly sliced
2 large garlic cloves, minced
1 medium green bell pepper, thinly sliced
1 medium red bell pepper, thinly sliced
2 medium tomatoes, peeled and cut into ½-inch wedges
¼ teaspoon freshly ground black pepper
½ teaspoon hot Hungarian paprika
¼ cup chopped parsley
2 tablespoons chopped fresh basil
2 cups grated Italian Fontina cheese (about 6 ounces)
½ cup freshly grated Parmesan cheese
1 egg white, lightly beaten

1. Make the crust: In a medium bowl, combine 1½ cups of the flour with the sugar and salt. Cut in the vegetable shortening until the mixture resembles coarse crumbs. Lightly beat the egg yolk with the vinegar, water and basil. Mix the liquid into the flour mixture just until a soft dough forms. Sprinkle with the remaining 1 teaspoon flour, cover with plastic wrap and refrigerate for 1 hour.

2. Prepare the ratatouille: Halve the eggplant lengthwise and cut crosswise into ¼-inch-thick slices. Put the eggplant in a colander, sprinkle with ½ teaspoon of the salt and let stand for 30 minutes. Pat the eggplant dry with damp paper towels to remove the salt.

3. In a large heavy skillet, heat 2 tablespoons of the oil over moderately high heat. Add the eggplant slices in batches and sauté, tossing, until golden, about 2½ minutes per side. Repeat with the remaining eggplant, adding up to 5 tablespoons more oil as needed. Drain the eggplant on paper towels, then transfer to a bowl.

4. Add the remaining 1 tablespoon of oil to the same skillet, add the zucchini and cook over high heat, tossing constantly, until golden, 2 to 3 minutes. Reduce the heat to moderate and cook until soft, about 3 minutes longer. Drain and add to the eggplant in the bowl.

5. Wipe the skillet clean. Add the butter and melt over moderate heat. Add the onion and cook for 2 minutes. Stir in the garlic and green and red bell peppers, cover and cook until the peppers are soft, about 5 minutes. Stir in the tomatoes, remaining ¼ teaspoon salt, the black pepper and paprika. Increase the heat to high and cook until all the liquid has evaporated, 2 to 3 minutes. Remove from the heat and add the eggplant, zucchini, parsley and basil; mix well.

6. Preheat the oven to 425°. Spoon one-third of the ratatouille into a buttered 10-inch glass or ceramic quiche dish. Sprinkle with one-third each of the Fontina and Parmesan cheeses. Continue layering until all the ingredients are used, ending with the Parmesan cheese.

7. Roll out the pastry on a lightly floured board and lay it over the dish. Trim and flute the edges. Cut a slash in the center of the pie. Brush with the beaten egg white and bake for 25 minutes, or until golden brown. Let stand for 5 minutes before serving.
—*Phillip Stephen Schulz*

PIPERADE PIE

I like to serve stewed corn made the southern way with this delicious, homey pie. Simply scrape off the kernels, add a pat of butter, salt and white pepper to taste, and just enough milk or cream (a few tablespoons) to make it liquid. Bring to a simmer and cook over moderately low heat for three to four minutes.

6 Servings

¼ cup extra-virgin olive oil
1 small red bell pepper, coarsely chopped
1 sweet onion, coarsely chopped
¼ cup coarsely chopped scallions
2 large garlic cloves, minced
6 plum tomatoes—peeled, seeded and coarsely chopped
1 tablespoon coarsely chopped fresh basil or 1 teaspoon dried
1 teaspoon salt
1 teaspoon freshly ground black pepper
4 dashes of hot pepper sauce
1 tablespoon unsalted butter
5 eggs, lightly beaten
1 prebaked Savory Cheese Crust (recipe follows)

1. Preheat the oven to 350°. In a large skillet, heat the olive oil over moderate heat. Add the red pepper, onion and scallions. Sauté, stirring occasionally, until the vegetables are softened

but not browned, about 10 minutes. Add the garlic, tomatoes, basil, salt, black pepper and hot sauce. Bring to a boil, reduce the heat and simmer, stirring frequently, until almost all the liquid evaporates, about 10 minutes. Stir in the butter and remove from the heat. Transfer to a bowl and let cool for 15 minutes.

2. Add the eggs to the cooled vegetable mixture and blend well. Spread the filling into the prebaked Savory Cheese Crust and bake for 30 minutes, or until the filling is set but not dry. Let stand for 15 minutes before slicing. Serve warm, at room temperature or slightly chilled.
—Lee Bailey

SAVORY CHEESE CRUST
Makes a Single 10-Inch Crust

1 cup all-purpose flour
Pinch of salt
4 tablespoons frozen unsalted butter, cut into small cubes
2 tablespoons frozen vegetable shortening, cut into small cubes
1 cup coarsely grated Cheddar cheese
2 tablespoons ice water

1. In a medium bowl, mix together the flour and salt. Cut in the butter, shortening and cheese until the mixture resembles coarse meal. Stir in the water, gather the dough into a ball and flatten into a 6-inch disk. Cover with plastic wrap and refrigerate until chilled, at least 30 minutes.

2. Preheat the oven to 400°. On a lightly floured surface, roll out the pastry into a 12-inch circle. Fit into a 10-inch pie pan. Trim the dough to ½ inch, turn under and crimp the edges. Prick the pastry all over with a fork. Cover with plastic wrap and refrigerate for 15 minutes.

3. Line the pastry with foil and fill with pie weights or dried beans. Bake for 15 minutes, or until set. Remove the pie weights and continue baking until golden, about 10 minutes. Let cool completely on a rack before filling.
—Lee Bailey

NANCY SILVERTON'S POTATO PIE
My kids are usually very fussy eaters; they like mild flavors. This recipe contains three of their favorite foods—bacon, potatoes and eggs.
8 Servings

5 medium boiling potatoes
½ pound thinly sliced pancetta or bacon, cut crosswise into thin strips
2 medium onions, thinly sliced
Salt
Sugar
¼ teaspoon freshly ground pepper
2 tablespoons olive oil
1½ tablespoons fresh rosemary or 2 teaspoons dried
3 garlic cloves, minced
5 eggs
2 cups heavy cream
Pinch of freshly grated nutmeg
3 cups shredded Gruyère cheese (about ½ pound)

1. Bring a large saucepan of water to a boil. Add the potatoes and cook over high heat for 5 minutes. Drain and pat dry; let cool slightly.

2. In a 12-inch ovenproof skillet, cook the pancetta over moderate heat, stirring until almost all the fat has been rendered, about 12 minutes. Using a slotted spoon, transfer the pancetta to paper towels to drain, then finely chop.

3. Set aside 2 tablespoons of the rendered fat. Leave 1 tablespoon of fat in the skillet and add the onions. Season with a pinch each of salt and sugar. Cover and cook over moderate heat, stirring occasionally, until softened and browned, about 25 minutes. Remove the onions from the skillet and set aside.

4. Meanwhile, peel the potatoes and then grate them on the coarse side of a grater or in a food processor fitted with a shredding disk. Season the potatoes with the pepper and 1 teaspoon of salt.

5. Preheat the oven to 350°. Add the reserved 2 tablespoons of pancetta fat to the skillet and rub it over the bottom and sides generously. Add the grated potatoes and evenly press them into the bottom and sides of the pan to form a crust. Cook over moderately low heat until the potatoes begin to brown on the bottom, about 15 minutes.

6. Meanwhile, in a small skillet, heat the olive oil. Add the rosemary and garlic and cook over moderate heat, stirring, until fragrant, about 3 minutes. Remove from the heat.

7. In a bowl, beat the eggs with the cream until blended. Season with ¼ teaspoon salt, the nutmeg and two-thirds of the herbed olive oil.

8. When the potato crust has begun to brown, drizzle the remaining herbed oil over the surface. Spread the reserved onions over the potatoes and sprinkle the pancetta and Gruyère on top. Pour the custard mixture into the potato crust and bake the pie for about 50 minutes, rotating the pan once, until the custard is set and golden on top. Remove the pie from the oven and let cool for about 10 minutes before serving in wedges.
—Nancy Silverton

Green Pea Soup (p. 78).

Bacon, Potato, White Bean and Red Pepper Soup (p. 72).

Clockwise from left: Red Tomato Salsa (p. 262), Yellow Tomato Salsa (p. 263) with flour tortillas, Tomatillo Salsa (p. 262) with blue corn chips.

PARMENTIER PIE

Potage Parmentier was the first recipe in the first chapter of Julia Child's first cookbook, *Mastering the Art of French Cooking*, and it was the first soup I ever made. As she promised, it was simplicity itself, but it made me feel like a real country cook from some little French farm town. I later learned that the French have a similar affection for this dish. One evening, after a very special and luxurious multi-course meal prepared by Daniel Boulud, former chef at Le Cirque in New York, I asked what he had eaten for dinner: Boulud sighed with contentment when he said, "Potage Parmentier, just as my mother used to make it. I still love it so." I've never stopped cooking this soup and just recently started making a pie based on the same satisfying ingredients, with the addition of mushrooms. I think of it as a solid rendition of the treasured soup.

4 to 6 Servings

1 ounce dried porcini mushrooms
4 Idaho potatoes (about 2¼ pounds), peeled and quartered
¼ cup plus 3 tablespoons olive oil
¼ cup plus 3 tablespoons warm milk
¾ teaspoon salt
½ teaspoon freshly ground pepper
Cayenne pepper
3 small leeks (white and tender green), thinly sliced
1 medium onion, chopped
½ pound fresh mushrooms, thinly sliced
2 garlic cloves, minced
2 tablespoons dry vermouth
½ teaspoon herbes de Provence, or substitute ⅛ teaspoon each of marjoram, oregano, tarragon and thyme
¼ cup freshly grated Parmesan cheese

1. In a small saucepan, gently simmer the dried mushrooms in 2 cups of water over moderate heat until soft, about 20 minutes. Drain and rinse the mushrooms and reserve the cooking liquid. Cut off and discard any tough bits from the mushrooms. Halve the mushrooms if the pieces are very large. Strain the liquid.

2. Meanwhile, boil the potatoes in a large pot of salted water over moderate heat until very tender, about 30 minutes. Drain well. Return the potatoes to the pot and shake over very low heat for 30 seconds to dry them out.

3. Place the potatoes in a large mixer bowl and beat with a paddle until smooth (or pass them through a food mill or ricer). Mix in ¼ cup of the olive oil, ¼ cup of the warm milk, ½ teaspoon of the salt, ¼ teaspoon of the black pepper and a pinch of cayenne.

4. Preheat the oven to 425°. In a large skillet, heat the remaining 3 tablespoons olive oil. Add the leeks, onion, fresh mushrooms and garlic. Cook over moderately high heat, stirring occasionally, until the vegetables are softened, about 10 minutes.

5. Mix in the porcini mushrooms and their reserved liquid. Cook, stirring occasionally, until the liquid is almost evaporated, about 8 minutes.

6. Add the vermouth and cook for 1 minute. Season with the remaining ¼ teaspoon salt, ¼ teaspoon black pepper, a pinch of cayenne and the *herbes de Provence*.

7. Spread half the mashed potatoes over the bottom of a lightly oiled 10-inch pie dish. Spread the mushroom mixture over the potatoes. Spread the rest of the potatoes evenly over the mushrooms. Brush the surface with the remaining 3 tablespoons milk. Run the tines of a fork over the potatoes to form a crisscross pattern and sprinkle the cheese on top. Bake in the upper third of the oven for 20 minutes, or until golden.
—*Dorie Greenspan*

POTATO PIE A L'ALSACIENNE

Says André Soltner, "My mother, Mimi Soltner, made this potato tart—the favorite of my father, my sister and my brother. It is not nouvelle cuisine."

8 Servings

2 egg yolks
1¾ cups all-purpose flour
¾ teaspoon salt
1 stick (4 ounces) plus 1 tablespoon unsalted butter, cut into pieces
1¼ pounds boiling potatoes (about 3 medium), peeled and thinly sliced
¼ cup chopped parsley
¼ teaspoon freshly ground pepper
5 ounces sliced mild-smoked bacon, cut crosswise into ¼-inch-wide strips
5 hard-cooked eggs, peeled and thinly sliced
½ cup crème fraîche or heavy cream

1. The day before you make the pie, beat 1 egg yolk with enough ice water to make ¼ cup.

2. Mix the flour and ¼ teaspoon of the salt in a bowl. Rub in the butter with your fingertips until the mixture resembles coarse meal. Pour in the egg yolk beaten with ice water and work with the fingers until just moistened. Gather the dough into a ball. Immediately roll out the dough into a rectangle on a floured work surface. Fold in thirds, wrap in plastic and refrigerate overnight.

3. Preheat the oven to 400°. Divide the dough in half. Roll one half into a 13-inch round. Fit into a 9-inch pie pan or tart pan with a removable bottom. Refrigerate for 10 minutes.

4. Rinse the sliced potatoes in cold water to remove any starch; drain and pat dry on paper towels. Transfer to a bowl and toss with the remaining ½ teaspoon salt, the parsley and pepper.

5. In a medium skillet, sauté the bacon over moderately high heat, stirring,

until browned on the edges, 1 to 2 minutes. Drain the bacon on paper towels.

6. Arrange a layer of overlapping slices of potato in the bottom of the prepared pie shell. Cover with the bacon. Arrange the egg slices over the bacon and top with the remaining potato slices. Spoon the crème fraîche over the potatoes and spread smooth with a spatula.

7. Preheat the oven to 400°. Beat the remaining egg yolk with 1 teaspoon water to make an egg glaze. Roll the remaining pastry into a round large enough to cover the pie. Brush the edges of the lower crust with some of the egg glaze and cover the pie with the top crust. Trim the edges, then crimp to seal.

8. Brush the top crust lightly with more egg glaze. Cut steam vents in the top of the pie with a small sharp knife. Bake in the middle of the oven for 20 minutes.

9. Reduce the oven temperature to 350° and bake for 1 hour. Reduce the oven temperature to 300° and bake for 10 minutes. Let the pie rest for 10 minutes before serving.
—André Soltner

AVOCADO FRITTATA
1 Serving

½ of an avocado, cut into
   1-inch chunks
½ teaspoon fresh lemon juice
⅛ teaspoon salt
1 ounce goat cheese, such as
   Montrachet or Bucheron,
   crumbled (about ¼ cup)
¼ teaspoon rosemary, crushed
4 oil-cured black olives, pitted and
   cut into slivers
3 eggs
¼ teaspoon coarsely ground pepper
1½ teaspoons olive oil

1. Preheat the broiler. In a small bowl, toss the avocado with the lemon juice and salt. Add the goat cheese, rosemary and olives; toss gently.

2. Beat the eggs with the pepper and a pinch of salt.

3. Set an ovenproof omelet pan or heavy medium skillet over moderately high heat and add the oil. When the oil is hot but not smoking, add the eggs. Cook for about 1½ minutes, stirring once or twice, until the bottom is set and the top is still slightly runny. Remove from the heat and sprinkle the avocado mixture evenly over the eggs.

4. Broil about 4 inches from the heat until the top of the frittata is set and the edges slightly browned, about 2 minutes.
—Anne Disrude

PEPPER AND SPAGHETTI FRITTATA
🍷 Beaujolais or California Gamay
6 to 8 Servings

6 tablespoons extra-virgin olive oil
2 medium onions, chopped
5 medium garlic cloves, minced
10 bell peppers, chopped
   (about 4 cups)
⅛ teaspoon sugar
½ teaspoon oregano
1 teaspoon salt
¼ teaspoon freshly ground
   black pepper
6 ounces spaghetti, cooked
   and drained
8 eggs
½ cup freshly grated Parmesan
   cheese
¼ cup minced parsley

1. In a large ovenproof skillet, heat the olive oil. Add the onions, garlic and bell peppers, and sauté over moderately high heat, stirring frequently, until the vegetables begin to brown, about 30 minutes. Season with the sugar, oregano, ½ teaspoon of the salt and ⅛ teaspoon of the black pepper.

2. Preheat the oven to 350°. Add the cooked spaghetti to the skillet and toss well. Cook, stirring occasionally, until the pasta is lightly browned, about 10 minutes.

3. In a medium bowl, beat the eggs with ¼ of water and the remaining ½ teaspoon salt and ⅛ teaspoon black pepper. Stir in ¼ cup of the Parmesan cheese and 2 tablespoons of the parsley.

4. Pour the egg mixture over the pasta and stir with a fork to distribute evenly. Cook without stirring until the eggs are set around the edges. Place in the oven and bake until the eggs are set, about 10 minutes.

5. Slide the frittata onto a platter and cut into wedges. Toss together the remaining ¼ cup Parmesan cheese and 2 tablespoons minced parsley and pass separately.
—Anne Disrude

POTATO FRITTATA SANDWICHES WITH ROASTED RED PEPPERS
With a glass of wine, these peasant sandwiches become a meal fit for a king.
4 Servings

4 large red bell peppers
6 eggs
2 tablespoons heavy cream
¼ teaspoon salt
¼ teaspoon freshly ground
   black pepper
2 medium baking potatoes (about 6
   ounces each), peeled and sliced
   ¼ inch thick
½ cup extra-virgin olive oil
1 medium onion, thinly sliced
Garlic-Olive Mayonnaise
   (recipe follows)
8 large slices of Portuguese bread
   or round peasant bread,
   lightly toasted

1. Roast the peppers directly over a gas flame or under the broiler as close to the heat as possible, turning until charred all over. Place the peppers in a bag and let steam for 10 minutes. Peel the peppers and remove the cores, seeds and ribs. Cut the peppers into ½-inch slices and set aside. *(The peppers can be roasted several days ahead and stored whole, but cored and seeded, in the refrigerator.)*

2. In a medium bowl, combine the eggs with the cream, salt and black pepper. Beat lightly until blended.

3. In a medium saucepan, cook the potatoes in boiling water over moderately high heat until tender, about 10 minutes; drain well.

4. Preheat the oven to 400°. In a large ovenproof skillet, heat ¼ cup of the olive oil over moderate heat. Add the onion, reduce the heat to low and cook until it is softened and golden, about 12 minutes.

5. Increase the heat to high and add the remaining ¼ cup olive oil. Stir in the potatoes. Add the beaten eggs and tilt the pan to evenly distribute.

6. Transfer to the oven and bake the frittata until firm to the touch, about 12 minutes. Return the skillet to high heat for about 30 seconds, shaking to loosen the frittata. Slide onto a very large plate or cutting board and let cool slightly. Cut the frittata into 4 lengthwise strips, then halve the strips crosswise.

7. Spread a tablespoon or two of the Garlic-Olive Mayonnaise on 4 of the toast slices. Top the remaining toast with the frittata slices and the roasted pepper strips and put the sandwiches together. Serve the sandwiches warm or at room temperature.
—*Marcia Kiesel*

GARLIC-OLIVE MAYONNAISE
Makes About ½ Cup

1 large garlic clove, minced
1 teaspoon chopped drained capers
1 teaspoon chopped oil-cured olives
1 anchovy fillet, mashed
½ cup mayonnaise

Stir the garlic, capers, olives and anchovy into the mayonnaise.
—*Marcia Kiesel*

ASPARAGUS PANCAKE WITH MOZZARELLA
For serving, treat this vegetable pancake as you would an omelet. It makes a fine brunch or light lunch dish.
1 or 2 Servings

10 to 12 large asparagus spears, trimmed and peeled
1 egg, beaten
¼ teaspoon salt
Pinch of freshly ground pepper
2 teaspoons olive oil
2 thin slices of mozzarella cheese

1. In a large pot of boiling salted water, cook the asparagus until tender but still bright green, about 3 minutes. Drain and rinse under cold running water; drain well.

2. Reserve 2 asparagus tips for garnish and coarsely chop the remainder. Place the pieces in a clean towel and press out the moisture. Stir the asparagus into the beaten egg. Add the salt and pepper.

3. Warm a large skillet over moderately high heat for 1 minute. Add the olive oil and let heat for 30 seconds. Scrape in the asparagus mixture and press it into an even layer. Reduce the heat to low and cook until the pancake is almost set, about 5 minutes.

4. Slide the pancake onto a plate. Cover the plate with the skillet and invert so that the pancake slides back into the pan, uncooked-side down. Place the mozzarella on top of the pancake, cover the skillet and cook until the cheese melts, about 1 minute. Slide the pancake onto a serving plate. Garnish with the asparagus tips.
—*Anne Disrude*

HUEVOS RANCHEROS NEW YORK STYLE
In this variation on the Mexican classic, thin egg crêpes, spiked with scallions and cayenne, replace the traditional fried eggs and tortillas.
❦ Full-flavored beer, such as Samuel Adams
4 Servings

6 tablespoons olive oil
1 large onion, chopped
2 large garlic cloves, minced
½ teaspoon oregano
½ teaspoon chili powder
¾ teaspoon cumin
¾ teaspoon salt
1 teaspoon Dijon mustard
2 cans (15 ounces each) pinto beans, rinsed and drained
1 can (14 ounces) Italian peeled tomatoes, drained and chopped
½ cup sour cream
2 large fresh plum tomatoes— peeled, seeded and chopped
½ cup coarse fresh bread crumbs (made from 1½ slices of firm-textured white bread)
8 eggs
Pinch of cayenne pepper
¼ cup minced scallion greens
Fresh Tomato Salsa (recipe follows)

1. In a large skillet, heat 3 tablespoons of the olive oil over moderate heat. Add the onion and sauté until golden, about 10 minutes. Add the gar-

lic, oregano, chili powder, ½ teaspoon of the cumin and ¼ teaspoon of the salt. Sauté for 1 minute longer.

2. Add the mustard, beans and canned tomatoes to the skillet and cook over low heat, stirring occasionally, until the beans form a puree, about 20 minutes. Stir in ¼ cup of the sour cream and the chopped fresh plum tomatoes. Remove from the heat and set aside. *(The recipe can be made to this point several days ahead.)*

3. In a medium skillet, heat 2 tablespoons of the olive oil over moderate heat. Add the bread crumbs and cook, tossing frequently, until golden and very crunchy, about 15 minutes. Toss with ¼ teaspoon of the salt and the remaining ¼ teaspoon cumin.

4. Beat the eggs with 2 tablespoons of water, the remaining ¼ teaspoon salt, the cayenne and the minced scallion greens.

5. Preheat the oven to 250°. Heat a 7-inch crêpe pan or nonstick skillet over moderate heat. Brush with some of the remaining olive oil to grease lightly. Add about 2 tablespoons of the beaten egg mixture and immediately tilt and swirl the skillet to form a thin, even crêpe. Cook until the egg is set and only slightly wet on top. Turn the crêpe and cook for 5 seconds on the second side. Transfer the egg crêpe to a cookie sheet and place in the oven to keep warm. Continue cooking the remaining eggs in the same manner, lightly oiling the skillet each time, to make 8 egg crêpes.

6. To assemble the dish, reheat the bean mixture over moderate heat until warm. Spread about 1 heaping tablespoon of beans onto each crêpe. Fold into a triangle and place 2 on each of 4 warmed plates. Spoon about 3 tablespoons of Fresh Tomato Salsa, a sprinkling of bread crumbs and a small dollop of sour cream over each serving.

—*Anne Disrude*

FRESH TOMATO SALSA

This sauce can be spooned over your favorite meats or served simply as a dip with tortilla chips.

Makes About 1½ Cups

5 scallions, coarsely chopped
6 large plum tomatoes—peeled, seeded and chopped
1 tablespoon olive oil
½ teaspoon salt
¼ teaspoon cumin
¼ teaspoon oregano
Pinch of cayenne pepper

In a large bowl, combine the scallions and tomatoes. Season with the olive oil, salt, cumin, oregano and cayenne. Mix well. Cover and refrigerate for up to 3 hours before serving.

—*Anne Disrude*

HOMINY AND RED CHILE STEW

Hominy is dent corn (so called for its characteristic indentation at the crown of the kernel), which has been treated with slaked lime or ash. Its flavor is nutty and nice, and the chewy texture of the large plump kernels is very satisfying. You can buy it dried and treat it as you would beans, but it's far easier to use the canned variety that's available at supermarkets.

You can easily vary this recipe by adding tomatoes or by using spices like cumin, cinnamon and clove. For the richest flavor possible, use pure ground dried chiles from New Mexico. The amount called for below will make a spicy dish, but the sour cream will soften the heat considerably. So will serving thick tortillas alongside to mop up the sauce. As with most stews, the flavor of this one improves upon sitting. It also freezes well.

▼ The earthy flavors and heat of the chile call for something equally spicy—California Zinfandel, preferably Burgess or Nalle.

4 Servings

2½ tablespoons light olive, safflower or peanut oil
1 medium onion, coarsely chopped
1 large garlic clove, finely chopped
¾ pound butternut or banana squash, peeled and cut into ½-inch cubes
1 teaspoon Greek or Mexican oregano
½ teaspoon salt
2 tablespoons pure chile powder (see headnote)
1 tablespoon all-purpose flour
2 cans (16 ounces each) hominy, drained and rinsed
1 green bell pepper, cut into ½-inch dice
½ cup sour cream
Chopped fresh coriander (cilantro)

1. In a large casserole, heat the oil over moderate heat. Add the onion, garlic, squash, oregano and salt and cook, stirring frequently, for 5 minutes. Stir in the chile powder and flour until blended.

2. Add the hominy and enough water to cover generously, about 3 cups. Cook over moderate heat, stirring occasionally, until the squash is barely tender, about 20 minutes.

3. Add the green pepper and cook, stirring occasionally, until the vegetables are tender, about 15 minutes longer. *(The recipe can be prepared ahead to this point and refrigerated, covered, for up to 3 days or frozen for 1 month. Reheat slowly and thoroughly before proceeding.)*

4. Transfer the stew to a serving bowl and swirl in the sour cream. Sprinkle with coriander.

—*Deborah Madison*

THREE-BEAN CASSEROLE

The next time you have a leftover ham bone, try this casserole. During the long baking, the flavors of the ham and vegetables permeate the three varieties of dried beans. Since the beans must soak overnight, plan your time accordingly.

6 to 8 Servings

½ pound dried baby lima beans
½ pound dried small black beans
½ pound dried red kidney beans or small red beans
1 tablespoon minced garlic
5 medium onions, thinly sliced and separated into rings
3 celery ribs, finely diced
2 to 3 cups chicken broth
1 tablespoon basil
½ cup malt vinegar
¼ cup molasses
2 teaspoons salt
½ teaspoon freshly ground pepper
1 large meaty ham bone or 2 smoked ham hocks

1. The day before you wish to bake the casserole, rinse all three varieties of beans and combine them in a large bowl. Cover them with cold water and soak overnight.

2. Drain the beans, place them in a large pot and add cold water to cover the beans by 1 inch. Place the pot over moderate heat and bring the water to a boil. Reduce the heat to low and simmer the beans for 30 minutes. Drain well, reserving the liquid.

3. Place the beans in a large bowl. Add the garlic, onion rings and celery, tossing well. Measure the bean cooking liquid and add enough chicken broth to make 3 cups; add to the beans and stir to blend.

4. Preheat the oven to 400°. In a small bowl, combine the basil, vinegar, molasses, salt and pepper and pour over the beans.

5. Place the ham bone in the bottom of a 4- or 5-quart casserole. Pour the bean mixture over the ham bone, cover the casserole and bake for 30 minutes. Reduce the temperature to 300° and bake, covered, for 2 hours.

6. Uncover the casserole and bake until only about 1½ cups of liquid remain. Let the casserole rest at room temperature at least 30 minutes before serving.

—F&W

MILLET-AND-EGGPLANT CASSEROLE

This spicy casserole, served with cold yogurt, also makes a delicious accompaniment to a roast leg of lamb. To make it a completely vegetarian dish, substitute 1 cup tomato juice and ⅓ cup water for the chicken broth in the ingredient list.

4 Servings

⅓ cup plus 2 tablespoons olive oil
1 cup millet
1⅓ cups chicken broth
⅔ cup dry vermouth
1 medium eggplant, cut into ½-inch cubes
1 large red bell pepper, cut into ½-inch squares (2 cups)
1 medium onion, minced
2 garlic cloves, minced
¼ teaspoon coriander
¾ teaspoon cumin
2 teaspoons dill
1½ teaspoons dry mustard
2 teaspoons mint
½ teaspoon ground cardamom
¼ teaspoon cayenne pepper
½ teaspoon salt
2 cups plain yogurt, for serving

1. Preheat the oven to 400°. In a medium skillet warm 1 tablespoon of the oil over moderately high heat until hot but not smoking. Add the millet and stir constantly until toasted and golden, about 4 minutes. Add the chicken broth and vermouth and bring just to a boil. Transfer the millet and liquid to a shallow 1½-quart baking dish, cover tightly, and bake for 20 minutes.

2. Meanwhile, in a large heavy skillet, warm ⅓ cup of the oil over moderately high heat. When the oil is almost smoking, add the eggplant and bell pepper and stir constantly for 6 minutes.

3. Add the remaining 1 tablespoon oil and the onion and garlic to the skillet; continue cooking and stirring over high heat for 3 minutes. Remove from the heat and stir in the coriander, cumin, dill, mustard, mint, cardamom, cayenne and salt. Stir until blended.

4. Add the baked millet and any remaining broth to the skillet. Stir the ingredients together, making sure to incorporate the brown bits that cling to the pan.

5. Turn the entire mixture back into the casserole, cover, and bake for 30 minutes, or until the eggplant is just tender and the peppers still firm. Serve with a side dish of cold yogurt.

—F&W

GARLIC-BRAISED EGGPLANT AND CHICKPEA CASSEROLE

Here's a savory Indian meatless main dish. Serve it with a chunk of crusty bread and a nice green salad.

❦ Given the Indian spices in this hearty dish, a light but hop-flavored lager, such as Corona or Steinlager, would harmonize with the flavors better than wine.

6 to 8 Servings

¼ cup plus 2 tablespoons vegetable oil
1½ teaspoons cumin seeds
½ teaspoon fennel seeds
½ teaspoon black peppercorns, cracked

2 medium onions, sliced
12 large garlic cloves, thickly sliced
2 teaspoons dry mustard
1 teaspoon crushed red pepper
1 teaspoon turmeric or curry powder
¾ teaspoon salt
1 small eggplant (about ¾ pound), unpeeled, cut into ½-by-½-by-2-inch sticks
5 plum tomatoes (about ¾ pound), quartered lengthwise
1 can (19 ounces) chickpeas, rinsed and drained
2 tablespoons chopped fresh coriander (cilantro) or mint

1. In a large skillet or flameproof casserole, heat the oil over high heat. Add the cumin seeds and cook until dark brown, about 15 seconds. Add the fennel seeds and black pepper and cook for 5 seconds. Add the onions and garlic and reduce the heat to moderately high. Cook, stirring frequently, until the onions and garlic are lightly browned, about 5 minutes.

2. Stir in the mustard, crushed red pepper, turmeric and salt. Add the eggplant. Reduce the heat to moderate and cook, stirring gently, until the eggplant is limp, about 5 minutes.

3. Add the tomatoes and cook, stirring constantly, until softened, about 5 minutes.

4. Gently stir in the chickpeas, cover and simmer over low heat until the liquid thickens to a gravy and the flavors have blended, about 5 minutes. Season with additional salt to taste. Sprinkle the coriander over the top and serve.
—*Julie Sahni*

SIMPLE STEW OF LIMA BEANS, CORN AND TOMATOES

Frozen baby lima beans are simply frozen fresh beans with no preservatives or added salt.

4 to 6 Servings

1 package (10 ounces) frozen baby lima beans
¼ of a medium onion
1 parsley sprig plus 2 tablespoons chopped
1 small bay leaf
4 whole black peppercorns
Salt
1¾ cups fresh corn kernels (from 4 to 6 ears) or 1 box (10 ounces) frozen corn
2 tablespoons unsalted butter
6 medium scallions (white and tender green), thinly sliced
1 tablespoon minced basil
½ teaspoon chopped fresh mint
1 large tomato—peeled, seeded and coarsely chopped
½ teaspoon freshly ground pepper
3 tablespoons sour cream
3 tablespoons chopped fresh coriander (cilantro)

1. In a medium saucepan, combine the lima beans, onion, parsley sprig, bay leaf, peppercorns and a pinch of salt. Add 2 cups of water and bring to a boil over high heat. Reduce the heat to moderate and simmer until the beans are tender, about 15 minutes. Drain the beans in a colander set over a bowl and reserve the broth. Discard the bay leaf, parsley, onion and peppercorns.

2. If using fresh corn, boil the kernels in lightly salted boiling water until tender, 3 to 5 minutes. Drain and rinse briefly; set aside.

3. In a large saucepan, melt the butter over moderate heat. Add the scallions and cook, stirring occasionally, until softened, about 2 minutes. Add the lima beans, chopped parsley, basil, mint, fresh or frozen corn, reserved bean broth and ½ teaspoon of salt. Bring the mixture to a simmer and cook for 3 minutes.

4. Add the tomato and cook until the tomato is soft and the broth just coats the vegetables, about 5 minutes. Remove from the heat. Stir in the ground pepper, sour cream and coriander. Season with salt to taste and serve warm.
—*Deborah Madison*

CHICKPEA STEW WITH GREENS AND SPICES

Chickpeas are probably the best of the canned beans. They are generally firm, and once the salty liquid is washed away, they taste quite good. Since chickpeas take several hours to cook, the canned variety can come in very handy.

Served with rice or couscous, this stew makes a good meatless meal. It is even better the next day, so it's a good dish to make ahead.

4 Servings

1½ pounds swiss chard, tough ribs removed—or a mixture of greens, such as chard, spinach, kale, tatsoi, mustard and arugula—torn into large pieces (8 to 10 cups total)
1 can (16 ounces) chickpeas, drained and rinsed
6 medium garlic cloves, coarsely chopped
1 teaspoon coarse (kosher) salt
2 teaspoons sweet paprika
1½ teaspoons cumin
½ teaspoon turmeric
1 teaspoon whole black peppercorns
3 tablespoons plus 1 teaspoon extra-virgin olive oil
¼ cup chopped fresh coriander (cilantro)

2 tablespoons minced parsley
1 medium onion, chopped
1 green bell pepper or Anaheim chile, cut into ½-inch dice
1 small dried red chile
2 large tomatoes—peeled, seeded and chopped, juice reserved—or 2 cups drained canned Italian plum tomatoes, coarsely chopped

1. Place the greens in a steamer, cover and steam over boiling water until wilted, about 5 minutes. When the greens are cool enough to handle, coarsely chop them and set aside.

2. Place the chickpeas in a medium bowl and add water to cover. Rub the beans between your hands to loosen the skins. Discard the skins.

3. In a mortar, pound the garlic with the salt until it begins to break down. Add the paprika, cumin, turmeric and peppercorns and pound until the peppercorns are well broken up. Add 1 teaspoon of the olive oil to moisten the mixture. Add 2 tablespoons of the coriander and the parsley and continue to pound for a few minutes until a rough paste forms.

4. In a large nonreactive skillet, heat the remaining 3 tablespoons olive oil over moderate heat. Add the onion and green pepper and crumble in the dried chile. Cook, stirring occasionally, for 4 minutes. Stir in the garlic paste, chickpeas and ½ cup of water. Bring to a simmer and cook until the onion is soft, about 5 minutes.

5. Stir in the tomatoes and their juice, the greens and ½ cup of water. Cook over moderately low heat for 20 minutes. Stir in the remaining 2 tablespoons coriander and serve hot.
—*Deborah Madison*

SANTA FE FARMERS' MARKET STEW

This stew was inspired by the bustling Santa Fe farmers' market one day last fall. The green chiles were in, as well as pinto beans and dried red chiles; and squash, tomatoes, peppers and onions were everywhere. With some cooked greens, a tomato salad and tortillas or corn bread, this stew makes a fine meal.
❣ Beer might be an obvious choice, but if choosing wine, a big red with forthright flavors would be a good match for this hearty, thick, chili-style stew. An Australian Shiraz, such as Taltarni, or a California Petite Sirah, such as Louis M. Martini Reserve, comes to mind.
4 Servings

1 cup dried Anasazi or pinto beans (7 ounces), picked over
3 medium tomatoes (about 1½ pounds)
5 large, unpeeled garlic cloves
1 tablespoon sunflower oil
1 small onion, finely chopped
1 medium zucchini or yellow summer squash, cut into ⅓-inch dice
1 green bell pepper, cut into ¼-inch dice
2 to 3 teaspoons pure ground red chile, such as New Mexican
1 can (16 ounces) white or yellow hominy, rinsed and drained
½ teaspoon salt
½ cup chopped fresh coriander (cilantro)
1 cup sour cream, for serving
1½ cups shredded sharp Cheddar cheese (4 ounces), for serving

1. In a large saucepan, soak the beans in plenty of cold water for at least 6 hours or overnight. Pour off the water, re-cover the beans with fresh water and bring to a boil. Boil the beans vigorously for 5 minutes. Drain the beans in a colander and rinse well to remove any scum.

2. Return the beans to the saucepan, add 8 cups of water and bring to a boil over high heat. Reduce the heat to moderately low and simmer until the beans are tender, about 1¼ hours. Drain the beans, reserving the broth.

3. Heat a heavy, medium, nonreactive skillet over moderate heat. Place the tomatoes in the skillet and roast, turning frequently, for about 20 minutes; the skins will brown lightly and the tomatoes will soften partially. Peel the tomatoes and transfer them to a blender.

4. In the same pan, roast the garlic cloves, turning frequently, until browned and the garlic is soft, 12 to 15 minutes. Peel the garlic cloves and add them to the blender. Puree until smooth.

5. In a medium casserole, warm the oil over moderately high heat. Add the onion, zucchini and bell pepper and cook for 3 minutes. Stir in the ground chile and 1 cup of the reserved bean broth. Bring to a rapid simmer and cook until the liquid is reduced by half, about 5 minutes.

6. Stir in the beans, hominy, salt, tomato-and-garlic puree and 1 cup of the bean broth. Bring to a boil over moderately high heat. Reduce the heat to moderately low and cook for 20 minutes. Season with salt to taste. Stir in the coriander and serve the stew in bowls. Pass the sour cream and Cheddar cheese separately.
—*Deborah Madison*

MAIN COURSES: VEGETABLES & GRAINS

BULGUR WHEAT, PEPPERS AND BEAN CURD

Although completely vegetarian, this colorful dish has enough complementary protein to be as nutritious as meat.
4 Servings

1 cup coarse or medium bulgur wheat
4 tablespoons unsalted butter
3 garlic cloves, minced
1 teaspoon cumin
2 red bell peppers, seeded and cut into julienne strips
3 tablespoons cider vinegar
10 ounces spinach, stemmed and coarsely chopped
4 small firm Chinese tofu cakes* (about 12 ounces), cut into cubes
½ teaspoon salt
½ teaspoon freshly ground pepper
*Available at Asian markets

1. Place the bulgur in a small bowl and rinse several times to remove any impurities. Add cold water to cover by 1½ inches and let soak for 45 minutes, or until tender. Drain in a fine sieve, pressing to remove as much moisture as possible.

2. In a large skillet, melt the butter over moderately low heat. Add the garlic and sauté for 30 seconds, until fragrant but not browned. Stir in the cumin and add the peppers, cover and cook for 5 minutes.

3. Add the vinegar and bulgur and cook uncovered for 5 minutes, stirring frequently to prevent scorching. Add the spinach and tofu, cover and simmer for about 5 minutes longer, or until the spinach wilts. Season with the salt and pepper.
—*Michèle Urvater*

RAGOUT OF FALL VEGETABLES

At Mondrian in Manhattan, chef Tom Colicchio varies the vegetables he adds to a base of braised artichokes according to what's fresh and in season at the market.
6 Servings

6 large artichokes
1 lemon, halved
⅓ cup plus 2 tablespoons olive oil
1 medium carrot, coarsely chopped
1 medium onion, coarsely chopped
1 leek (white and tender green), thinly sliced crosswise
3 garlic cloves, lightly crushed
1 sprig of fresh thyme
1 sprig of fresh rosemary
1 bay leaf
1 cup dry white wine
¼ pound chanterelle mushrooms, cut into 1-inch pieces
¼ pound shiitake mushrooms, stemmed, caps cut into 1-inch pieces
Salt and freshly ground pepper
3 small parsnips, peeled and cut into ½-inch dice
1 large tomato—peeled, seeded and cut into ¼-inch dice

1. Break off the stems of the artichokes. Trim the artichokes by snapping off the tough outer leaves near the base. Using a sharp stainless steel knife, cut off all the leaves. Trim the base of any remaining dark green parts. Rub the base well with the cut lemon to prevent discoloration. Trim the base to an even round shape and scoop out the hairy chokes with a spoon. Rub the surfaces again with the cut lemon and set aside.

2. In a large nonreactive skillet, heat 1½ tablespoons of the oil over moderately low heat. Add the carrot, onion and leek and cook, stirring, until softened, about 3 minutes. Add the artichoke bottoms, garlic, thyme and rosemary sprigs, bay leaf, wine, ⅓ cup of the olive oil and 3 cups of water. Cook, covered, over low heat, turning the artichokes once, until they are easily pierced, about 20 minutes.

3. Meanwhile, in a large skillet, heat the remaining 1½ teaspoons olive oil over moderately high heat. Add the chanterelle and shiitake mushrooms and season with salt and pepper. Cover and cook, stirring once, until softened, about 3 minutes.

4. Bring a small saucepan of water to a boil over high heat. Add the parsnips and cook until tender, about 4 minutes. Drain and set aside.

5. Cut each artichoke bottom in half and then slice crosswise ¼ inch thick. Return to the pan and add the chanterelles, shiitakes, parsnips and tomato. Warm over moderately high heat until heated through, about 3 minutes. Spoon the ragout onto 6 serving plates and serve hot.
—*Baba S. Khalsa*

POTATOES BAKED IN PARCHMENT

These delicious herb-flavored potatoes make a great focal point for a vegetarian meal. They're also a good accompaniment to poultry and beef. For a dramatic presentation, serve the vegetables in their parchment pouches.
2 Servings

8 large sprigs of parsley, cut into 2-inch lengths
6 large sprigs of thyme, cut into 2-inch lengths, or ½ teaspoon dried
6 large sprigs of marjoram or oregano, cut into 2-inch lengths, or ½ teaspoon dried

6 sage leaves, or ¼ teaspoon dried
6 large garlic cloves
6 ounces small red potatoes, unpeeled and whole, or large red potatoes, cut into 1¼-inch cubes
8 medium mushrooms, sliced
2 tablespoons olive oil, preferably extra-virgin
½ teaspoon salt
½ teaspoon freshly ground pepper

1. Preheat the oven to 400°. In a large bowl, combine the parsley, thyme, marjoram, sage, garlic and potatoes.

2. Cut a 32-by-24-inch sheet of parchment paper crosswise in half. Lay the halves on a workspace with the longest sides facing you. Crease each piece of paper in the center.

3. Divide the mushrooms and the potatoes, garlic and herbs equally between the 2 sheets of parchment, centering them on the right half of the paper, nearest the crease.

4. Drizzle 1 tablespoon of olive oil over each set of vegetables and season with the salt and pepper.

5. Fold the parchment in half, enclosing the ingredients. Seal the packet by folding and crimping the edges to create a semicircular shape.

6. Place the 2 packets on a baking sheet and bake in the middle of the oven for 20 to 25 minutes, until the potatoes are tender when poked with a skewer.
—Annie Somerville

EGGPLANT CREPES
In preparing the eggplant, if you only have table salt, use half as much.
4 Servings

2 medium eggplants (about 2 pounds), peeled and sliced lengthwise ⅛ inch thick
1 teaspoon coarse (kosher) salt
¼ cup all-purpose flour
1½ cups peanut oil
1 tablespoon olive oil
1 medium onion, thinly sliced
8 garlic cloves, minced
½ cup dry white wine
1 can (35 ounces) crushed Italian tomatoes, with their juice
¼ teaspoon sugar
¾ teaspoon table salt
¼ teaspoon freshly ground pepper
⅓ cup chopped fresh basil
1 teaspoon chopped fresh thyme or ½ teaspoon dried
1 teaspoon chopped fresh oregano or ½ teaspoon dried
2 pounds spinach, stemmed
1 cup ricotta cheese
½ cup freshly grated Parmesan cheese
1⅓ cups shredded Gruyère cheese (about 4 ounces)

1. Arrange the eggplant slices in a single layer on a large nonreactive jelly-roll pan. Sprinkle with the coarse salt. Set a large cutting board or baking sheet on top of the eggplant and weigh down with canned goods. Cover with plastic wrap and set aside at room temperature for 3 hours or refrigerate overnight.

2. Drain the liquid from the eggplant and pat the slices dry. Dredge in the flour; shake off any excess. In a large, heavy nonreactive skillet, heat the peanut oil over moderately high heat. Add the eggplant in batches and fry until golden brown, about 2 minutes per side; drain on paper towels. Pour off the oil and wipe out the skillet.

3. Add the olive oil to the skillet and warm over moderate heat. Add the onion and cook until softened, about 6 minutes. Add half the garlic and cook until fragrant, about 2 minutes. Add the wine and boil until it reduces to ¼ cup, 3 to 5 minutes. Stir in the tomatoes and their juice, the sugar, ½ teaspoon of the table salt and ⅛ teaspoon of the pepper. Simmer over moderate heat until slightly thickened, about 15 minutes. Stir in the basil, thyme and oregano and simmer for 5 minutes longer to blend the flavors.

4. Meanwhile, in a large pot, bring 2 quarts of water to a boil. Add the spinach and cook until just wilted, about 1 minute. Drain, rinse under cold water and squeeze dry.

5. In a food processor, combine the spinach, ricotta, Parmesan and remaining minced garlic, ¼ teaspoon salt and ⅛ teaspoon pepper. Process until the spinach is finely chopped, about 1 minute.

6. Preheat the oven to 400°. Butter a large, shallow nonreactive baking dish. Spread an eggplant slice with about 2 tablespoons of the spinach mixture and roll up. Place the roll, seam-side down, in the baking dish. Repeat with the remaining eggplant slices and filling. Spoon the tomato sauce over the eggplant rolls and sprinkle the Gruyère cheese over the top. Bake until heated through and golden brown on top, 25 to 30 minutes.
—Adam Esman

POLENTA WITH PEPPERS, MUSHROOMS AND ONIONS

This delicious entrée can be made completely vegetarian by substituting vegetable broth (or water and 1 teaspoon of salt) for the chicken stock.

4 to 6 Servings

¼ cup olive oil
1 medium onion, sliced
1 garlic clove, minced
1 medium green bell pepper, sliced
1 medium red bell pepper, sliced
½ pound mushrooms, sliced
½ teaspoon salt
¼ teaspoon freshly ground black pepper
2 teaspoons chopped fresh oregano or ⅔ teaspoon dried
3 cups chicken stock or 1½ cups canned broth diluted with 1½ cups water
1 cup instant polenta
¼ cup boiling water

1. In a large skillet, heat the oil. Add the onion and garlic and cook over moderately high heat until the onion is softened and slightly browned, about 5 minutes.

2. Add the green and red peppers and the mushrooms. Season with the salt and black pepper. Reduce the heat to moderate and continue to cook, stirring occasionally, until the peppers are tender, about 15 minutes. Stir in the oregano, remove from the heat and cover to keep warm.

3. In a medium saucepan, bring the chicken stock to a boil over high heat. Gradually stir in the polenta and cook, stirring constantly, for 5 minutes. Add the boiling water and cook for 5 minutes longer. Season with additional salt and black pepper to taste.

4. Spoon the polenta into a serving dish, smoothing it into an even layer with a wet spatula and mounding the edges slightly. Pour the sautéed vegetables into the center and serve.

—John Robert Massie

POBLANOS RELLENOS WITH CORIANDER-CORN SAUCE

These stuffed chiles require a bit of work, but they're well worth the effort. The large, deep-green poblano peppers are incredibly sweet and only mildly hot. Brie works well as a filling because it melts easily, and its nutty, buttery flavor remains, even when the cheese is heated.

4 Servings

8 large poblano peppers, stems intact
1 tablespoon unsalted butter
1 small onion, chopped
1 cup corn kernels, fresh or frozen
½ cup heavy cream
2 tablespoons sour cream
1 large tomato, seeded and diced
¼ cup plus 2 teaspoons chopped fresh coriander (cilantro)
½ teaspoon salt
¼ teaspoon freshly ground black pepper
1 pound Brie cheese, chilled
3 eggs, separated
½ cup corn oil
½ cup all-purpose flour
2 tablespoons toasted pine nuts

1. Roast the peppers directly over a gas flame or under the broiler as close to the heat as possible, turning, until charred all over. Place the peppers in a paper bag to steam for 5 minutes. Gently remove the blackened skin without tearing the pepper or removing the stem. Rinse briefly under cold water. Using kitchen scissors, cut a slit lengthwise to the core on one side of each pepper. Cut the seed core from the base of the stem, keeping the pepper intact. Remove and rinse all seeds from the cavity. Pat the peppers dry.

2. In a medium saucepan, melt the butter over low heat. Add the onion and corn and cook, stirring occasionally, until the onion is tender, about 5 minutes.

3. Stir in the heavy cream and increase the heat to moderate. Bring to a boil and cook until thickened and slightly reduced, about 3 minutes.

4. Remove from the heat and stir in the sour cream, tomato, 2 tablespoons of the chopped coriander, the salt and black pepper. Remove from the heat; set the sauce aside.

5. Cut the rind off the Brie and divide the cheese into 8 equal pieces. Place a piece of cheese plus 1 teaspoon of the remaining chopped coriander inside each pepper and gently press it closed.

6. In a medium bowl, beat the egg whites with a pinch of salt until stiff. Beat in the egg yolks until well blended.

7. In a large skillet, heat the corn oil over high heat until it begins to shimmer. Dredge the stuffed peppers in the flour; shake off any excess. Beat the eggs again briefly. One at a time, dip the peppers into the egg, coating them all over. Fry 4 peppers at a time over high heat until browned on one side, about 2 minutes. Turn the peppers over and cook until the cheese is melted and the peppers are browned on the other side, about 2 minutes. Drain on paper towels.

8. Briefly warm the sauce over moderate heat. Pour onto a large platter or individual serving plates. Place the peppers on the sauce, sprinkle the pine nuts on top and serve.

—Marcia Kiesel

SWEET-AND-SOUR VEGETABLE STIR-FRY

Not everything in an Indonesian *rijsttafel* is chile-hot and spicy. This lightly seasoned vegetarian dish provides a contrast to the more vivid ones commonly found in this multi-dish meal.

12 Servings

2 tablespoons corn oil
3 garlic cloves, chopped
3 small shallots, halved
4 pounds Chinese long beans* (asparagus beans) or standard green beans, cut into 2-inch pieces
2 medium carrots, cut into thin julienne strips
1 large red bell pepper, cut into long slender strips
1 package (10 ounces) frozen baby corn, thawed
1 small head of cauliflower, cut into 1-inch florets
½ teaspoon minced fresh hot chile pepper (green or red) or ¾ teaspoon crushed red pepper
2 tablespoons cider vinegar
3 tablespoons kecap manis,* preferably homemade (p. 39)
2 teaspoons (packed) brown sugar
1 large tomato, cut into ½-inch dice
*Available at Asian markets

1. In a wok or large heavy skillet, heat the oil. Add the garlic and shallots and stir-fry over moderate heat until softened but not browned, about 1 minute. Add the long beans, carrots, bell pepper, baby corn, cauliflower and hot pepper and continue to stir-fry for 5 minutes.

2. Add the vinegar, *kecap manis* and brown sugar and stir-fry for 3 minutes; the vegetables should still be crunchy. (The recipe can be prepared to this point up to 3 hours ahead. Set the vegetables aside at room temperature.)

3. Shortly before serving, reheat the vegetables, if desired. Add the tomato and toss gently to mix. Serve warm or at room temperature.
—Copeland Marks

BLACK-EYED-PEA CAKES WITH SALSA

At Carolina's, a popular restaurant in Charleston, these cakes are served as an appetizer (squirted decoratively with a mixture of sour cream thinned with milk). But they're hearty enough for a light lunch or supper.

4 Servings

1 cup dried black-eyed peas, rinsed and picked over to remove any grit
1 small smoked ham hock (about ½ pound)
5 cups rich unsalted chicken stock or 2 cans (13¾ ounces each) chicken broth mixed with 1½ cups water
1 tablespoon unsalted butter
¼ cup minced red onion
2 large garlic cloves, minced
2 tablespoons minced red bell pepper
1 jalapeño pepper, minced
1 egg yolk
About ½ cup fresh bread crumbs
½ teaspoon hot pepper sauce
2 tablespoons chopped fresh coriander (cilantro)
½ teaspoon cumin
¼ teaspoon freshly ground black pepper
⅓ cup yellow cornmeal, for dredging
¼ cup vegetable oil, for frying
Salsa Vinaigrette (p. 262)

1. In a large heavy saucepan, combine the black-eyed peas, ham hock and chicken stock. Bring to a boil over moderate heat. Reduce the heat to low, cover and simmer until the peas are very tender, about 1¼ hours.

2. Drain into a large sieve over a heatproof bowl; reserve the ham hock and the cooking liquid for making soup later, if desired. Transfer the peas to a large mixing bowl and let them cool to room temperature.

3. Meanwhile, in a small heavy skillet, melt the butter over moderately low heat. Add the red onion, garlic, red bell pepper and jalapeño pepper; cook until limp and golden, about 5 minutes. Add to the black-eyed peas and let cool.

4. Add the egg yolk, ½ cup bread crumbs, hot sauce, fresh coriander, cumin and black pepper, and mix well with your hands or a potato masher to squash the peas. When the mixture is well blended, test to see if it is stiff enough to shape. If not, add a few more tablespoons of bread crumbs. Refrigerate uncovered for at least 1 hour. (The recipe can be prepared up to 1 day in advance. Cover and refrigerate until ready to proceed.)

5. Shape the black-eyed-pea mixture into 12 small cakes about 2½ inches across and ½ inch thick. Dredge in the cornmeal, shaking off any excess.

6. In a large heavy skillet, heat the oil over high heat until it ripples. Add half the black-eyed-pea cakes and fry over high heat, turning once, until browned, 1 to 1½ minutes on each side. Drain on paper towels. Repeat with the remaining cakes.

7. Arrange 3 cakes on each of 4 heated plates. Spoon about ½ cup of the Salsa Vinaigrette onto each plate. Pass the remainder separately.
—Carolina's, Charleston, South Carolina

FRESH TOMATOES STUFFED WITH RICE AND CHEESE

These elegantly creamy stuffed tomatoes can be prepared in the cool of morning, then left at room temperature until lunch or dinner. Or take them on a picnic; they are equally good hot or at room temperature.

8 Servings

8 large, ripe tomatoes
Coarse (kosher) salt
¼ cup olive oil
½ cup chopped shallots or scallions
1 garlic clove, minced
1 cup converted rice
2 tablespoons dry white wine or dry vermouth
1½ cups chicken stock or canned broth
½ cup chopped fresh basil, or ½ cup minced parsley mixed with 1 teaspoon dried basil
8 ounces mozzarella cheese, cut into ½-inch cubes
2 tablespoons fresh lemon juice
Salt and freshly ground pepper
⅓ cup freshly grated Parmesan cheese

1. If the tomatoes do not stand straight, cut a small slice off the bottom of each. Carefully core the tomatoes. Scoop out and reserve the tomato flesh, leaving a firm wall on all sides. Sprinkle a little coarse salt into each tomato, invert onto a rack and allow to drain while you prepare the filling.

2. In a saucepan, heat 2 tablespoons of the olive oil. Sauté the shallots and garlic over moderately high heat for about 2 minutes, then add the rice. Toss for a minute or so, then add the wine. Reduce for a couple of minutes, until almost dry, then add the broth. Bring to a boil, then lower the heat and partially cover the pot. Simmer until the rice is almost cooked but still quite firm, about 15 minutes. Remove from the heat and transfer the rice to a large bowl and set aside to cool slightly.

3. Meanwhile, remove the seeds and squeeze excess juice from the reserved tomato pulp. Chop the pulp coarsely and add to the rice in the bowl.

4. Preheat the oven to 350°. Toss together the rice, basil (or parsley and dried basil), mozzarella, lemon juice and salt and pepper to taste.

5. Stuff the tomato shells with the rice mixture, mounding it slightly on top. (Use any surplus rice for a rice salad, adding oil and vinegar to taste.)

6. Place the stuffed tomatoes in a shallow baking pan (preferably not aluminum) and sprinkle the Parmesan over each. Drizzle ¾ teaspoon of the remaining olive oil over each tomato. Add hot water to the pan to a depth of about ½ inch, and bake until the tomatoes are tender but still intact, 15 to 25 minutes. Serve hot or at room temperature.
—F&W

THAI JASMINE RICE

This is a fried-rice dish from Thailand. It is made with an extremely fragrant long-grain variety from Thailand called jasmine rice. If you cannot find jasmine rice, substitute another aromatic rice type, such as basmati, or one of the domestic aromatics.

6 Servings

2 tablespoons oyster sauce*
1 teaspoon sugar
½ pound thick-sliced bacon
2½ tablespoons peanut oil
5 eggs, beaten
Pinch of freshly ground white pepper
3 medium shallots, minced
4 large broccoli stalks, peeled and cut into ¼-inch dice (about 1¼ cups)
Basic Cooked Rice (recipe follows)—made with jasmine rice* and chicken stock or broth—cooled, uncovered
6 medium scallions, thinly sliced
Salt
*Available at Asian markets and specialty food shops

1. In a bowl, combine the oyster sauce and sugar. Set aside.

2. Using a small sharp knife, cut the fat from the bacon. Cut the meat into ½-inch pieces and the fat into ¼-inch dice. Set aside separately.

3. Heat a wok over high heat for 40 seconds. Add the diced bacon fat and stir-fry until golden brown and crisp, about 2 minutes. Remove the wok from the heat. Using a slotted spoon, transfer the cracklings to paper towels to drain. Pour off the fat. Rinse and dry the wok and spatula.

4. Heat the wok over high heat for 40 seconds. Add 1½ tablespoons of the peanut oil and swirl the oil around the wok with the spatula until a wisp of white smoke appears, about 1 minute. Add the eggs and the pepper and scramble until firm, about 2 minutes. Remove the wok from the heat and, using the spatula, cut the eggs into ½-inch pieces. Transfer to a plate and set aside. Again, rinse and dry the wok and spatula.

5. Heat the wok over high heat for 40 seconds. Add the remaining 1 tablespoon peanut oil and swirl the oil around with the spatula until a wisp of white smoke appears, about 1 minute. Add the meaty bacon and spread in a single layer. Cook for 1 minute, flip over and cook until well browned, 1 to 1½ minutes. Stir in the shallots and cook until softened, about 2 minutes. Add the broccoli stalks, mix well and cook for 2 minutes more.

6. Stir in the rice and cook until heated through, about 2 minutes. Add the reserved oyster sauce mixture and stir until the rice is thoroughly coated. Stir in the reserved eggs. Add the scallions and mix well; season to taste with salt. Transfer the rice to a warm platter, sprinkle the reserved fat cracklings on top and serve.
—*Eileen Yin-Fei Lo*

BASIC COOKED RICE
In this formula, washing and soaking the rice accounts for the small amount of water used in cooking it.
6 Servings

2 cups rice
1¾ cups plus 2 tablespoons cold water, chicken stock or canned broth

1. In a large saucepan, combine the rice and enough water to cover. Rub the rice between your hands a few times. Drain well in a colander. Repeat this procedure 2 more times. Return the rice to the pan and then add the 1¾ cups plus 2 tablespoons water (or chicken stock). Set aside to soak for 2 hours before cooking.
2. Place the saucepan over high heat and bring to a boil. Boil, stirring, until most of the water evaporates, about 4 minutes. The rice will still be quite hard. Reduce the heat to very low, cover and cook, stirring occasionally, until tender, about 8 minutes.
3. Fluff the rice with a fork. Serve at once or cover tightly until ready to use.
—*Eileen Yin-Fei Lo*

WILD MUSHROOM RISOTTO
Although risotto is commonly served as a first course, this satisfying version makes a good main course. In smaller portions, it could also be a side dish.
❗ With the risotto, try the rich and buttery Simi Chardonnay.
6 Servings

5½ cups chicken stock or canned low-sodium broth
6 tablespoons unsalted butter
½ pound mixed fresh wild mushrooms, such as shiitake, chanterelles, porcini or oyster mushrooms, trimmed and sliced
1 garlic clove, minced
3 tablespoons olive oil
1 small onion, minced
1½ cups arborio rice
½ cup minced flat-leaf parsley
2 ounces asiago or Parmesan cheese, finely grated
Salt and freshly ground pepper

1. In a medium saucepan, bring the stock to a simmer over moderate heat. Reduce the heat to very low and cover until ready to use.
2. In a large heavy saucepan, melt 3 tablespoons of the butter over moderately high heat. Add the mushrooms and cook until wilted, about 3 minutes. Add the garlic and cook, tossing, for 1 minute. Transfer the mushrooms and garlic to a plate or bowl and set aside.
3. Reduce the heat to moderate. Add the remaining 3 tablespoons butter and the olive oil to the saucepan. When the butter is melted, add the onion and cook until softened, about 5 minutes. Add the rice and stir to coat well.
4. Increase the heat to moderately high. Add about ½ cup of the hot stock and stir constantly until the mixture comes to a simmer. Add 2 more cups of the hot stock, ½ cup at a time, stirring constantly until the stock is absorbed after each addition, about 10 minutes.
5. Toss in the reserved mushrooms and garlic and the parsley. Continue adding the remaining stock, about ½ cup at a time, stirring constantly until the stock is absorbed after each addition and the rice is tender, but still firm, and creamy but not soupy, 8 to 10 minutes longer.
6. Remove the risotto from the heat and stir in the cheese. Season to taste with salt and lots of pepper and serve at once.
—*Mary Evely*

INDONESIAN RICE WITH RAISINS
On the island of Sumatra, this savory, perfumed rice main dish is accompanied occasionally with pickled vegetables.
6 to 8 Servings

3 pounds beef soup bones, rinsed
1 teaspoon whole cloves
1 teaspoon coriander seeds
1 teaspoon cumin seeds
1 whole nutmeg
2 cinnamon sticks
3 ounces fresh ginger (a 2-by-½-inch piece), peeled and coarsely chopped
1 stalk of lemon grass,* rinsed and cut into 1-inch pieces
1 tablespoon salt
1½ pounds London broil, rinsed
2½ cups extra-long-grain rice
2½ tablespoons peanut oil
5 large shallots, finely chopped
2 large garlic cloves, minced
⅔ cup raisins
*Available at Asian markets

1. In a large pot, place the beef bones and 10 cups of water. Add the cloves, coriander seeds, cumin seeds, nutmeg, cinnamon sticks, ginger, lemon grass and 2 teaspoons of the salt. Cover and bring to a boil over high heat. Reduce the heat to low and cover, leaving a slight opening. Simmer the stock for 2¼ hours, skimming occasionally.

2. Add the London broil and bring to a boil over high heat. Reduce the heat to low and cover, leaving a slight opening. Simmer until the meat is tender, about 1 hour. Remove the meat and set aside on a platter to cool. Wrap the meat in plastic wrap and refrigerate overnight.

3. Strain the stock and discard the solids. Return the stock to the pot and bring to a boil over high heat; boil until reduced to 4 cups, about 20 minutes. Remove from the heat, let cool and refrigerate overnight.

4. Skim the fat from the stock and discard. Set aside at room temperature. Remove the London broil from the refrigerator and set aside at room temperature. In a medium bowl, combine the rice and enough water to cover. Rub the rice between your hands a few times. Drain well in a colander. Repeat this procedure 2 more times and set aside.

5. In a large saucepan, heat the peanut oil over high heat until hot, about 1 minute. Add the shallots and cook, stirring, until beginning to brown, about 2 minutes. Add the garlic and cook, stirring, until browned, 1 minute longer. Add the rice and stir until well coated.

6. Pour in the reserved stock and bring to a boil over moderately high heat; boil for 3 minutes. Stir, reduce the heat to very low, cover and cook until tender, about 15 minutes. Remove from the heat. Stir in the raisins and the remaining 1 teaspoon salt. Cover and let stand until the raisins are plump, about 5 minutes.

7. Meanwhile, slice the London broil against the grain ½ inch thick, then cut the slices into ½-inch dice. Lightly toss the meat with the rice and serve hot. *(The recipe can be prepared up to 2 days ahead; cover and refrigerate. Reheat over moderate heat, stirring.)*
—Eileen Yin-Fei Lo

BASIC AND BEAUTIFUL FRIED RICE

Just about the only requirement for good fried rice is that the rice be neither gummy nor dry. Rice that has been refrigerated overnight is ideal. It will break apart easily and will not stick to the pan. Overly dry rice requires a bit of stock or water, which can be added to the pan just after the rice is tossed in. Cover and steam over low heat until the liquid is absorbed.

2 Servings

2 to 6 tablespoons corn or peanut oil
1 small onion, cut into ¼-inch dice
1 small red bell pepper, cut into
 ¼-inch dice
1 small carrot, cut into ¼-inch dice
3½ cups cold cooked rice
½ pound cooked string beans or
 asparagus, cut into ½-inch
 pieces, or other cooked vegetable
¼ pound cooked chicken, beef,
 pork, fish or shellfish, slivered
 or cubed
½ teaspoon salt
3 medium scallions, thinly sliced on
 the diagonal

1. Heat a wok or large heavy skillet over high heat for 30 seconds. Add 1½ tablespoons of the oil and swirl to glaze the pan. Reduce the heat to moderately high, add the onion and stir-fry until partially softened, about 1 minute. Add the red pepper and carrot and stir-fry until crisp-tender, about 2 minutes. Drizzle a bit more oil down the side of the wok as needed to prevent sticking. Adjust the heat to maintain a sizzle without scorching the vegetables.

2. Add the rice, toss to blend and stir-fry until heated through, about 3 minutes. Add a bit more oil if needed to prevent sticking. Add the string beans and toss briskly for 30 seconds. Add the chicken and toss to combine and heat through, about 1 minute. Season with the salt and fold in the scallions.
—Barbara Tropp, China Moon Cafe, San Francisco

TAIWANESE RICE WITH PINEAPPLE

If you wish to serve this in an authentic Taiwanese fashion, buy the largest pineapples you can find, lay each pineapple on its side and slice off one-quarter along the length. Scoop out the flesh and use the pineapple shell as the serving vessel.

6 to 8 Servings

2 tablespoons oyster sauce*
3 teaspoons white wine
2 teaspoons Oriental sesame oil
1 teaspoon light (thin) soy sauce*
1½ teaspoons sugar
1 teaspoon cornstarch
1 teaspoon salt
Freshly ground white pepper
1 pound medium shrimp—shelled,
 deveined and cut crosswise
 into thirds
1 small ripe pineapple
3 tablespoons chicken stock or
 canned broth
2 teaspoons soy sauce
4 Chinese pork sausages*
About 2 tablespoons peanut oil
2 teaspoons minced fresh ginger
1½ teaspoons minced garlic

Basic Cooked Rice (p. 109),
  made with 2½ cups short-grain
  rice and 2¼ cups plus 1
  tablespoon cold water
3 scallions, thinly sliced
*Available at Asian markets

1. In a medium bowl, combine 1 tablespoon of the oyster sauce, 2 teaspoons of the wine, 1 teaspoon of the sesame oil, the light soy sauce, 1 teaspoon of the sugar, the cornstarch, ½ teaspoon of the salt and a pinch of pepper. Add the shrimp, toss to coat and set aside to marinate until ready to use.

2. Quarter and core the pineapple. Using a grapefruit knife, cut enough of the fruit into ⅓-inch pieces to make 1 cup, drained; set aside. Reserve the remainder for another use.

3. In a small bowl, combine the chicken stock, soy sauce and remaining 1 tablespoon oyster sauce, 1 teaspoon wine, 1 teaspoon sesame oil, ½ teaspoon salt, ½ teaspoon sugar and a pinch of pepper. Mix well and set the sauce aside.

4. Rinse the sausages and pat dry. Slice on the diagonal ¼ inch thick. Heat a wok over high heat for 30 seconds. Add 1 tablespoon of the peanut oil, swirl it around with a metal spatula and heat until a wisp of white smoke appears, about 1 minute. Add the sausages and stir-fry for 2 minutes. Using a slotted spoon, transfer the sausages to a plate and set aside.

5. Reheat the wok over high heat for 20 seconds. Stir in the ginger, then stir in the garlic and cook until golden brown, about 30 seconds. Add the shrimp and its marinade, spreading the shrimp in a single layer. Cook for 10 seconds, then flip the shrimp and cook until pink, another 10 seconds. Add the Basic Cooked Rice and mix well until very hot, about 2 minutes.

6. Stir in the reserved pineapple pieces; if the rice sticks to the wok, add the remaining 1 tablespoon peanut oil and mix well. Add the sausages and stir well. Stir the reserved sauce, then drizzle it over the rice, stirring with a wooden spoon. Toss thoroughly to coat all the rice. Add the scallions and mix well.

—Eileen Yin-Fei Lo

WARM CAESAR SALAD

All the elements of a classic Caesar Salad—anchovy, garlic, Parmesan cheese and Worcestershire sauce—are even zestier when the salad is served warm. Guests are always delighted with this adaptation of an old favorite.

6 Servings

2 medium heads of romaine lettuce
2 tablespoons unsalted butter
¾ cup olive oil
8-ounce loaf of Italian or Viennese bread, cut into ¾-inch cubes
2 hard-cooked eggs
1½ tablespoons anchovy paste
1 tablespoon Worcestershire sauce
1 garlic clove, crushed through a press
3 tablespoons white wine vinegar
¼ cup freshly grated Parmesan cheese
Freshly ground pepper

1. Trim away the ends and any wilted outer leaves from the romaine. Separate the leaves, wash them well and dry. Tear the leaves into bite-size pieces, wrap and refrigerate. *(The romaine can be prepared to this point up to 1 day ahead.)*

2. In a medium skillet, melt the butter in 2 tablespoons of the oil over moderate heat. When the foam subsides, add the bread cubes and toss to coat. Reduce the heat to moderately low and sauté the bread cubes, stirring often, until crisp and golden brown, 5 to 7 minutes. Set aside.

3. Meanwhile, coarsely chop the eggs; then force them through a sieve into a small bowl. Put the prepared romaine in a large salad bowl.

4. In a small nonreactive saucepan, combine the anchovy paste, Worcestershire sauce, garlic and vinegar. Whisk to blend well. Whisk in the remaining olive oil. Set the pan over moderate heat and bring just to a boil.

5. Remove from the heat and immediately pour the hot dressing over the lettuce; toss well. Add the Parmesan to the bowl and toss again.

6. Divide the salad among 6 plates. Sprinkle the croutons on top and spoon a small mound of sieved egg into the center of each salad. Season generously with pepper and serve at once.

—*Michael McLaughlin*

LAYERED SALAD WITH SPICY PEANUT DRESSING (GADO-GADO)

Gado-gado is the premier Indonesian salad. It can serve as the centerpiece of an all-vegetable meal, as one of several dishes in an Asian-style meal or as a colorful addition to a buffet. Though the salad is traditionally layered, you can compose it to achieve a more dramatic effect.

6 Servings

½ pound red potatoes
2 cups cauliflower florets (1-inch)
1 medium carrot, cut into 2-by-¼-inch julienne strips
¼ pound green beans, cut into 2-inch pieces
2 cups shredded cabbage (about 5 ounces)
1 cup jicama, sliced into 2-by-¼-inch julienne strips (optional)
1 firm Chinese tofu cake
1 cup canned unsweetened coconut milk
⅓ cup crunchy peanut butter
1 large garlic clove, minced
3 tablespoons kecap manis* or homemade Indonesian Sweet Soy Sauce (p. 39)
3 tablespoons fresh lemon juice
1 square inch of lemon zest
2 tablespoons brown sugar
1 to 2 teaspoons crushed red pepper or 1 small fresh hot red chile, finely chopped
½ teaspoon salt
2 scallions, thinly sliced
2 hard-cooked eggs, sliced
6 cherry tomatoes

*Available at Asian markets

1. In a medium saucepan of boiling salted water, cook the potatoes until tender, 12 to 20 minutes, depending on size; drain. When cool enough to handle, peel and slice.

2. In another saucepan of boiling salted water, cook the cauliflower until just tender, about 3 minutes. Remove with a slotted spoon and rinse under cold water; drain well.

3. In the same boiling water, repeat this process with the carrot, then the green beans and finally the cabbage. (The jicama is not cooked.)

4. In a medium saucepan of simmering water, poach the tofu for 10 minutes. Remove with a slotted spoon, wrap in a kitchen towel and let drain for at least 15 minutes. Cut into ½-inch cubes.

5. To make the sauce, in a medium saucepan, combine the coconut milk, peanut butter, garlic, *kecap manis*, lemon juice, lemon zest, brown sugar, hot pepper, salt and 2 tablespoons of water. Bring to a boil over moderate heat, stirring frequently. Reduce the heat to moderately low and simmer until thickened, about 10 minutes. Keep warm over very low heat until ready to use.

6. Arrange the vegetables in layers in a large salad bowl. Scatter the bean curd and scallions on top. Garnish with the eggs and tomatoes. Pour the warm peanut butter sauce over all. Serve at room temperature.

—*Copeland Marks*

TUSCAN-STYLE WHILE BEAN SALAD

While this dish can serve as accompaniment to almost any simple meat, I like to offer it as the centerpiece of a compose-your-own salad for a long, lazy summer lunch. I present the beans, surrounded by ripe red tomato wedges, with an assortment of tasty and colorful components for each guest to choose according to individual taste: Italian tuna packed in olive oil, good black olives, crisp greens, scallions, sliced cucumbers and the like. Set out a cruet of extra-virgin olive oil, lemon wedges and a pepper mill for seasoning. Serve with crusty loaves of Italian bread and a chilled, light-bodied wine.

6 to 8 Servings

2 cups dried cannellini (about 12 ounces), Great Northern white or cranberry beans (see Note)
1 small yellow onion
2 garlic cloves
4 sprigs of fresh sage or thyme or ¼ teaspoon dried thyme
4 plum tomatoes
Salt
1 teaspoon freshly ground pepper
1 small red onion, chopped
1 small celery rib with leaves, cut into ¼-inch dice
¼ cup thinly sliced scallion greens
2 tablespoons shredded fresh basil
1 tablespoon chopped parsley
⅓ cup extra-virgin olive oil
3 tablespoons fresh lemon juice

1. Either soak the dried beans overnight in cold water to cover by 4 inches or place in a large saucepan with several inches of water to cover and bring to a boil. Boil, covered, for 2 minutes, then remove from the heat and let stand for 1 hour. Drain the beans.

2. Place in a large saucepan and cover with 4 inches of fresh cold water. Cover and bring to a boil over high heat. Tuck the yellow onion, garlic and 2 sprigs of the sage in with the beans. Simmer over low heat until the beans are tender but not mushy, about 1½ hours. Drain the beans; discard the onion, garlic and sage.

3. Core and seed one of the tomatoes and cut it into ½-inch dice. Cut the remaining 3 tomatoes lengthwise into 6 wedges each and set aside. Finely chop the remaining 2 sprigs of sage.

4. Place the beans in a large mixing bowl and season with the salt to taste and the pepper. Add the chopped tomato and sage, the red onion, celery, scallion greens, basil, parsley, olive oil and lemon juice. Toss gently with 2 rubber spatulas to combine the ingredients without crushing the beans. Cover and set aside for at least 1 hour at room temperature to blend the flavors. *(The recipe can be prepared 1 day in advance up to this point and refrigerated, covered.)*

5. Transfer to a serving bowl or platter and surround with the tomato wedges. Serve at room temperature.

NOTE: If fresh cranberry beans are available, use 3 pounds (weighed in the pod), shelled. Begin at Step 2, but cook the beans for only about 20 minutes, until tender. If you prefer to use canned beans, use 2 cans (19 ounces each) cannellini. Simply rinse and drain, then proceed to Step 3.
—Richard Sax

SINGAPORE NOODLE SALAD

One of my early-morning pleasures in Singapore and Thailand is to sample the spicy noodles available from almost every street vendor. This salad is excellent even when it is tossed up to a day in advance of serving and has been known to disappear during the silent hours of early morning.

6 to 8 Servings

¼ cup sesame seeds
½ pound dried Chinese spaghetti-style noodles*
2 tablespoons vegetable oil
1 bunch of broccoli, separated into 1-inch florets
1 pound asparagus, preferably pencil thin, tough ends snapped off, cut on the diagonal into 1-inch pieces
½ cup chicken stock or canned broth
½ cup peanut butter
¼ cup red wine vinegar
2 tablespoons soy sauce, preferably dark*
1 tablespoon Oriental sesame oil
1 tablespoon dry sherry
2 teaspoons sugar
1½ teaspoons Chinese chili sauce
¼ cup minced scallions
2 tablespoons finely minced fresh ginger
1 large garlic clove, minced
1 cup bean sprouts
1 cup thinly sliced button or shiitake mushrooms or separated enoki*
1 large red bell pepper, cut into thin julienne strips
2 tablespoons minced chives
*Available at Asian markets

1. In a small dry skillet, toast the sesame seeds over moderately high heat, tossing, until fragrant and golden brown, about 1 minute. Set aside.

2. In a large pot of boiling salted water, cook the noodles, separating them with a fork, until tender but still firm, about 2½ minutes. Drain and rinse under cold running water, drain well and toss with the vegetable oil.

3. Bring a large saucepan of salted water to a boil over high heat. Add the broccoli and cook until crisp-tender, 2 to 3 minutes. Using a slotted spoon, transfer to a bowl of ice water and chill until cold, about 5 minutes. Drain on paper towels. Repeat with the asparagus, cooking for only 1 to 2 minutes.

4. In a small saucepan, bring the stock to a boil over moderately high heat; remove from the heat. Stir in the peanut butter, vinegar, soy sauce, sesame oil, sherry, sugar, chili sauce, scallions, ginger, garlic and toasted sesame seeds.

5. In a large bowl, toss together the noodles, broccoli, asparagus, bean sprouts, mushrooms and red pepper. Add the dressing and toss to coat. Sprinkle the chives over the top just before serving.
—Hugh Carpenter

BEEFSTEAK TOMATOES WITH MINTED CRACKED WHEAT SALAD
4 Servings

¼ cup plus 1 tablespoon olive oil
¾ cup bulgur wheat
¼ cup fresh lemon juice
2 tablespoons fresh lime juice
½ teaspoon salt
½ teaspoon freshly ground pepper
2 scallions, minced
2 tablespoons minced fresh mint
4 medium beefsteak tomatoes

1. In a medium skillet, heat ¼ cup of the olive oil. Add the bulgur and toss over high heat until coated with oil, about 1 minute; remove from the heat. Stir in ¾ cup of hot water, cover and soak until the water is absorbed and the bulgur is tender, at least 1 hour.

2. In a medium bowl, combine the lemon and lime juices with the salt and pepper. Add the scallions, mint, remaining 1 tablespoon oil and the soaked bulgur; toss well. Cover and refrigerate for 1 hour or longer.

3. Cut the tops from the tomatoes and hollow out the centers. Toss the bulgur salad well and spoon it into the tomatoes.
—Molly O'Neill

THAI RICE NOODLE SALAD
South Chinese influence shows through in *pad thai*, yet the raw vegetables and the bite of the chile-vinegar dressing make this salad distinctively Thai.
6 to 8 Servings

3 bundles of medium-thin, flat rice noodles* (12 ounces)
2 cups rice vinegar*
1 tablespoon sugar
2 Anaheim chiles, thinly sliced
2 tablespoons peanut oil
6 scallions—3 whole, 3 chopped
4 garlic cloves, chopped
1 pound fresh bean sprouts
2 medium tomatoes, cut into small wedges
3 eggs, lightly beaten
1 tablespoon soy sauce
1 tablespoon nam pla*
1 head of leaf lettuce, leaves separated
2 tablespoons chopped roasted peanuts
2 tablespoons fresh coriander (cilantro) leaves
1 lime, thinly sliced
*Available at Asian markets

1. Soak the rice noodles in hot water until softened, 25 to 30 minutes. Drain.

2. For the dressing, combine the rice vinegar and sugar and stir until the sugar dissolves. Add the sliced chiles. *(The dressing can be made in advance and refrigerated.)*

3. Heat a wok over moderately high heat. Add the peanut oil and when hot, add the chopped scallions and garlic and stir-fry for 15 seconds. Add half the bean sprouts and stir-fry for 1 minute. Add the tomato wedges and stir-fry for 30 seconds; try not to mash the tomatoes. Push the contents of the wok slightly up the sides and add the eggs to the center. When the eggs begin to set, stir to break into large pieces.

4. Push the contents of the wok as high up the sides as you can and add the drained rice noodles. Stir-fry until heated through, 2 to 3 minutes. Try to use as much of the wok's surface as possible; the noodles have a tendency to bunch up, but keep spreading them out. Add the soy sauce and *nam pla*, then stir well to combine all the ingredients in the wok.

5. To serve, line a large flat platter with the lettuce leaves. Spoon the noodle mixture into the center. Sprinkle the chopped peanuts and the coriander leaves on top. Garnish the platter with the remaining bean sprouts, the lime slices and the whole scallions. Serve the chile-vinegar dressing alongside.
—Jeffrey Alford

DANDELION AND BACON SALAD WITH HARD CIDER VINAIGRETTE

Wilted dandelion greens are a southern classic. Served in spring, dandelion greens were a welcome addition to the diet after a long winter of root vegetables and sauerkraut. Here, hard cider and cider vinegar add a fruity note. This salad makes a wonderful lunch served with crusty French bread and sparkling cider.

6 Servings

¾ pound small boiling potatoes
½ cup hard cider
½ pound thickly sliced bacon, cut crosswise into ¼-inch strips
¼ cup cider vinegar
¾ pound young dandelion greens, or substitute arugula or chicory (about 5 cups)
¼ teaspoon salt
¼ teaspoon freshly ground pepper
Cider vinegar or hard cider, as accompaniment

1. Place the potatoes in a large saucepan of lightly salted water and bring to a boil. Cook until just tender, about 20 minutes; drain. While still warm but cool enough to handle, peel and quarter. Place the warm potatoes in a large bowl, toss with ¼ cup of the hard cider and cover to keep warm.

2. In a large nonreactive skillet, cook the bacon over moderate heat until crisp and lightly browned, about 7 minutes. Drain on paper towels and set aside. Measure out 7 tablespoons of the drippings into a small bowl and reserve in a warm place; discard the remaining fat.

3. Add the remaining ¼ cup hard cider and the cider vinegar to the skillet and boil over high heat until reduced to ⅓ cup, about 2 minutes. Remove from the heat and set aside.

4. Add the dandelion greens and the reserved bacon and warm bacon drippings to the warm potatoes and toss. Add the hot cider reduction and toss quickly. Season with the salt and pepper. Pass a cruet of cider vinegar or hard cider on the side.

—*Sarah Belk*

WARM EGG AND RICE SALAD

I like this dish so much that I often make a version of it just for myself when I'm at home on Sunday.

4 to 6 Servings

1 cup rice
2 cups chicken stock or canned low-sodium broth
8 cups torn, crisp salad greens, such as romaine, radicchio or Belgian endive
1½ tablespoons balsamic vinegar
2½ tablespoons olive oil
¼ cup minced red bell pepper
½ pound cured ham, sliced ¼ inch thick and cut into ¼-inch strips
1 tablespoon safflower oil
3 medium scallions, very thinly sliced
1 tablespoon milk
4 eggs
Pinch of salt
Pinch of freshly ground black pepper
4 ounces Fontina cheese, cut into ½-inch dice
Black olives, for garnish

1. In a medium saucepan, combine the rice and the chicken stock and bring to a boil over moderately high heat. Cover and reduce the heat to low. Cook until the rice is tender and the liquid has been absorbed, about 20 minutes.

2. Meanwhile, place the greens in a medium bowl. Pour the vinegar into a small bowl and gradually whisk in the olive oil until incorporated. Pour the dressing over the greens and toss to coat. Arrange the greens on a platter and sprinkle the red bell pepper on top.

3. Heat a large heavy skillet over moderately high heat. Add the ham and cook, stirring, until lightly browned, about 3 minutes. Transfer the ham to a plate and cover to keep warm.

4. Add the safflower oil to the skillet and heat over moderately high heat. Add the scallions and cook, stirring constantly, until softened, about 2 minutes.

5. In a medium bowl, whisk the milk into the eggs until well blended. Season with the salt and black pepper and stir in the cooked rice. Transfer the mixture to the skillet and cook, stirring with a fork, until the mixture is soft and still somewhat wet, about 2 minutes. Add the cheese and cook, stirring, until the eggs are just set, 1 to 2 minutes longer.

6. Spoon the egg and rice mixture over the greens. Garnish with the reserved ham strips and black olives and serve hot.

—*Lee Bailey*

WARM POTATO SALAD WITH SALMON

Freshly poached salmon and new potatoes are the key to this rich and refreshing salad.

8 Servings

½ teaspoon salt
½ teaspoon freshly ground pepper
1 garlic clove, minced
1 tablespoon anchovy paste
1 tablespoon Dijon mustard
½ cup olive oil
¼ cup fresh lemon juice
2 tablespoons white wine vinegar
2 large cucumbers—peeled, halved lengthwise, seeded and cut crosswise into thin slices

½ cup minced fresh parsley leaves
2½ pounds (about 20 small) waxy potatoes, unpeeled
1 pound center-cut salmon steaks
3 tablespoons capers, rinsed and drained
Boston lettuce leaves, for lining the platter
3 tablespoons diced sweet pickle

1. In a small bowl, combine the salt, pepper and garlic. Using the handle of a wooden spoon or a pestle, pound the garlic into the salt and pepper to form a paste. Add the anchovy paste and mustard, stirring until blended.
2. Alternating the olive oil, lemon juice and vinegar, slowly add these ingredients to the bowl, beating them in with a whisk.
3. Place the cucumbers in a large bowl and pour in the vinaigrette. Add the parsley and toss the ingredients well. Let the mixture marinate while you prepare the remaining ingredients.
4. Cook the potatoes in a large saucepan of boiling salted water until just tender, about 20 minutes; do not overcook. Remove the potatoes from the pan and allow them to cool slightly while you prepare the salmon.
5. Bring 2 cups of water to a full boil in a heavy skillet. Add the salmon steaks, reduce the heat to low, cover the pan and simmer until the salmon is cooked through and flakes easily when pierced with a fork, about 8 minutes. Transfer the fish to a cutting board. Remove and discard the skin and bones. Flake the salmon into 1-inch pieces and add to the bowl of marinated cucumbers; toss gently.
6. Peel the potatoes; cut them into slices ¼ inch thick and then into ¼-inch-wide strips. Add the potatoes to the salmon and cucumbers as you cut them. Toss the salad gently to coat the potatoes with the dressing, using two large spoons.

7. Add the capers and carefully toss the salad to distribute them. Taste the salad for seasoning and add more salt if desired.
8. Line a serving platter or individual plates with the lettuce leaves and arrange the salad on top. Sprinkle with the diced pickle and serve warm.
—Jim Fobel

TUNA AND WHITE
BEAN SALAD
8 Servings

12 ounces (1½ cups) dried white kidney or navy beans
½ cup olive oil
3 tablespoons fresh lemon juice
½ teaspoon salt
⅛ teaspoon freshly ground pepper
½ cup thinly sliced scallions
¼ cup finely minced red onion
1 can (7 ounces) tuna fish in olive oil
2 tablespoons chopped parsley
2 lemons, each cut into 8 wedges

1. Rinse the beans in a colander under cold water, place in a 4-quart pot and add cold water to cover by an inch. Bring the water to a boil over high heat and cook for 2 minutes. Remove from the heat and let the beans soak for 1 hour. Return the pot to high heat and bring the liquid to a boil. Reduce the heat to low and simmer, partially covered, until the beans are just tender, (adding more water if the liquid cooks away), about 1½ hours. Drain.
2. In a large bowl, combine the oil, lemon juice, salt and pepper. Add the still-warm beans and gently toss until coated. Stir in the scallions and onion and set aside for 1 hour. Taste for seasoning and add more salt and pepper if necessary.

3. Drain the tuna fish, flake it into bite-size pieces and fold it into the beans. Serve the salad garnished with the chopped parsley and lemon wedges.
—F&W

GRILLED FRESH
TUNA AND LIMA
BEAN SALAD
The combination of grilled tuna, tender white beans, fresh herbs and tomatoes makes this a sunny Provençal-style salad. This substantial main-course salad could be served on its own, or as the centerpiece of a multi-dish buffet or picnic.
6 to 8 Servings

2 cups baby lima or other dried white beans
¾ cup extra-virgin olive oil
1 tablespoon minced fresh thyme
1 tablespoon minced fresh oregano
1 tablespoon minced parsley
2 tablespoons grainy mustard
1 tablespoon red wine vinegar
¼ cup plus 2 tablespoons fresh lemon juice
½ teaspoon freshly ground pepper
1½ teaspoons salt
1 pound fresh tuna steak, cut ½ inch thick
4 large tomatoes—peeled, seeded and coarsely chopped

1. Soak the beans overnight in cold water to cover by 2 inches.
2. The next day, drain and rinse the beans. In a medium saucepan, combine the beans with 6 cups of cold water and bring to a boil. Reduce the heat to moderately low and simmer the beans until tender, 20 to 25 minutes.
3. Meanwhile, in a small bowl, whisk together the olive oil, thyme, oregano, parsley, mustard, vinegar, lemon juice, pepper and ½ teaspoon of the salt.
4. Place the tuna in a shallow glass dish. Pour two-thirds of the dressing

over the tuna. Cover and let marinate at room temperature for 30 minutes, turning once.

5. Light a grill or preheat the broiler. Drain the beans and place in a bowl. Add the tomatoes, the remaining 1 teaspoon salt and the remaining dressing. Toss well. Set aside to cool.

6. Grill the tuna 4 to 5 inches from the heat, turning once and basting occasionally with the marinade, for 8 to 10 minutes or until just cooked through. Let cool to room temperature.

7. Break the tuna into bite-size chunks and toss with the beans. Cover and refrigerate until serving time.
—Bob Chambers

FRESH TUNA SALAD NICOISE

This recipe is a variation on the classic *salade niçoise*. The tuna can be grilled over a charcoal fire instead of baked, and a garnish of tiny black olives adds a nice touch.

❦ Thyme finds an echo in the flavor of Chardonnay, and the richness of the tuna needs an oaky wine, such as Wente Chardonnay.

3 to 4 Main-Course Servings

3 tablespoons chopped fresh thyme
3 garlic cloves, minced
½ teaspoon freshly ground pepper
¾ cup extra-virgin olive oil
1 tuna steak (about 1 pound), cut 1¼ inches thick
2 medium shallots, minced
1 teaspoon salt
¼ cup balsamic vinegar
½ pound green beans
¾ pound small red potatoes, quartered
1 pint cherry tomatoes, halved

1. In a shallow baking dish, combine 2 teaspoons of the thyme, 1 clove of the garlic, ⅛ teaspoon of the pepper and ¼ cup of the olive oil. Place the tuna in the marinade and turn to coat well. Set aside at room temperature, turning occasionally, for 1 hour.

2. Preheat the oven to 350°. In a small bowl, combine the shallots, salt and vinegar with the remaining 2 tablespoons plus 1 teaspoon thyme, 2 garlic cloves and ⅜ teaspoon pepper. Whisk in the remaining ½ cup olive oil.

3. Bring a medium pot of salted water to a boil. Add the green beans and cook until crisp-tender and bright green, 3 to 5 minutes. Using a slotted spoon, remove the green beans to a bowl of ice water; drain. In the same boiling water, cook the potatoes over moderately high heat until tender, 8 to 10 minutes; drain. Place the green beans, potatoes and tomatoes in separate bowls.

4. Cover the baking dish holding the tuna with aluminum foil, crimping the edges to seal. Bake for 10 minutes. Remove from the oven and let stand, covered, for 10 minutes. Transfer the tuna to a shallow dish. Drizzle the vinaigrette evenly over the tuna and the three vegetables. Toss the vegetables to coat. Marinate, turning the tuna occasionally, for at least 1 hour at room temperature or up to 4 hours in the refrigerator.

5. To assemble, place the tuna in the center of a large serving platter. Arrange the marinated vegetables around the fish and season with additional salt and pepper if desired. Serve the salad at room temperature.
—Marcia Kiesel

CORIANDER SWORDFISH SALAD WITH JICAMA AND THIN GREEN BEANS

Fresh jalapeño and coriander give this salad its south-of-the-border flavor.
❦ Chilled lager beer
4 Servings

¼ cup plus 2 tablespoons pine nuts
2 limes
½ cup (packed) fresh coriander (cilantro)
1 large jalapeño pepper, stemmed but not seeded
1 cup olive oil
½ teaspoon salt
⅛ teaspoon freshly ground black pepper
Hot pepper sauce (optional)
1½ pounds swordfish steaks, cut 1 inch thick
1 pound thin green beans
1 pound jicama, peeled and cut into thin julienne strips
1 small red onion, thinly sliced
1 large tomato—peeled, seeded and cut into thin strips

1. Preheat the oven to 350°. In a small baking dish, toast the pine nuts in the oven until golden, about 10 minutes.

2. Grate the zest from the limes. Squeeze the limes to extract their juice. In a food processor or blender, combine ¼ cup of the pine nuts with the coriander, jalapeño pepper, olive oil, lime zest, lime juice, salt and black pepper; puree until smooth, about 1 minute. Add hot pepper sauce to taste.

3. Place the swordfish in a nonreactive baking dish and sprinkle with ½ cup of the dressing. Let marinate for 1 hour at room temperature.

4. Preheat the oven to 400°. Bake the swordfish until just opaque throughout, about 15 minutes; let cool. Remove and discard the skin. Cut the fish into 1-inch cubes.

5. Bring a medium saucepan of salted water to a boil over high heat. Add the green beans and cook until crisp-tender, about 2 minutes. Drain and refresh under cold running water; drain well.

6. In a medium bowl, mix together the green beans, jicama and red onion. Add ½ cup of the dressing and toss well. Spoon the salad onto 4 plates. Arrange the swordfish cubes on top and garnish with the remaining 2 tablespoons pine nuts and the tomato strips. Pass the remaining dressing separately.

—Mimi Ruth Brodeur

GRILLED SWORDFISH AND AVOCADO SALAD BAJA-STYLE

Warm flour tortillas and sweet butter go well with this easy summertime salad.

6 Servings

1½ pounds swordfish steak, cut about 1 inch thick
6 tablespoons fresh lemon juice
2 tablespoons fresh lime juice
1 tablespoon distilled white vinegar
2 tablespoons hot red or green chile salsa
2 tablespoons minced onion
1 tablespoon minced fresh coriander (cilantro)
½ teaspoon salt
½ cup plus 2 tablespoons olive oil, preferably extra-virgin
¾ pound tomatoes (about 5 plum)— peeled, seeded and coarsely chopped
1½ cups cooked corn kernels, fresh or frozen
3 large avocados
1 head of romaine lettuce

1. Place the fish in a shallow bowl. Sprinkle with 3 tablespoons of the lemon juice, the lime juice, vinegar, chile salsa, onion, minced coriander and salt. Let marinate at room temperature, turning occasionally, for 30 minutes.

2. Light the charcoal or preheat the broiler. Remove the steaks, reserving the marinade. Brush both sides of the fish liberally with about 2 tablespoons of the olive oil. Grill or broil 3 to 4 inches from the heat for 8 to 10 minutes on the first side and 4 to 8 minutes on the second, or until just opaque throughout. (The recipe may be prepared ahead to this point. Wrap and refrigerate overnight. Let return to room temperature before continuing.)

3. Gradually whisk the remaining ½ cup oil into the reserved marinade to make a dressing. Cut the cooked fish into 1-inch cubes and toss with half of the dressing. Let marinate at room temperature for 30 minutes.

4. To serve, toss the fish with the tomatoes, corn and remaining dressing. Halve, pit and peel the avocados. Paint the avocado halves with the remaining 3 tablespoons of lemon juice. On a large serving platter, arrange the romaine leaves and top with the avocado halves. Spoon the swordfish salad in and around the avocados.

—Beverly Cox

CURRIED SALMON SALAD WITH LEMON GRASS VINAIGRETTE

This Thai-inspired salad is from Roy Yamaguchi of Roy's in Honolulu.

4 Servings

2 tablespoons Thai Mussaman curry paste*
1 tablespoon canned unsweetened coconut milk*
1 teaspoon finely grated fresh ginger
1 teaspoon minced garlic
1 teaspoon Oriental sesame oil
7 tablespoons vegetable oil
3½ teaspoons soy sauce
Four 5-ounce center-cut salmon fillets
2 stalks of fresh lemon grass,* white part only, minced
2 tablespoons strained fresh lemon juice
¼ cup light olive oil
Salt
½ head of romaine lettuce, torn into bite-size pieces
½ head of bronze-leaf lettuce, torn into bite-size pieces
2 medium Belgian endives, cored and sliced crosswise ½ inch thick
1 package enoki mushrooms or radish sprouts†
½ of a ripe medium papaya— peeled, seeded and cut into ½-inch dice
⅓ cup macadamia nuts, coarsely chopped (1½ ounces)
2 ounces pickled red ginger,† cut into matchsticks (see Note)

*Available at Thai markets
†Available at Asian markets

1. In a medium bowl, combine the curry paste, coconut milk, grated fresh ginger, garlic, sesame oil, vegetable oil and 2 teaspoons of the soy sauce. Whisk well. Pour half of the marinade into a 9-by-11-inch glass dish. Add the salmon and spoon the remaining marinade over the fish. Let marinate at room temperature for 1 hour, turning occasionally.

2. Meanwhile, in a small saucepan, combine the lemon grass with 1 cup of water and bring to a simmer over moderate heat. Simmer until only 1 teaspoon of liquid remains, about 30 minutes. In a small nonreactive bowl, whisk the lemon grass and its liquid with the lemon juice, olive oil, ½ teaspoon salt and the remaining 1½ teaspoons soy sauce. Set the dressing aside.

3. Keeping as much of the marinade solids clinging to the fish as possible, transfer the salmon to a rack set over a platter and let drain for 20 minutes, turning once.

4. Light a grill. When the fire is medium-hot, grill the salmon for 3 minutes, then turn and grill for 2 to 3 minutes longer, until the flesh, ½ inch from the center of a fillet, flakes with a fork. Lightly salt the salmon. Alternatively, preheat the broiler. Broil the fish for about 5 minutes, until the surface is firm and begins to sizzle. Turn and broil for 2½ to 3½ minutes longer.

5. While the salmon cooks, assemble the salad: In a large bowl, toss the romaine and bronze-leaf lettuces with the endives and three-quarters of the enoki mushrooms. Whisk the reserved dressing to blend. Toss the salad with the dressing and arrange on 4 plates.

6. Scatter the diced papaya over the salads and set a salmon fillet in the center of each one. Sprinkle with the macadamia nuts. Garnish the salads with the remaining enoki mushrooms and the pickled ginger.

NOTE: As a substitute, cut ½ of a medium red bell pepper into thin matchsticks and marinate overnight in 6 tablespoons unseasoned rice vinegar mixed with 2 tablespoons sugar. Drain well before using.
—Linda Burum

DILLED TABBOULEH
Fresh dill, smoked trout and bulgur make a luncheon salad that would be appealing packed into lettuce leaves, scooped up on endive leaves or stuffed into tomatoes.
6 Servings

1 smoked brook trout (about 8 ounces)
1 cup bulgur
¼ cup extra-virgin olive oil
⅓ cup fresh lemon juice
¼ cup minced fresh dill
¼ cup chopped red onion
2 tablespoons capers, rinsed and drained
½ teaspoon salt
¼ teaspoon freshly ground pepper
1 large tomato, cut into ½-inch dice

1. Remove the fillets from the smoked trout; peel off the skin. Break the fish into ½- to ¾-inch pieces.

2. In a large bowl, pour 2 cups of warm water over the bulgur, cover and let soak for about 1 hour, until the grains are puffed and softened.

3. Line a sieve with a double thickness of dampened cheesecloth and drain the bulgur into the sieve. Draw up the ends of the cheesecloth and gently squeeze the bundle until all the excess moisture is removed.

4. Place the bulgur in a large salad bowl and toss with the olive oil and lemon juice. Add the trout, dill, red onion, capers, salt and pepper; toss well. Cover and refrigerate for 2 to 3 hours, until chilled.

5. Before serving, add the tomato and toss into the salad.
—Dorie Greenspan

SMOKED SEAFOOD SALAD WITH ROASTED PEPPERS AND CAPERS
▼ The roasted peppers, capers and smoked fish call for wines that offer a contrast to their intense flavors. Straightforward whites such as Hugel Pinot Blanc Cuvée des Amours or Torres Viña Sol would work well here.
6 Servings

2 large red bell peppers
2 large green bell peppers
1 large yellow bell pepper
2 teaspoons drained capers
2 tablespoons olive oil
2 tablespoons fresh lemon juice
1 tablespoon chopped fresh basil
½ teaspoon freshly ground black pepper
¼ teaspoon salt
1½ pounds assorted smoked seafood, such as shrimp, scallops, mussels, trout, sturgeon, eel and bluefish, cut into large chunks

1. Roast the red, green and yellow bell peppers directly over a gas flame or under a preheated broiler as close to the heat as possible, turning until charred all over. Immediately place the peppers in a paper bag to steam for 5 minutes. Peel the peppers and remove the cores, seeds and membranes.

2. Cut the peppers into long, thin strips and place in a large bowl. Add the capers, olive oil, lemon juice, basil, black pepper and salt.

3. Add the smoked seafood and toss lightly but thoroughly. Mound the salad on individual serving plates. If desired, remove some of the pepper strips from the salad and use them as a garnish.
—*Steve Mellina*

SMOKED TROUT NICOISE
6 to 8 Servings

1½ pounds small red potatoes
½ pound snow peas
1 pound green beans
1 pint cherry tomatoes
1 red onion, thinly sliced
1 cup Niçoise olives
10 ounces skinned and boned smoked trout or salmon, cut into chunks (about 2½ cups)
4 hard-cooked eggs, halved lengthwise
⅓ cup white wine vinegar
⅔ cup olive oil
1 tablespoon minced shallot
1 tablespoon minced fresh dill
½ teaspoon salt
Freshly ground pepper
Sprigs of fresh dill, for garnish

1. Cook the potatoes in a large pot of boiling water until tender, 15 to 20 minutes. Drain and rinse under cold running water until they are cool enough to handle. Slip the skins off the potatoes and cut them into quarters.

2. Meanwhile, in another pot of boiling water, cook the snow peas for 2 to 3 minutes, until they swell slightly and turn bright emerald green. Remove them with a slotted spoon and rinse until cool. Cook the green beans in the same boiling water for 7 to 10 minutes, until tender but still firm. Drain the beans and rinse them under cold running water until cool.

3. Arrange the potatoes, snow peas, green beans, cherry tomatoes, and onion decoratively on a large platter. Scatter the olives over the vegetables. Mound the smoked trout in the center. Place the eggs decoratively around the salad. If not serving immediately, cover and refrigerate.

4. In a small bowl, whisk the vinegar and oil until blended. Whisk in the shallot, minced dill, salt and pepper to taste.

5. Just before serving, whisk the dressing and drizzle it over the salad. Garnish with the dill sprigs.
—*F&W*

JADE CRAB SALAD
The special ingredient in the dressing, red sweet ginger (preserved red ginger in syrup), is available at all Chinese markets (Koon Chun or Mee Chun are a couple of well-known brands).
6 Servings

½ cup pine nuts
3 tablespoons red wine vinegar
2 tablespoons syrup from the jar of preserved red ginger* (see below)
2 tablespoons safflower oil
1 tablespoon minced fresh ginger
1 teaspoon grated orange zest
½ teaspoon Chinese chili sauce
½ teaspoon salt
¾ pound spinach, stems removed, leaves shredded
½ pound lump crabmeat
1 large red bell pepper, cut into thin julienne strips
1 bunch of enoki mushrooms, separated, stem ends trimmed
2 scallions, thinly sliced on the diagonal
⅓ cup finely julienned preserved red ginger*
*Available at Asian markets

1. Preheat the oven to 350°. Spread out the pine nuts on a baking sheet and toast in the oven until golden brown, about 10 minutes.

2. In a bowl, whisk together the vinegar, red ginger syrup, oil, fresh ginger, orange zest, chili sauce and salt.

3. In a large bowl, combine the spinach, crabmeat, red pepper, mushrooms, scallions, preserved red ginger and toasted pine nuts. Add the dressing; toss well.
—*Hugh Carpenter*

CRAB AND AVOCADO SALAD WITH CORIANDER-TOMATILLO VINAIGRETTE
Tart, lemony tomatillos and coriander make a refreshing dressing for this juicy, summery dish.
4 Servings

1 cup dry white wine
2 large shallots, minced
½ pound sole fillet, cut crosswise into ¼-inch-thick strips
2 fresh tomatillos, finely chopped
1 garlic clove, minced
1 teaspoon grated lemon zest
½ teaspoon salt
½ teaspoon freshly ground pepper
½ cup chopped fresh coriander (cilantro)
2½ tablespoons white wine vinegar
⅓ cup extra-virgin olive oil
1 medium cucumber—peeled, seeded and cut into ½-inch dice
1 large tomato, seeded and cut into ¼-inch dice
4 cups shredded romaine lettuce
½ pound lump crabmeat, picked over to remove any cartilage
2 large avocados, preferably Hass, cut lengthwise in half and sliced crosswise ¼ inch thick

1. In a small saucepan, bring the wine and half of the minced shallots to a boil over high heat. Reduce the heat to low, add the sole and simmer until just opaque throughout, about 3 minutes. Using a slotted spoon, transfer the sole to a plate and set aside at room temperature to cool. Place the sole in the refrigerator to chill.

2. Boil the poaching liquid over high heat until it is reduced to 3 tablespoons, about 5 minutes. Pour the poaching liquid into a shallow bowl and let cool in the refrigerator.

3. In a medium bowl, combine the remaining minced shallots, the tomatillos, garlic, lemon zest, salt, pepper, coriander and vinegar. Whisk in the olive oil and the chilled poaching liquid.

4. In a bowl, combine the cucumber, tomato, lettuce and ½ cup of the vinaigrette and toss. In another bowl, gently combine the chilled sole, crabmeat and the remaining vinaigrette.

5. Arrange the sliced avocados on the outer rim of each of 4 plates. Dividing equally, place the lettuce mixture in the center. Spoon the crab-sole mixture over the lettuce, pouring any excess vinaigrette over the avocados.
—Marcia Kiesel

CRAB HOPPIN' JOHN

Though usually featured as a summer meal in itself, this salad can also be served with a number of other dishes in a buffet.

8 to 10 Servings

1 package (10 ounces) frozen black-eyed peas
1 pound lump crabmeat, picked over to remove any cartilage
Juice of 2 lemons
3 cups cooked rice
1 medium red onion, finely chopped
2 celery ribs, finely chopped
¼ cup chopped parsley
½ cup light vegetable oil
½ teaspoon salt
½ teaspoon freshly ground pepper
Hot pepper sauce

1. In a medium saucepan, bring 1 cup of water to a boil over high heat. Add the black-eyed peas and return to a boil. Reduce the heat to moderate and simmer until almost all the water has been absorbed, 20 to 25 minutes. The peas should no longer taste starchy. Set aside and let cool in the saucepan. Any remaining water will be absorbed during cooling.

2. In a large serving bowl, combine the crabmeat and lemon juice. Add the rice, onion, celery, parsley, oil, salt and pepper.

3. Add the cooled black-eyed peas and toss lightly but thoroughly to combine. Refrigerate, covered, until chilled. Serve with hot sauce on the side.
—John Martin Taylor

SOFT-SHELL CRAB AND HAZELNUT SALAD

4 Servings

1 cup hazelnuts (5 ounces)
¼ cup all-purpose flour
¾ teaspoon salt
½ teaspoon freshly ground pepper
8 cleaned soft-shell crabs
1 cup milk
3 tablespoons sherry wine vinegar
1 tablespoon Dijon mustard
2 tablespoons hazelnut oil
⅓ cup vegetable oil
½ cup olive oil
3 medium carrots, grated
2 large bunches of arugula, large stems removed
3 medium Belgian endives, cored and sliced crosswise ¼ inch thick
Lemon wedges, for serving

1. Preheat the oven to 350°. Spread the hazelnuts on a small baking sheet and roast in the middle of the oven for 10 to 12 minutes, until fragrant and the skins start to crack. Transfer the nuts to a kitchen towel and rub them against each other to remove the skins. Let the nuts cool completely. Reduce the oven temperature to 250°.

2. In a food processor, combine ½ cup of the cooled hazelnuts with the flour and ¼ teaspoon each of the salt and pepper. Pulse until the nuts are finely ground. Place on a plate. Coarsely chop the remaining ½ cup hazelnuts and set aside.

3. Place the crabs in a large shallow dish and pour the milk on top. Set aside to soak for 10 minutes. In a small bowl, combine the vinegar, mustard and the remaining ½ teaspoon salt and ¼ teaspoon pepper. Gradually whisk in the hazelnut oil, vegetable oil and ¼ cup of the olive oil. Set the dressing aside.

4. In a large skillet, preferably nonstick, heat 2 tablespoons of the olive oil over moderately high heat. Remove 4 crabs from the milk and shake gently. Dredge the crabs in the hazelnut flour, add to the skillet and fry, turning once, until browned and crisp, about 3 minutes per side. Transfer the crabs to a heatproof platter and place in the oven to keep warm. Repeat with the remaining 2 tablespoons olive oil and 4 crabs.

5. In a large bowl, combine the carrots, arugula and endives. Add all but ¼ cup of the dressing and toss well. Arrange the salad on plates and sprinkle the reserved chopped hazelnuts on top. Set the crabs on the salads and spoon some of the reserved dressing over each crab. Serve immediately, with lemon wedges on the side.
—Bob Chambers & Carl Parisi

WILD RICE AND MUSSEL SALAD

🍷 California Chardonnay, such as Sebastiani Proprietor's Reserve
4 Servings

¼ cup pine nuts
1 cup wild rice (about 8 ounces), well rinsed
1½ teaspoons salt
6 tablespoons unsalted butter
2 medium shallots, minced
½ cup dry white wine
1 bouquet garni: 4 parsley stems, 1 bay leaf and ¼ teaspoon thyme tied in a double thickness of cheesecloth
3 pounds mussels—well scrubbed, soaked and debearded
1 bunch of scallions, thinly sliced
½ teaspoon freshly ground pepper
4 to 6 large leaves of butter lettuce, for garnish

1. Preheat the oven to 375°. Spread the pine nuts in a small baking pan. Bake, shaking the pan occasionally, until the nuts are toasted and lightly browned, 3 to 4 minutes. *(The nuts can be toasted ahead of time.)*

2. In a medium saucepan, combine the rice and 1 teaspoon of the salt with 4 cups of cold water. Bring to a boil over high heat, reduce to a simmer, cover and cook for about 25 minutes, until the rice is barely tender and still slightly chewy. Drain and set aside.

3. Meanwhile, in a large flameproof casserole, melt 3 tablespoons of the butter over moderate heat. Add the shallots and sauté until softened but not browned, about 3 minutes. Add the wine, bouquet garni and mussels. Increase the heat to high, cover and cook, shaking the pan occasionally, for about 3 minutes, or until the mussels open. Remove from the heat.

4. When cool enough to handle, remove the mussels from their shells and place in a small bowl. Strain the cooking liquid through a sieve lined with a double thickness of dampened cheesecloth into a saucepan and boil over high heat until reduced to ½ cup, about 5 minutes.

5. In a large saucepan, melt the remaining 3 tablespoons butter over moderate heat. Add the wild rice, scallions and reduced mussel liquid. Toss gently to combine. Add the pepper, mussels, toasted pine nuts and remaining ½ teaspoon salt. Cook, tossing, until warmed through, about 3 minutes. Arrange each portion on a leaf of lettuce, and serve warm.

—John Robert Massie

MUSSEL AND POTATO SALAD

🍷 Ice cold American lager beer
6 Servings

2 pounds small red potatoes
5 pounds mussels
½ cup dry vermouth
½ cup olive oil
1 tablespoon plus 1 teaspoon Dijon mustard
1½ teaspoons finely chopped shallot
1 garlic clove, minced
¼ teaspoon salt
¼ teaspoon freshly ground pepper
Belgian endive or Romaine lettuce leaves, sprigs of watercress and strips of roasted red bell pepper, for garnish

1. In a medium saucepan of boiling salted water, cook the potatoes until tender, about 15 minutes. Drain and then rinse under cold running water until cool; drain well. Place in a bowl, cover and refrigerate until chilled.

2. Meanwhile, scrub the mussels under cold water; trim off the beards. Place in a large nonreactive pot. Add the vermouth, cover and bring to a boil over moderately high heat. Steam until the shells open, about 5 minutes. Remove from the heat; discard any unopened mussels. As soon as they are cool enough to handle, remove the mussels from their shells. Place in a large bowl, cover and refrigerate until chilled. Strain and reserve the cooking liquid.

3. Peel the cold potatoes and cut into thin slices. Add to the mussels. In a small bowl, combine ⅓ cup of the strained cooking liquid, the oil, mustard, shallot, garlic, salt and pepper. Pour over the mussels and potatoes and toss gently to coat. Cover and refrigerate for at least 2 hours.

4. To serve, mound the mussel and potato salad on individual plates or on a platter. Garnish with Belgian endive leaves, sprigs of watercress and strips of roasted red bell pepper.

—Rick Ellis

MUSSEL AND RED BEAN SALAD

This salad makes a perfect light meal. All you need to add is a loaf of bread and a good bottle of wine.

🍷 California Sauvignon Blanc, such as Cakebread Cellars
6 Servings

3 pounds mussels, preferably cultivated, scrubbed and debearded
2 tablespoons unsalted butter
3 medium shallots, minced
½ cup dry white wine
2 cans (19 ounces each) red kidney beans, rinsed and drained
¼ cup olive oil
2 tablespoons sherry wine vinegar
½ teaspoon Dijon mustard
¼ teaspoon freshly ground pepper
Salt
¼ cup chopped parsley

1. Soak the mussels in a bowl of cold salted water to remove any sand, about 10 minutes.

2. In a large heavy saucepan, melt the butter over moderate heat. Add the shallots and sauté until softened, 3 to 4 minutes. Add the wine, increase the heat to high and bring to a boil. Add the mussels, cover and cook, shaking the pan occasionally, until the mussels have opened, about 5 minutes.

3. Remove the mussels from the pan with a slotted spoon. Discard any that have not opened. Increase the heat to high and boil the cooking liquid until reduced by half, about 10 minutes.

4. Remove the mussels from their shells and place in a large bowl. Add the beans and toss.

5. In a small bowl, whisk together the oil, vinegar, mustard, pepper and ½ cup of the reduced mussel cooking liquid. Season with salt to taste. Pour the dressing over the beans and mussels, add the parsley and toss to combine. Serve at room temperature.
—John Robert Massie

WARM MUSSEL SALAD WITH LEMON GRASS AND FRESH CORIANDER

Lemon grass adds a subtle lemon perfume to the mussel cooking liquid.
4 Servings

3 pounds mussels, scrubbed and debearded
⅓ cup dry white wine
2 stalks of lemon grass* (bottom third only), finely minced
2 teaspoons minced fresh ginger
¼ cup fresh lime juice
1½ teaspoons sugar
½ teaspoon salt
⅓ cup plus 1 tablespoon peanut oil
½ teaspoon freshly ground black pepper
½ teaspoon crushed red pepper
½ cup finely chopped red onion
1 cup chopped fresh coriander (cilantro)
*Available at Asian markets

1. In a stockpot, place the mussels, wine, lemon grass and ginger. Cover and cook over high heat, shaking the pan once or twice, until the mussels steam open, 3 to 5 minutes. With a slotted spoon, transfer the mussels to a large bowl. Discard any that do not open.

2. In a small jar, combine the lime juice, sugar and salt. Cover tightly and shake to blend well. Add the oil, black pepper and red pepper and shake until blended.

3. Pour the dressing over the mussels. Add the red onion and coriander and toss. Transfer to a serving bowl and serve warm or at room temperature.
—Bruce Cost

STIR-FRIED SCALLOP AND RED PEPPER SALAD

I can't tell you how much leftover Chinese food I have eaten in my life. But knowing that if Chinese food is rewarmed, the vegetables will become soggy and the meat overcooked, I always eat the remains at room temperature. One day I put some leftover scallops in ginger sauce on a bed of lettuce leaves and discovered that the cooled scallops made a great salad.

The following dish is traditionally served hot, but if you make it a few hours in advance and let it cool to room temperature, you'll enjoy an exciting main-course salad.
4 to 6 Servings

1 tablespoon water chestnut powder* or cornstarch
3 tablespoons medium-dry sherry
1 egg white
1 pound bay scallops or quartered sea scallops
1½ tablespoons dark soy sauce*
1 tablespoon chicken stock or water
1 tablespoon demiglace (optional)
1 teaspoon red wine vinegar
1 teaspoon sugar
½ to 1 teaspoon Szechuan chili paste with garlic*
1 cup peanut oil
2 tablespoons minced fresh ginger
2 scallions, sliced
1 garlic clove, minced
1 red bell pepper, cut into 1-inch squares or triangles
2 ounces snow peas, strings removed and cut in half crosswise on the diagonal (about ¾ cup)
¼ pound peeled water chestnuts, preferably fresh,* thinly sliced (about ¾ cup)
2 teaspoons Oriental sesame oil
*Available at Asian markets

1. In a medium bowl, dissolve the water chestnut powder in 1 tablespoon of the sherry. Whisk in the egg white until blended. Add the scallops and toss to coat evenly. Cover and refrigerate for at least 1 hour, or up to 12 hours.

2. In a small bowl, mix the remaining 2 tablespoons sherry with the soy sauce, chicken stock, demiglace, vinegar, sugar and chili paste. Set the seasoning sauce aside.

3. Place a wok over high heat until it smokes; this will take about 2 minutes. Add the peanut oil and heat to 325°, or until shimmering.

4. Stir the scallops and marinade. Carefully add to the hot oil all at once, stirring with a pair of chopsticks in a circular motion for about 1 minute, or until the scallops turn white. Turn off the heat and drain the oil and scallops into a colander. Shake to remove as much oil as possible.

5. Return the wok to high heat. In the oil that sticks to the wok, stir-fry the ginger, scallions, garlic and bell pepper for 1 minute. Return the scallops to the wok and add the snow peas and water chestnuts. Toss briefly over the heat to mix.

6. Add the seasoning sauce and stir-fry for 30 seconds. Turn off the heat. Add the sesame oil. Toss to mix. Transfer to a platter and let cool to room temperature. *(The salad can be made up to 1 day ahead and refrigerated, covered. Let return to room temperature before serving.)*
—Karen Lee & Alaxandra Branyon

THAI PAPAYA SCALLOP SALAD

Cool papaya, fresh bay scallops, sweet red pepper and crisp cucumber mingle with buttery pine nuts in a Thai sweet and sour lime dressing.

4 to 6 Servings

¾ cup pine nuts
1 tablespoon unsalted butter
1 tablespoon peanut oil
1 small garlic clove, minced
4 teaspoons finely minced fresh ginger
1 pound bay scallops
3 tablespoons fresh lime juice
2 tablespoons (packed) light brown sugar
1 tablespoon light soy sauce or 2 tablespoons fish sauce (nuoc mam)*
½ teaspoon Chinese chili sauce*
2 scallions, minced
1 tablespoon minced fresh coriander (cilantro)
1½ ripe papayas (about 2 pounds total)—peeled, seeded and cut into ½-inch dice

1 medium red bell pepper, cut into ½-inch dice
1 small cucumber, seeded and cut into ½-inch dice
1 pound spinach—stemmed, washed and patted dry
*Available at Asian markets

1. Preheat the oven to 325°. On a small baking sheet, toast the pine nuts until golden, about 10 minutes.

2. In a large skillet, melt the butter in the oil with the garlic and 2 teaspoons of the ginger over moderately high heat. Add the scallops and cook, tossing, until firm when pressed with your finger and just opaque throughout, about 2 minutes. Transfer the scallops to a plate and let cool for 15 minutes. Cover and refrigerate until ready to use.

3. In a medium jar, combine the lime juice, brown sugar, soy sauce, chili sauce, scallions, coriander and the remaining 2 teaspoons ginger. Cover the jar tightly and shake vigorously to blend and dissolve the sugar.

4. In a large bowl, combine the papayas, red pepper and cucumber. Add the scallops, all but 2 tablespoons of the reserved pine nuts and the dressing. Toss well to coat.

5. Divide the spinach among chilled dinner plates. Spoon the salad on top. Sprinkle the remaining 2 tablespoons pine nuts over the salads and serve.
—Hugh Carpenter

LEMON SHRIMP SALAD

Here's a twist on the usual pairing of chicken and lemon. I find shrimp particularly appealing on a hot summer night.

4 Servings

1 pound medium shrimp, shelled and deveined
1 tablespoon minced lemon zest
¼ cup fresh lemon juice
2 tablespoons ginger syrup (from preserved stem ginger)*
1 tablespoon soy sauce
½ teaspoon Chinese chili sauce*
½ teaspoon salt
2 tablespoons finely minced fresh ginger
1 garlic clove, minced
3 cups safflower or other vegetable oil
8 wonton skins,* cut into ¼-inch-wide strips
2 ounces rice sticks,* separated into small bundles
8 large romaine lettuce leaves
1 cup shredded carrots
¼ cup thinly sliced scallions
*Available at Asian markets

1. In a large saucepan of boiling water, cook the shrimp until opaque throughout, 1½ to 2 minutes. Drain and transfer to a bowl of ice water. Split the cooled shrimp in half lengthwise, pat dry, cover and refrigerate.

2. In a small jar, combine the lemon zest, lemon juice, ginger syrup, soy sauce, chili sauce, salt, fresh ginger, garlic and 2 tablespoons of the safflower oil. Cover tightly and shake vigorously.

3. In a large skillet, heat the remaining oil over moderate heat until it just begins to smoke. Cook the wonton strips in several batches until light golden brown, about 1 minute. Using tongs or a slotted spoon, transfer them to a tray lined with paper towels. Cook the rice

stick bundles in the same oil until they puff, about 10 seconds; drain on paper towels. *(The recipe can be prepared to this point up to several hours ahead.)*

4. Stack the romaine leaves and roll up from a long side into a tight cylinder. Slice the roll at ⅛-inch intervals and transfer the shredded lettuce to a large mixing bowl. Add the carrots, scallions and cooked shrimp. Shake the dressing, pour it over the salad and toss. Gently fold in the fried wonton skins and rice sticks. Serve immediately.

—Hugh Carpenter

SHRIMP SALAD
WITH
CAULIFLOWER
❡ California white wine, such as Wente Grey Riesling
6 to 8 Servings

2½ pounds medium shrimp, in the shell
1 medium head of cauliflower, cut into 1-inch florets
1 cup mayonnaise
1 cup plain yogurt
⅓ cup grainy mustard
6 tablespoons chopped fresh dill
1 cucumber, scored and thinly sliced, for garnish
1 large carrot, peeled and shredded, for garnish

1. Cook the shrimp in a large pot of salted boiling water, uncovered, until they are pink and slightly firm to the touch, 1 to 2 minutes. Pour into a colander and rinse under cold running water to stop the cooking; drain well. Shell the shrimp, and devein if desired. Place in a large bowl. Add the cauliflower, cover and refrigerate.

2. In a small bowl, combine the mayonnaise, yogurt, mustard and dill; blend well. Add to the shrimp and cauliflower; toss well. Cover and refrigerate for at least 2 hours or preferably overnight.

3. Arrange the cucumber slices decoratively around the salad on individual plates or a serving dish and garnish with shredded carrot.

—Rick Ellis

SPICY SOUTHEAST
ASIAN SHRIMP
SALAD
6 Servings

2 tablespoons fresh lemon juice
1 large garlic clove, crushed
1 teaspoon salt
¼ cup peanut oil
½ teaspoon Oriental sesame oil
¼ teaspoon Chinese hot oil or cayenne pepper
1½ pounds medium shrimp, shelled and deveined
½ cup plus 1 tablespoon rice vinegar or distilled white vinegar
2 tablespoons honey
3 tablespoons finely chopped unsalted dry-roasted peanuts
3 heads of Boston lettuce, separated into leaves
4 carrots, shredded (about 2 cups)
1 small cucumber, seeded and thinly sliced
12 radishes, thinly sliced
¼ cup coarsely chopped fresh mint
⅓ cup chopped fresh coriander (cilantro) or flat-leaf parsley
4 scallions, thinly sliced

1. In a medium bowl, combine the lemon juice, garlic, salt, peanut oil, sesame oil and hot oil. Add the shrimp and toss to coat. Let marinate at room temperature, tossing occasionally, for at least 30 minutes or up to 1 hour. Meanwhile, soak about 1 dozen 6-inch bamboo skewers in water.

2. Light the charcoal or preheat the broiler. Thread the shrimp onto the skewers. Grill or broil about 4 inches from the heat for 1½ to 2 minutes on each side, or until loosely curled and just opaque throughout. *(The recipe may be prepared 1 day ahead to this point. Remove the shrimp from the skewers and let cool to room temperature. Wrap and refrigerate overnight. Let return to room temperature before continuing.)*

3. In a small bowl, make a dressing by mixing together the vinegar and honey until blended. Stir in the peanuts.

4. Cover a large serving platter with some of the lettuce leaves. Arrange the shrimp, bunches of the remaining lettuce and mounds of the carrots, cucumber, radishes, mint, coriander and scallions on top. To serve, drizzle with some of the dressing and serve the rest on the side, or suggest that guests make a package of a lettuce leaf filled with shrimp, vegetables, a sprinkle of herbs and a drizzle of the dressing.

—Beverly Cox

SHRIMP AND
ENDIVE SALAD
4 Servings

½ pound large shrimp, shelled and deveined
4 small heads Belgian endive
1 can (8 ounces) water chestnuts, drained and sliced
1 tablespoon soy sauce
2 teaspoons cornstarch
1 teaspoon sugar
½ teaspoon salt
1 jar (3 ounces) pimientos, drained and cut into 2-inch strips
1 large grapefruit, halved and sectioned, juice reserved
1 teaspoon grated fresh ginger

1. Cut each shrimp into three pieces and set them aside. Separate the leaves from the heads of endive; cut the largest leaves in half crosswise. Wash them and pat them dry with paper towels.

2. Toss the water chestnuts with the soy sauce and allow them to rest for 3

minutes; discard any soy sauce that has not been absorbed.

3. In a small bowl, mix the cornstarch with 1 teaspoon water, the sugar, and the salt; set aside.

4. In a large bowl, combine the shrimp, endive, water chestnuts, pimientos and the grapefruit sections along with any reserved juice; toss gently. Turn the mixture into a nonstick or nonreactive skillet and place over moderately high heat. Cook, stirring gently but constantly, for 3 to 4 minutes, or until the shrimp have turned pink and the other ingredients are hot.

5. Using a slotted spoon, quickly transfer the salad to a bowl. Add the cornstarch mixture to the liquid in the pan, and, stirring constantly, cook until the liquid thickens, about 1 minute. Toss the salad with the dressing and ginger; arrange on a platter and serve immediately.
—*Jim Fobel*

CABBAGE SALAD WITH SHRIMP (ACHAR UDANG)
Tart and crisp, this salad makes a refreshing light luncheon dish.
2 Servings

2 tablespoons fresh lime juice
2 tablespoons cider vinegar or distilled white vinegar
2 teaspoons salt
1 tablespoon plus 1 teaspoon sugar
2 teaspoons minced fresh ginger
2 cups finely shredded cabbage (about 5 ounces)
2 Kirby cucumbers, unpeeled and sliced
1 red bell pepper, cut into julienne strips
½ pound medium shrimp—cooked, shelled and deveined

1. In a large bowl, mix together the lime juice, vinegar, salt, sugar, ginger and 2 tablespoons of water.

2. Add the cabbage, cucumbers, bell pepper and shrimp. Toss with the dressing until well mixed. Let marinate at room temperature, tossing occasionally, for up to 2 hours. Serve at room temperature or slightly chilled.
—*Copeland Marks*

COOL AND SPICY SALAD WITH SHRIMP AND PORK
Here is an example of the classic combination of pork and shrimp that the Vietnamese do so well. Note that the pork is cooked in water to eliminate any fat or strong taste.
6 to 8 Servings

½ pound boneless pork loin, trimmed of excess fat
½ pound medium shrimp, shelled and deveined
1 large cucumber—peeled, seeded and cut into very thin 2-inch-long strips
2 teaspoons salt
6 tablespoons sugar
6 tablespoons distilled white vinegar
3 celery ribs, cut into thin 2-inch strips
2 large carrots, cut into long, thin strips
1 medium onion, thinly sliced
½ cup chopped fresh coriander (cilantro)
3 tablespoons chopped fresh mint
½ cup Nuoc Cham Dipping Sauce (p. 133)
3 cups vegetable oil, for deep-frying
24 dried shrimp chips*
½ cup chopped, unsalted dry-roasted peanuts
*Available at Asian markets

1. Put the pork into a small saucepan with enough cold water to cover. Bring to a simmer over moderate heat and cook until tender, about 25 minutes. Drain, cool slightly and cut into thin 2-inch-long strips. Let the pork cool to room temperature.

2. In a medium saucepan of boiling water, cook the shrimp until just opaque throughout, 2 to 3 minutes. Drain and cool slightly, then halve each shrimp lengthwise. Set the shrimp aside to cool to room temperature. *(The shrimp and pork can be cooked 1 day ahead and refrigerated.)*

3. Toss the cucumber strips with 1 teaspoon of the salt and set aside in a colander for 5 minutes. Then squeeze to remove any excess moisture.

4. In a measuring cup, combine the remaining 1 teaspoon salt with the sugar and vinegar. Put the celery, carrots and onion in 3 separate bowls. Toss each vegetable with one-third of the vinegar mixture. Let stand for 5 minutes.

5. Drain the vegetables and squeeze dry. Put them in a large bowl. Add the cucumber, pork, shrimp, ¼ cup of the coriander and 2 tablespoons of the mint. Add the Nuoc Cham Dipping Sauce and toss well.

6. In a large saucepan or a wok, heat the oil to 350°. Fry the shrimp chips, 5 or 6 at a time, until they expand fully and float on the surface, about 12 seconds. Try not to let the chips turn brown. Drain on paper towels.

7. Spoon the salad onto a large platter. Sprinkle the top with the remaining coriander, mint and the chopped peanuts. Scatter the shrimp chips around the salad.
—*Marcia Kiesel*

Collard Greens and Black-Eyed-Pea Soup with Cornmeal Croustades (p. 65).

(Relatively) Quick Greens in Potlikker (p. 164).

Sweet Potato Vichyssoise (p. 60).

NUOC CHAM DIPPING SAUCE

This indispensable dipping sauce is served with every Vietnamese meal.
Makes About 2½ Cups

1 teaspoon crushed red pepper
1 tablespoon distilled white vinegar
½ cup fish sauce (nuoc mam)*
¼ cup fresh lime juice
1 small carrot—finely shredded, rinsed and squeezed dry
2 small garlic cloves, minced
½ cup sugar
*Available at Asian markets

1. In a small dish, soak the red pepper in the vinegar for 2 minutes.
2. In a small bowl, combine the fish sauce, lime juice, carrot, garlic and sugar. Stir in 1½ cups warm water and the hot pepper-vinegar mixture. Stir until the sugar dissolves. Serve at room temperature. Store the sauce in a jar in the refrigerator for up to 3 days.
—Marcia Kiesel

SHRIMP, PORK AND WATERCRESS SALAD

Thais, like the Vietnamese, frequently combine shrimp and pork, and the effect is always satisfying.
6 Servings

1 pound small shrimp, shelled and deveined
1 tablespoon peanut oil
¼ cup chopped shallots
4 garlic cloves, chopped
½ pound lean ground pork
1 large bunch of watercress, large stems removed
½ cup roasted peanuts, chopped
2 tablespoons nam pla*
3 tablespoons fresh lime juice
1 teaspoon sugar
2 large tomatoes, thinly sliced
2 teaspoons crushed red pepper
*Available at Asian markets

1. In a medium saucepan, bring 5 cups of water to a boil. Add the shrimp and cook until loosely curled and opaque throughout, 2 to 3 minutes. Drain and rinse under cold water. Drain well, pat dry and set aside.
2. In a small skillet, heat the peanut oil over moderate heat. Add the shallots and garlic and cook until softened but not browned, about 30 seconds. Add the ground pork and cook, stirring, until no traces of pink remain, 5 to 6 minutes. Remove from the heat. If there is any excess fat in the skillet, drain it off. Let the pork cool slightly.
3. In a large bowl, combine the watercress, reserved shrimp and the peanuts. Add the pork mixture and toss well.
4. In a small bowl, stir together the *nam pla*, lime juice and sugar. Pour the dressing over the salad and toss. Arrange the tomato slices on a serving platter, overlapping them slightly. Spoon the salad on top and garnish with the crushed red pepper.
—Jeffrey Alford

SQUID SALAD WITH LEMON AND FRESH MINT

Squid can become firm and elastic if overcooked for even a few seconds. I use an off-the-stove cooking method to keep the squid tender and moist.
6 to 8 Servings

¼ cup Shellfish Boil (recipe follows)
2 pounds cleaned small whole squid, with tentacles
1 cup (packed) fresh mint leaves
1 teaspoon dry mustard
½ cup fresh lemon juice
½ of a jalapeño pepper, minced
1 teaspoon salt
1½ teaspoons sugar
½ teaspoon freshly ground black pepper
⅔ cup olive oil
Lemon slices and mint sprigs, for garnish

1. Bring a large saucepan of water to a boil. Stir in the Shellfish Boil and the squid. After 1 minute, remove the pan from the heat and set it on a rack. Let the squid cool in the liquid until tepid, about 30 minutes. Drain well.
2. Meanwhile, in a food processor, combine the mint leaves, dry mustard, lemon juice, jalapeño pepper, salt, sugar and black pepper. Process until pureed. With the machine on, pour in the olive oil in a thin stream until blended.
3. Thinly slice the squid bodies crosswise. Leave the tentacles whole. Place the squid in a large bowl and toss with the dressing. Arrange on a serving platter and garnish with lemon slices and mint sprigs.
—Steve Mellina

SHELLFISH BOIL
Makes About ⅔ Cup

1 tablespoon fresh thyme or 1 teaspoon dried
3 bay leaves, crumbled
1 teaspoon white peppercorns
1 teaspoon black peppercorns
1 teaspoon celery seeds
½ teaspoon mustard seeds
⅔ cup coarse (kosher) salt

Mix together all the ingredients and store in a covered jar for up to 1 year in a cool, dark place.
—Steve Mellina

SEAFOOD SALAD WITH WATERCRESS, RED PEPPER AND FRESH HERBS

As you cook the seafood, be sure to transfer it to a platter large enough for it to cool in a single layer.

❢ The dressing paired with the pasta and seafood calls for an assertive light wine. Either Domaines Ott Côtes de Provence rosé or Trimbach Gewürztraminer from Alsace fit the bill: Both are bone-dry and have a hint of bitterness in the aftertaste, which contrasts with the flavors of the dish.

6 Servings

¾ cup plus 1 tablespoon extra-virgin olive oil
2 large shallots, minced
7 garlic cloves, minced
1 cup dry white wine
1 pound bay scallops
1 pound medium shrimp, shelled and deveined
1 pound small squid, cleaned and sliced into ¼-inch rings, large tentacles cut in half
1 large tomato—peeled, seeded and minced, juices reserved
1 tablespoon fresh lemon juice
1 tablespoon chopped capers plus 2 tablespoons liquid from the jar
2 tablespoons chopped fresh dill
1½ tablespoons chopped fresh tarragon or 2 teaspoons dried
2 tablespoons chopped parsley
1 teaspoon chopped fresh thyme or ¼ teaspoon dried
½ teaspoon salt
½ teaspoon freshly ground black pepper
3 cups fusilli or other curly pasta
2 medium red bell peppers, thinly sliced
2 large celery ribs, finely diced
2 bunches of watercress, large stems removed, torn into bite-size pieces

1. In a large nonreactive saucepan, heat ¼ cup of the olive oil over high heat. Add the shallots and 3 of the minced garlic cloves. Reduce the heat to low and cook, stirring occasionally, until softened and translucent, about 3 minutes.

2. Increase the heat to high, add the white wine and bring to a boil. Add the scallops and cook for 1 minute. With a slotted spoon, transfer the scallops to a large shallow dish or platter to cool.

3. Add the shrimp to the saucepan and cook until opaque, about 1½ minutes. Remove the shrimp to the platter to cool. Add the squid to the saucepan and cook until tender, about 1½ minutes. Remove to the platter to cool.

4. Boil the shellfish poaching liquid until syrupy and reduced to ½ cup, about 5 minutes. Pour into a medium bowl and stir in the remaining 4 minced garlic cloves, the tomato and its juice, the lemon juice, capers and their liquid, dill, tarragon, parsley, thyme, salt and pepper. Whisk in ½ cup of the olive oil in a thin stream and set aside. (The recipe can be prepared to this point up to 1 day ahead. Cover the seafood and dressing separately and refrigerate.)

5. In a large pot of boiling salted water, cook the fusilli until tender but still firm, about 10 minutes. Drain and rinse under cold running water; drain well.

6. In a large bowl, toss the fusilli with the remaining 1 tablespoon olive oil to lightly coat. Add the seafood, red peppers and celery. Stir in the dressing. Just before serving, add the watercress and toss well.

—Marcia Kiesel

MOLDED SEAFOOD SALAD WITH SHIITAKE MUSHROOMS AND AVOCADO SAUCE

❢ A crisp, herbaceous white will underscore this salad's rich flavors. A California Fumé Blanc, such as Grgich Hills or Robert Mondavi, would be ideal.

4 Servings

6 ounces small shrimp, shelled and deveined
¼ pound bay scallops
½ cup plus 1 tablespoon olive oil
1 pound fresh shiitake mushrooms, stems removed, caps quartered if large
½ teaspoon salt
¼ teaspoon freshly ground white pepper
1 tablespoon plus 1 teaspoon balsamic vinegar
¼ pound cooked lobster meat, cut into small dice (from a 1- to 1½-pound lobster)
½ pound lump crabmeat, picked over to remove any cartilage
1 tablespoon mixed chopped fresh herbs, such as basil, parsley, tarragon and chives
½ of a medium Hass avocado, finely diced
1 small garlic clove, minced
1 teaspoon chopped fresh coriander (cilantro)
1 teaspoon fresh lemon juice
2 drops of hot pepper sauce

1. In a medium saucepan of boiling water, cook the shrimp and scallops together until just opaque throughout, about 1½ minutes. Drain in a colander and refresh under cold running water. Transfer to paper towels to drain. Halve the shrimp crosswise and set the shrimp and scallops aside.

2. In a large skillet, heat ¼ cup of the olive oil over high heat. When the oil is

hot, add the mushrooms in an even layer. Season with the salt and white pepper and cook, without turning, until browned on one side, about 2 minutes. Stir and cook the mushrooms until tender, about 1 minute longer. Add 1 tablespoon of the vinegar and cook until evaporated, about 10 seconds. Transfer the mushrooms to a large plate and let cool.

3. In a large bowl, combine the lobster, crabmeat and the reserved shrimp and scallops. Stir in the cooled mushrooms, 1 tablespoon of the olive oil, the remaining 1 teaspoon vinegar and the chopped fresh herbs. Toss well to combine. Tightly pack equal amounts of the seafood salad into four 3½- to 4-inch-wide ramekins or small bowls. Cover with plastic wrap and refrigerate for 1 hour.

4. In a blender, combine the avocado, garlic, coriander, lemon juice and hot sauce. Add ¼ cup of water and blend until smooth. With the machine on, pour in the remaining ¼ cup olive oil in a thin stream and blend until thoroughly incorporated.

5. To serve, unmold the salads by turning each ramekin over onto a plate and giving it a sharp tap. Spoon a heaping tablespoon of the avocado sauce on either side of each salad and serve immediately.
—Steve Mellina

COUSCOUS SALAD
6 to 8 Servings

2 cups chicken stock or canned broth
¾ teaspoon cinnamon
½ teaspoon ground ginger
½ teaspoon cumin
¼ teaspoon turmeric
3 tablespoons extra-virgin olive oil
½ pound skinless, boneless chicken breast or leftover cooked chicken
1 cup couscous
1 medium carrot, cut into ¼-inch dice
1 small red bell pepper, cut into ¼-inch dice
1 small cucumber or zucchini, cut into ¼-inch dice
1 small red onion, cut into ¼-inch dice
1 small Granny Smith apple, cut into ¼-inch dice
⅓ cup currants or raisins
1 cup canned chickpeas, rinsed and drained
¼ cup fresh lemon juice
½ teaspoon salt
¼ teaspoon freshly ground black pepper

1. In a heavy medium saucepan, whisk together the chicken stock, cinnamon, ginger, cumin, turmeric and 1½ tablespoons of the olive oil. Bring to a boil, reduce to a bare simmer and add the chicken breast if you are using uncooked chicken. Poach until white throughout but still moist, about 15 minutes. Remove the chicken and set aside to cool.

2. Return the stock to a boil. Add the couscous in a slow steady stream, stirring constantly, and continue to boil, stirring, for 1 minute. Cover the pot tightly, remove from the heat and let stand for 15 minutes.

3. Fluff the couscous grains with a fork, transfer to a large mixing bowl and let cool. Then fluff again, rubbing with your fingers to break up any lumps.

4. Cut the chicken (poached or leftover) into ½-inch dice. Add the chicken to the couscous. Add the carrot, bell pepper, cucumber, onion, apple, currants and chickpeas and toss.

5. In a small jar with a lid, shake the remaining 1½ tablespoons olive oil with the lemon juice, salt and pepper until well mixed. Pour over the salad and toss well. Cover and refrigerate for several hours or up to 3 days. Season with additional salt, pepper and lemon juice to taste before serving.
—Dorie Greenspan

CURRIED TURKEY SALAD WITH GRAPES AND ALMONDS
8 Servings

1 turkey breast, about 5 pounds
2 medium carrots, cut into chunks
1 medium onion, quartered
2 celery ribs with leaves, cut into large pieces
2 bay leaves
6 black peppercorns, coarsely crushed
4 sprigs of parsley
2 teaspoons thyme
1½ cups mayonnaise
1½ tablespoons curry powder
2 tablespoons mango chutney, finely chopped
¼ teaspoon cumin
3 tablespoons chopped basil leaves or 1 teaspoon dried
⅛ teaspoon cayenne pepper
2 tablespoons dry white wine
2 tablespoons fresh lemon juice
Salt and freshly ground pepper
8 ounces slivered, blanched almonds (about 2 cups)
2 pounds seedless green grapes
1 to 2 bunches of watercress, for serving

1. Poach the turkey breast: Place the turkey breast in a stockpot with 5 quarts of water, the carrots, onion and celery. Make a bouquet garni by tying the bay leaves, peppercorns, parsley and thyme in a square of cotton cheesecloth and add it to the stockpot. Simmer over moderate heat for 1½ to 2 hours, or until the juices run clear when the turkey is pierced with a fork, or when a meat thermometer registers an internal temperature of 165°. Let the turkey cool to

room temperature in the stock to help keep it moist.

2. Meanwhile, make the curry mayonnaise: In a medium bowl, mix the mayonnaise with the curry powder, chutney, cumin, basil and cayenne. Add the white wine and lemon juice and mix well. Season with salt and black pepper to taste, cover and refrigerate until needed.

3. In a large ungreased skillet, toast the almonds over moderate heat until golden brown.

4. Remove the turkey breast from the stock (discard the bouquet garni, strain the stock and reserve for another use). Remove the turkey meat from the bones and tear or cut the meat into strips about ¼ inch wide and 2 to 3 inches long. Mix the strips with 1¼ cups of the curry mayonnaise, adding more if the mixture seems dry. Add the almonds and grapes and gently incorporate them. Cover and refrigerate for at least 30 minutes.

5. Place the salad in a mound in the center of a large, chilled serving dish and surround with the watercress.
—F&W

CHINESE-STYLE CHICKEN SALAD
4 to 6 Servings

1 pound bean sprouts
3 tablespoons Oriental sesame oil
3 tablespoons soy sauce
7 tablespoons rice vinegar
¼ cup peanut oil
2 tablespoons minced fresh ginger
1 teaspoon minced garlic
¼ pound snow peas, cut lengthwise into ⅛-inch-wide strips
3 cups shredded, cooked chicken breasts (about 2 whole breasts)
2 celery ribs, thinly sliced on the diagonal
6 large mushrooms, thinly sliced
½ cup thinly sliced scallions
2 bunches of watercress, stemmed
2 tablespoons toasted sesame seeds
Roasted red bell peppers or pimientos, for garnish

1. Bring a large saucepan of water to a boil and drop the bean sprouts in; blanch them for 1 minute, then drain them in a colander. Refresh under cold running water and drain well.

2. Place the bean sprouts in a large bowl and add 2 tablespoons of the sesame oil, 1 tablespoon of the soy sauce and 2 tablespoons of the vinegar; toss gently to combine. Set the sprouts aside while you prepare the rest of the salad.

3. In a blender or food processor, combine the peanut oil, ginger, garlic, 2 tablespoons of the vinegar and 1 tablespoon of the soy sauce. Process the dressing until the garlic and ginger are pureed.

4. In a wok or large skillet, heat the dressing until very hot but not smoking. Add the snow peas and stir-fry for 1 minute. Remove half of them with a slotted spoon and set aside. Add the shredded chicken to the wok and stir-fry for 1 minute. Add the celery and stir-fry for 1 minute.

5. Remove the wok from the heat and stir in the mushrooms, scallions, and remaining 3 tablespoons vinegar, 1 tablespoon soy sauce and 1 tablespoon sesame oil; toss gently.

6. Arrange the watercress around the edge of a large platter. Just inside the watercress ring, form the bean sprouts into a nest over the rest of the platter. Arrange the reserved snow peas in a ring inside the bean sprouts and mound the chicken salad in the center of the snow pea ring. Sprinkle with the sesame seeds. Cut the roasted peppers into strips and arrange in a starburst pattern on top of the salad.
—Jim Fobel

CRUNCHY CHICKEN SALAD
Be sure the noodles and sesame seeds are not added until the last minute; otherwise they will get soggy.
10 Servings

2½ tablespoons olive oil
1½ pounds skinless, boneless chicken breasts
1 teaspoon salt
¾ teaspoon freshly ground pepper
1 cup fresh bean sprouts (about 3 ounces)
1 large celery rib, cut into ½-inch dice
¾ cup mayonnaise
2 tablespoons fresh lemon juice
1½ tablespoons soy sauce
1 head of romaine lettuce, torn into bite-size pieces
1½ cups Chinese fried noodles
2 tablespoons sesame seeds (optional)

1. Preheat the oven to 350°. Coat the bottom of a large shallow casserole or baking sheet with ½ tablespoon of the oil. Sprinkle the chicken breasts with ¾ teaspoon of the salt and the pepper. Place the chicken in a single layer in the casserole, turn to coat in the oil and bake for about 20 minutes, just until opaque throughout. Set aside until cool to the touch. Using your fingers, shred the meat into bite-size pieces.

2. In a large bowl, toss the chicken with the bean sprouts, celery and the remaining ¼ teaspoon salt. In a small bowl, mix the mayonnaise with the lemon juice, soy sauce and the remaining 2 tablespoons olive oil. Spoon on top of the chicken and toss to combine. Refrigerate for 15 minutes.

3. In a large salad bowl, combine the lettuce, chicken salad, Chinese noodles and sesame seeds. Toss and serve.
—Annie Gilbar

TRIPLE-MUSTARD CHICKEN SALAD

Toasted mustard seeds make this chicken salad extra special. This dish will feed large crowds during the holidays, but it is also great for warm-weather picnics.

❦ Serve a clean, crisp, dry, but assertive white. Try Bordeaux Blanc Sec, such as Maître d'Estournel, which is refreshingly light.

16 Servings

8 pounds skinless, boneless chicken breasts
2 cups mayonnaise
¼ cup Pommery mustard
½ cup extra-sharp Dijon mustard
½ cup fresh lemon juice
½ teaspoon salt
1 teaspoon freshly ground pepper
12 large celery ribs, peeled and thinly sliced on the diagonal
¾ cup mustard seeds
¼ cup extra-virgin olive oil
4 large bunches of watercress, tough stems removed

1. Put the chicken in a large heavy pot and add enough cold salted water to cover by a least 1 inch. Bring to a simmer over moderate heat. Reduce the heat to low and simmer until the chicken is juicy but no longer pink in the center, about 20 minutes. Remove and cut the chicken into 1-inch dice. *(The recipe can be prepared to this point up to 1 day in advance. Cover and refrigerate.)*

2. In a medium bowl, combine the mayonnaise, Pommery and Dijon mustards, lemon juice and salt and pepper; whisk to blend well.

3. In a large bowl, combine the chicken, celery and mustard mayonnaise. Toss to coat well.

4. In a medium skillet, combine the mustard seeds and olive oil. Cook, covered, over moderate heat, shaking the pan, until the seeds begin to pop, 1 to 2 minutes. Immediately remove from the heat and continue shaking the pan until the seeds are toasted and fragrant and have stopped popping. Scrape the mustard seeds and oil over the salad and fold to combine.

5. Transfer the salad to a large serving bowl or a platter and surround with the watercress. Serve slightly chilled or at room temperature.

—Susan Wyler

BROILED THAI CHICKEN SALAD

The special taste of *yam gai* is in the marinade. Try this same marinade with shark fillets or thin slices of pork for an unusual and delicious barbecue.

❦ Although cold beer is an obvious choice for this salad, a fruity, off-dry white such as Folie à Deux Dry Chenin Blanc or Hogue Cellars Chenin Blanc would also work nicely.

6 Servings

3½ ounces cellophane noodles (wun sen)
4 large garlic cloves, chopped
2 teaspoons peppercorns
¼ cup chopped fresh coriander (cilantro)
¼ cup plus 1 tablespoon soy sauce
2 tablespoons plus 1 teaspoon nam pla*
1¼ pounds skinless, boneless chicken breasts, cut into 1-inch cubes
4 cups chicken stock or canned broth
1½ teaspoons sugar
3 tablespoons fresh lime juice
1 head of leaf lettuce, leaves separated
4 medium tomatoes, cut into wedges
3 scallions, chopped
1 European cucumber, cut into 1-inch dice
1 tablespoon chopped roasted peanuts
1 lime, thinly sliced
*Available at Asian markets

1. Place the cellophane noodles in a bowl with hot water to cover. Let soak for 30 minutes to soften.

2. In a blender or food processor, combine the garlic, peppercorns and 2 tablespoons of the coriander. Blend or process to a paste. Scrape down the sides of the bowl and add ¼ cup of the soy sauce and 2 tablespoons of the *nam pla*. Pour the marinade into a shallow pan and add the chicken cubes. Toss well and let marinate for 30 minutes.

3. Preheat the broiler. In a medium saucepan, bring the chicken stock to a boil. Drain the cellophane noodles. Add the noodles to the pan and boil until softened, about 5 minutes. Drain well. Cut the noodles into 4 pieces. Reserve the stock for another use.

4. Remove the chicken from the marinade and place on a baking sheet in a single layer. Broil for 4 to 7 minutes, stirring the pieces from time to time, until browned and cooked through.

5. In a small bowl, stir together the sugar, lime juice and the remaining 1 tablespoon soy sauce and 1 teaspoon *nam pla*, to make the dressing.

6. To assemble the salad, line a large platter with the lettuce leaves. Spread the noodles over the lettuce and top with the tomato wedges, chopped scallions and diced cucumber. Arrange the broiled chicken on the salad and pour on the dressing. Sprinkle the remaining 2 tablespoons coriander and the peanuts on top. Garnish with the lime slices.

—Jeffrey Alford

AVOCADO AND VEAL SALAD WITH WALNUTS

🍷 California Petite Sirah

6 to 8 Servings

3 pounds boneless veal shank or shoulder
3 medium carrots, cut into 2-inch lengths
2 celery ribs, cut into 2-inch lengths
3 medium onions, unpeeled, quartered
12 parsley stems (from about ½ bunch)
4 medium garlic cloves, unpeeled
10 black peppercorns
1 teaspoon thyme
1¼ teaspoons salt
½ cup walnut oil
⅓ cup walnut pieces
2 tablespoons sherry wine vinegar
1 teaspoon green peppercorns, finely chopped
2 ripe avocados

1. Put the veal, carrots, celery, onions, parsley stems, garlic, black peppercorns, thyme and ½ teaspoon of the salt into a stockpot. Add water to cover and simmer, skimming, until the meat is very tender, about 2½ hours. Remove the meat (see Note) and when cool enough to handle, shred into pieces about ½ by 2 inches (there will be about 3½ packed cups).

2. Toss the veal with ½ teaspoon of the salt and 2 tablespoons of the walnut oil; set aside.

3. In a small skillet, heat 1 tablespoon of the walnut oil. Add the walnut pieces and cook, tossing, over moderate heat until toasted, about 3 minutes.

4. In a medium bowl, whisk together the remaining 5 tablespoons walnut oil, the vinegar, green peppercorns and remaining ¼ teaspoon salt to make a dressing.

5. Pit and peel the avocados and cut each half lengthwise into 8 slices. Add to the dressing, tossing gently to coat.

6. To assemble, mound the veal in the center of a large platter. Remove the avocado slices from the dressing and reserve the dressing. Arrange the avocado around the veal. Sprinkle the walnuts over the avocado. Drizzle the reserved dressing over the salad.

NOTE: Strain the cooking liquid, which is delicious, and use as a stock or soup base. Discard the solids.

—Anne Disrude

PAILLARD OF VEAL WITH BITTER GREENS

🍷 Rosé of Cabernet Sauvignon, such as Simi

4 Servings

1 small head of chicory (curly endive)
1 small head of escarole
1 small head of romaine lettuce
4 veal rib chops (1½ inches thick)— boned, trimmed and pounded to ¼ inch (to save time, ask your butcher to do this)—or 4 large veal scallops, pounded ¼ inch thick
1 teaspoon salt
½ teaspoon freshly ground pepper
3 tablespoons unsalted butter
9 tablespoons olive oil
2 garlic cloves, crushed through a press
½ teaspoon Dijon mustard
2 tablespoons red wine vinegar

1. Remove and discard the tough outer leaves from the chicory, escarole and romaine. Separate the heads into individual leaves, remove the tough central ribs and tear the leaves into bite-size pieces. Combine the greens, rinse, dry and set aside in a large bowl.

2. Preheat the oven to 200°. Season the veal on both sides with ½ teaspoon of the salt and ¼ teaspoon of the pepper. In a large skillet, melt the butter in 3 tablespoons of the oil over moderately high heat. Working in batches, sauté the veal paillards, turning once, until well browned and tender, about 5 minutes. As they cook, place them on a heatproof platter, cover loosely with foil and keep warm in the oven.

3. Whisk together the remaining 6 tablespoons oil, the garlic, mustard and remaining ½ teaspoon salt and ¼ teaspoon pepper until blended. Pour over the mixed greens and toss to coat evenly with dressing.

4. When all the paillards are cooked, transfer them to 4 warmed large plates. Pour out the fat in the pan. Return the pan to moderate heat and add the vinegar. Bring to a boil, scraping up any brown bits from the bottom of the pan. Add the mixed greens to the hot pan and toss to mix in the vinegar. Divide evenly among the 4 plates, serving the salad on top of the veal.

—John Robert Massie

VEAL TONNATO SALAD

Serve this zesty salad with plenty of warm, crusty Italian bread. The grilled veal scallops may be served warm or at room temperature.

🍷 Orvieto, such as Antinori

6 Servings

2 tablespoons Dijon mustard
¼ cup fresh lemon juice
2 tablespoons minced red onion plus 1 small red onion, thinly sliced and separated into rings
2 tablespoons drained capers
1 tablespoon minced parsley
¾ teaspoon salt
1 cup plus 2 tablespoons olive oil, preferably extra-virgin

1 can (3½ ounces) tuna packed
　　in olive oil, drained and
　　finely chopped
6 veal scallops, pounded ⅛ inch
　　thick
¼ teaspoon freshly ground pepper
1½ teaspoons minced fresh thyme or
　　½ teaspoon dried
1 head of romaine lettuce
1 lemon, thinly sliced
3 hard-cooked eggs, sliced
6 flat anchovy fillets (optional)

1. In a small bowl, combine the mustard, lemon juice, minced onion, capers, parsley and ¼ teaspoon of the salt. Gradually whisk in 1 cup of the oil in a steady stream. Stir in the tuna and set aside.

2. Light the charcoal. Brush the veal scallops on both sides with the remaining 2 tablespoons olive oil. Combine the remaining ½ teaspoon salt, the pepper and thyme and sprinkle over both sides of the scallops.

3. Grill the veal scallops as close to the hot coals as possible until they just lose their pink color, 1½ to 2 minutes on each side. If no grill is available, sauté the veal in a heavy skillet over moderately high heat for about 2 minutes on each side.

4. Place the veal scallops in a large shallow bowl and add half the tuna dressing, turning the veal to coat well. Set the remaining dressing aside. *(The recipe may be prepared ahead to this point. Let the veal cool to room temperature, cover and refrigerate overnight. Let return to room temperature before continuing.)*

5. To serve, place the romaine leaves on a large serving platter and arrange the veal on top. Garnish with the onion rings, lemon slices, eggs and anchovies. Pour the reserved dressing over the veal or serve on the side.
—*Beverly Cox*

KENTUCKY PORK SALAD

This hearty dish is beautiful to behold as well as exceedingly tasty. Serve it with biscuits hot from the oven. For authentic regional flavor, add applewood or hickory chips to the coals before grilling. A grill with a cover allows the cook more heat control and helps capture the smoky taste imparted by the wood chips.
❢ Chilled lager beer
6 Servings

5 tablespoons Dijon mustard
6 tablespoons unsulphured molasses
5 tablespoons bourbon
¼ teaspoon hot pepper sauce
6 loin pork chops, cut 1 inch thick
¾ pound yams (about 3), peeled
　　and cut into ¼-inch slices
½ cup fresh lime juice
3 tablespoons minced shallots
1 tablespoon minced parsley
¾ teaspoon salt
½ teaspoon coarsely cracked
　　black pepper
⅔ cup peanut oil
1 head of chicory (curly endive),
　　torn into bite-size pieces
1 small red cabbage (about 1
　　pound), shredded
2 tart apples, such as Granny
　　Smith, cored and cut into wedges

1. In a large baking dish, combine 2 tablespoons of the mustard, 3 tablespoons of the molasses, 2 tablespoons of the bourbon and ⅛ teaspoon of the hot sauce. Add the pork chops and sliced yams and turn to coat. Cover and let marinate at room temperature for at least 30 minutes or up to 4 hours.

2. Light the charcoal or preheat the broiler. Remove the pork chops and yams from the marinade, scraping off most of it. Grill the chops, covered with a tent of aluminum foil, or broil about 3 to 4 inches from the heat, turning once, until glazed and browned but white and moist throughout, about 6 minutes on each side. Grill the yam slices for 8 to 10 minutes on each side, until soft and browned. *(The recipe may be prepared 1 day ahead to this point. Cool to room temperature, then wrap and refrigerate overnight. Let return to room temperature before continuing.)*

3. In a medium bowl, combine the remaining 3 tablespoons mustard, molasses and bourbon and ⅛ teaspoon hot sauce with the lime juice, shallots, parsley, salt and pepper. Gradually whisk in the oil in a steady stream to make a dressing.

4. Bone the pork chops and cut the meat into ½-inch cubes. Place the pork in a shallow bowl, add the yams and half the dressing. Toss to coat well.

5. Line a platter with the chicory tossed with a little dressing. Add the shredded cabbage, also tossed with dressing, in a ring on the platter. Toss the apple wedges in dressing and arrange in spokes around the platter, then mound the dressed pork and yam slices in the center. (Or, if you prefer, toss all the ingredients with the remaining dressing at one time.)
—*Beverly Cox*

VIETNAMESE SALAD WITH CHILE-MINT DRESSING

This salad is composed in the Vietnamese style. It contains a small amount of meat, lots of vegetables and a generous sprinkling of peanuts. If you want to double the amount of meat, cook it in two pans and double the dressing. This dish makes a delicious light lunch.
❢ The sharp-flavored, spicy yet slightly sweet components of this salad will not marry well with wine, but would be complemented by a fine beer, such as Ringnes Export Lager or Anchor Steam.
4 Servings

½ pound boneless pork shoulder, trimmed of excess fat and cut into ¼-inch-thick strips
2 large garlic cloves, minced
2 tablespoons soy sauce, preferably mushroom soy*
1 tablespoon peanut oil
1 medium onion, thinly sliced
1½ tablespoons cider vinegar
1½ tablespoons fresh lime juice
2 teaspoons fish sauce (nuoc mam)*
1½ tablespoons chopped fresh mint plus 1½ tablespoons shredded mint
⅛ teaspoon minced hot chile pepper
¼ teaspoon sugar
¼ teaspoon salt
2 large white turnips, peeled
2 large carrots, peeled
3 cups thinly sliced romaine lettuce
3 cups thinly sliced cabbage
⅓ cup chopped unsalted peanuts
*Available at Asian markets

1. Preheat the oven to 350°. In a medium bowl, toss the pork strips with 1 clove of the minced garlic and 1 tablespoon of the soy sauce. Set aside to marinate at room temperature for about 1 hour.

2. In a medium ovenproof skillet, heat 2 teaspoons of the peanut oil. Add the onion slices and cook over moderately high heat, stirring occasionally, until the onions are evenly browned, about 5 minutes. Place the skillet in the oven and cook, stirring occasionally, until the onions are dry and deep brown, about 25 minutes. Remove from the skillet and set aside.

3. In a medium bowl, whisk the vinegar, lime juice, fish sauce, chopped mint, remaining minced garlic, hot pepper, sugar, salt and ⅓ cup of water to blend well.

4. Using a food processor fitted with the grating disk, shred the turnips. Place them in a small bowl and toss with 3 tablespoons of the dressing. Shred the carrots. Place them in another bowl and toss with 3 tablespoons of the dressing.

5. In a large skillet, heat the remaining 1 teaspoon peanut oil over high heat. Add the pork strips to the skillet in an even layer and cook, without stirring, until browned on the bottom, about 2 minutes. Turn and cook until browned on the second side, about 2 minutes longer. Add the remaining 1 tablespoon soy sauce and stir to coat the meat well. Remove the meat to a plate and set aside to cool.

6. In a large bowl, toss the lettuce, cabbage, shredded mint and pork with the remaining dressing and any accumulated dressing from the turnips and carrots. Arrange the turnips and carrots on opposite sides of each of 4 salad plates and place the pork and vegetable mixture in the center. Sprinkle the charred onions and chopped peanuts over all. Or place each of the ingredients in a separate bowl and allow guests to compose their own salads.
—Marcia Kiesel

WILD SOUTH SEA SALAD

This salad combines an ingredient never used in traditional Oriental food—wild rice—with barbecued pork and a ginger-shallot rice vinegar dressing.
4 to 6 Servings

½ pound pork tenderloin
1½ tablespoons hoisin sauce*
1 tablespoon plum sauce*
1½ teaspoons oyster sauce*
1½ teaspoons honey
¼ teaspoon Chinese chili sauce*
1½ tablespoons minced fresh ginger
1 large garlic clove, minced
¾ cup wild rice
1½ teaspoons salt
12 snow peas
2 ounces fresh shiitake mushrooms, stemmed, caps cut into thin julienne strips
1 small red bell pepper, cut into thin julienne strips
1 small yellow bell pepper, cut into thin julienne strips
1 bunch of arugula, shredded (optional)
½ cup fresh coriander sprigs, chopped
2 tablespoons rice vinegar
¼ cup safflower oil
1 shallot, minced
*Available at Asian markets

1. Put the pork tenderloin in a small baking dish. In a small bowl, combine the hoisin sauce, plum sauce, oyster sauce, honey, chili sauce, 1½ teaspoons of the ginger and the garlic. Spread this marinade all over the pork and let stand at room temperature for 1 to 2 hours.

2. Preheat the oven to 350°. Bake the pork tenderloin until the internal temperature measures 155°, 30 to 40 minutes. Let cool completely. Cut into thin julienne strips about 2 by ⅛ inch.

3. Put the wild rice in a strainer and rinse under cold running water; drain. In a medium saucepan, bring 3 cups of water to a boil over high heat. Stir in the wild rice and 1 teaspoon of the salt. Reduce the heat to low and simmer until the rice grains are puffed and tender, about 45 minutes. Drain and fluff with a fork or chopsticks. Let cool to room temperature and refrigerate until well chilled.

4. Bring a small saucepan of salted water to a boil over high heat. Add the snow peas and cook until brightly colored and crisp-tender, 6 to 8 seconds. Drain and rinse under cold running water until cool. Drain on paper towels and cut into thin julienne strips.

5. In a large serving bowl, combine the pork, wild rice, snow peas, mush-

rooms, red and yellow peppers, arugula and coriander. Toss to mix well.

6. In a small bowl, whisk together the vinegar, oil, shallot and remaining 1 tablespoon ginger and ½ teaspoon salt. Pour over the salad and toss to coat.

—*Hugh Carpenter*

PORK TENDERLOIN SALAD WITH APPLES AND APRICOTS

Because pork and fruit go very well together, I've included apricots and apples in this salad. Be sure to add the pork pan juices to the dressing for extra flavor.

8 Servings

1 tablespoon unsalted butter
½ cup plus 2 tablespoons extra-virgin olive oil
3 pounds pork tenderloin, trimmed
4 ounces dried apricots, cut into thin strips (about ⅔ cup)
2 celery ribs, thinly sliced
3 large shallots, thinly sliced
6 medium Granny Smith apples—peeled, cored and cut into ¾-inch chunks
2 tablespoons fresh lemon juice
2 tablespoons white wine vinegar
1 tablespoon Dijon mustard
½ teaspoon salt
¼ teaspoon freshly ground pepper
2 tablespoons chopped parsley

1. Preheat the oven to 450°. In a large skillet, melt the butter in 1 tablespoon of the olive oil over moderately high heat. Add the pork and sauté, turning, until well browned all over, 7 to 10 minutes.

2. Transfer the pork to a baking dish; set the skillet aside. Roast the pork in the oven for 15 to 20 minutes or until a meat thermometer inserted into the thickest part of the tenderloin registers 150°. Let cool for 15 minutes.

3. Meanwhile, in a small heatproof bowl, cover the apricots with 1 cup of very hot water. Let stand for 15 minutes to soften.

4. In the skillet used for the pork, heat 1 tablespoon of the olive oil. Add the celery and shallots and cook over moderately high heat, stirring, until the shallots are softened and translucent, 2 to 3 minutes. Place in a large bowl.

5. Add 1 tablespoon olive oil to the skillet. Add the apples and sauté over moderately high heat, tossing frequently, until just tender, 4 to 5 minutes. Add to the shallots and celery and toss well.

6. Drain the apricots and add to the bowl. Slice the pork ¼ inch thick and then cut each piece into ¼-inch strips. Add to the apple mixture.

7. In a small bowl, whisk the lemon juice with the vinegar and any juices from the baking dish. Whisk in the mustard, salt, pepper and remaining 7 tablespoons olive oil. Pour the dressing over the salad and toss well. Cover and refrigerate. Toss with the parsley before serving.

—*Bob Chambers*

TORTELLINI SALAD

❢ Hearty Italian red wine, such as Lungarotti Rubesco

6 Servings

1 pound cheese-filled tortellini
¾ cup olive oil
½ pound prosciuttino* or other ham, cut into ⅜-inch slices
½ pound smoked turkey, cut into ⅜-inch slices
1 cup cooked fresh peas or defrosted frozen peas
2 medium carrots, cut into 1½-by-⅛-inch julienne strips
1 large red bell pepper, cut into 1½-by-⅛-inch julienne strips
½ cup minced parsley
2 garlic cloves, minced
1 head of red leaf lettuce, for lining the platter
*Available at Italian markets

1. Cook the tortellini in a large pot of boiling salted water until just tender, about 5 minutes for fresh, 10 to 15 minutes for frozen or dried. Drain into a colander and rinse under cold running water until cool. Let drain for 5 to 10 minutes. Place in a bowl and toss with ¼ cup of the oil. Cover and refrigerate.

2. Cut the prosciuttino and smoked turkey into ⅜-inch cubes; add to the tortellini. Add the peas, carrots and red pepper; cover and refrigerate until about 30 minutes before serving.

3. Shortly before serving, whisk together the parsley, garlic and remaining ½ cup oil. Pour over the salad and toss well to coat. Arrange the lettuce leaves around the edge of a serving bowl or on individual plates and fill with the tortellini salad.

—*Rick Ellis*

BARLEY-HAM SALAD WITH HONEY MUSTARD DRESSING

I enjoy barley as much for its texture as for its taste. This salad highlights both. Make it at least several hours ahead, and you'll have a piquant blend of flavors from the honey mustard dressing and the smoked ham.

6 to 8 Servings

¼ cup plus 1 tablespoon peanut oil
1 cup pearl barley
2 cups chicken stock or canned broth
½ teaspoon salt
1½ tablespoons Dijon mustard
1½ tablespoons cider vinegar
1 tablespoon fresh lemon juice
2 tablespoons honey
¼ teaspoon freshly ground black pepper
Dash of hot pepper sauce (optional)

1 cup peas, fresh or frozen
3 carrots, shredded
½ cup finely diced red or yellow bell pepper
½ pound smoked ham, such as Black Forest, cut into ½-inch dice
Tomato wedges, for garnish

1. In a heavy medium saucepan, heat 1 tablespoon of the oil. Add the barley and sauté over moderate heat until the grains are coated with oil. Add the chicken stock, salt and 1 cup of water. Bring to a boil over high heat. Reduce the heat to moderately low, cover and simmer for 40 minutes, or until almost all of the liquid is absorbed. Remove from the heat and let stand, covered, for 10 minutes. Turn the barley into a bowl and let cool to room temperature.

2. In a blender or food processor, mix together the remaining ¼ cup oil, the mustard, vinegar, lemon juice, honey, black pepper and hot sauce. Process for 2 minutes, or until the dressing is thickened and creamy.

3. If using fresh peas, blanch in boiling water until just tender, about 2 minutes; drain and pat dry. If using frozen peas, thaw and pat dry.

4. In a large salad bowl, toss the barley with the dressing. Add the carrots, bell pepper, ham and peas. Toss well and refrigerate for several hours or overnight. *(The salad can be made up to 3 days ahead.)* Garnish with tomato wedges before serving.
—Dorie Greenspan

HAM SALAD WITH RASPBERRIES AND PEARL ONIONS
2 Servings

4 ounces tiny white pearl onions
2 cups dry red wine
⅔ cup fresh orange juice
¼ cup sugar
1 navel orange
12 ounces hickory smoked ham, or other fine smoked ham, sliced ¼ inch thick
16 small spinach leaves, well rinsed and dried
1 cup fresh raspberries
¼ teaspoon coarsely ground pepper

1. Cut the root ends off the pearl onions. In a large pot of boiling water, blanch the onions for 2 minutes. Drain and rinse under cold running water until cool enough to handle; slip off the skins.

2. In a large skillet, combine the wine, orange juice and sugar. Bring to a boil over high heat. Stir to dissolve the sugar; continue to boil for 6 minutes. Add the onions and cook for 6 minutes longer, until the onions are tender and the liquid is reduced to about ½ cup of syrup.

3. Meanwhile, remove 1 large piece of zest from the orange with a swivel-bladed vegetable peeler and cut into very fine slivers. Blanch for 30 seconds in boiling water, run cold water over to cool; reserve.

4. Using a sharp knife, cut away the remaining peel and all of the inner white membrane from the orange. Cut down on either side of the dividing membranes to section the orange. Halve each section crosswise and set aside in a strainer to drain off any juice.

5. Trim the ham of any fat and then cut into 2-by-½-by-¼-inch strips. Arrange the spinach leaves around the edges of each of two plates.

6. Add the ham to the hot wine sauce and heat through for about 1 minute, stirring to coat with the sauce. Add the orange sections, raspberries and pepper and toss again just to coat. Using a slotted spoon, divide the salad between the two spinach-lined plates. Pour the remaining sauce over the top. Garnish with the orange zest. Serve warm.
—F&W

HAM AND CHEESE SALAD
Slices of rosy, smoky ham are teamed with Gruyère cheese for a satisfying main-course salad.
8 Servings

1½ pounds baked ham, sliced ¼ inch thick and cut into 3-by-¼-inch matchsticks
1½ pounds Gruyère cheese, sliced ¼ inch thick and cut into 3-by-¼-inch matchsticks
3 scallions, thinly sliced
2 tablespoons chopped flat-leaf parsley
2 tablespoons white wine vinegar
2 tablespoons Dijon mustard
½ teaspoon salt
¼ teaspoon freshly ground black pepper
Pinch of cayenne pepper
½ cup extra-virgin olive oil

1. In a large bowl, combine the ham, cheese, scallions and parsley.

2. In a small bowl, whisk the vinegar, mustard, salt, black pepper and cayenne. Gradually whisk in the olive oil in a thin stream.

3. Pour the dressing over the ham and cheese and toss well. Cover and refrigerate. Serve at room temperature.
—Bob Chambers

COMPOSED SALAD OF SAUSAGE AND ORZO WITH FRESH PEAS AND MIXED GREENS

One of my main complaints with most composed salads is that either they are overdressed—literally swimming in vinaigrette—or they come to the table very artfully arranged and with dressing on the side, making it impossible to properly dress the salad. Thus, you will notice, in this recipe I lightly dress the greens and pasta ahead of time. This isn't complicated and certainly makes for a tastier salad.

❦ California sparkling rosé, such as Korbel

6 Servings

½ cup pine nuts
2 cups orzo (about 12 ounces)
2 cups shelled fresh green peas (about 2 pounds in the pod)
12 sweet Italian sausages (about 2 pounds)
1 large garlic clove, minced
¼ cup dry red wine
½ cup chopped scallions
2 tablespoons red wine vinegar
2 teaspoons Dijon mustard
1 tablespoon mayonnaise
1 teaspoon salt
¾ teaspoon freshly ground pepper
⅓ cup plus 1½ tablespoons olive oil
4 bunches of mâche (lamb's lettuce), about ½ pound
2 Belgian endives, cut into thin strips
8 small radicchio leaves, torn into pieces
2 teaspoons fresh lemon juice

1. Preheat the oven to 375°. Scatter the pine nuts over a small baking sheet and toast in the oven until golden brown, about 10 minutes.

2. Bring a large saucepan of salted water to a boil over high heat. Stir in the orzo and cook until the pasta is tender but still firm, about 10 minutes. Drain and rinse under cold running water; drain well.

3. Steam the peas over boiling water until just tender, about 6 minutes. Rinse under cold running water; drain well.

4. Prick the sausages all over with a fork. In a large skillet, cook the sausages over moderately high heat, turning, until browned, about 10 minutes. Pour off all of the fat. Add the garlic and wine to the skillet. Reduce the heat to low, cover and simmer for 20 minutes. Remove the sausages. Degrease the juices in the pan and set aside.

5. In a large bowl, mix together the orzo, peas, pine nuts and scallions. In a small bowl, whisk together the vinegar, mustard, mayonnaise, ½ teaspoon of the salt and ½ teaspoon of the pepper. Gradually whisk in ⅓ cup of the olive oil and the reserved pan juices until well blended. Add to the orzo and stir to coat.

6. In a large bowl, toss together the mâche, Belgian endives and radicchio. Drizzle the lemon juice and the remaining 1½ tablespoons olive oil over the salad. Season with the remaining ½ teaspoon salt and ¼ teaspoon pepper and toss.

7. Place a portion of the greens on each plate, top with a portion of the orzo mixture and arrange 2 sausages, sliced if you like, over the top of the salads.
—Lee Bailey

MARINATED LAMB SALAD WITH RADISHES AND CURRANTS

❦ Ordinarily lamb calls for red wine, but the sharp flavors and raw vegetables in a salad conflict with all but the most vigorous—the intense De Loach Zinfandel, for example. A better choice might be an elegant beer, such as Palm Ale from Belgium.

4 to 6 Servings

1½ pounds boneless leg of lamb, in 2 to 3 pieces cut about 2 inches thick
1 teaspoon salt
¾ teaspoon freshly ground pepper
⅔ cup plus 1 tablespoon olive oil
¼ cup plus ½ tablespoon sherry wine vinegar
3 tablespoons dried currants
2 large garlic cloves, minced
2 large shallots, minced
2 tablespoons soy sauce, preferably mushroom soy*
½ small head of chicory (curly endive), torn into bite-size pieces
½ medium head of romaine lettuce, torn into bite-size pieces
1 bunch of red radishes, halved and thinly sliced
½ pint cherry tomatoes, quartered
2 large scallions, thinly sliced
*Available at Asian markets

1. Season the lamb with ½ teaspoon of the salt and ¼ teaspoon of the pepper. In a large skillet, heat 1 tablespoon of the olive oil over high heat. When the oil is hot, add the lamb, reduce the heat to moderately high and cook for 5 minutes. Turn the meat over, reduce the heat to moderate and cook until medium-rare, about 7 minutes longer. Remove the meat from the skillet and let rest on a cutting board for 10 minutes.

MAIN COURSES: SALADS

2. Meanwhile, in a small nonreactive saucepan, combine 2 tablespoons of the sherry vinegar and the currants. Cook over moderate heat until the vinegar is hot. Remove from the heat.

3. In a small bowl, combine the garlic, shallots, soy sauce and remaining ½ teaspoon salt, ½ teaspoon pepper and 2½ tablespoons sherry vinegar. Whisk in the remaining ⅔ cup olive oil in a thin stream.

4. Slice the lamb into ¼-inch strips and place in a medium bowl. Add ¾ cup of the dressing and toss well. Cover and let marinate in the refrigerator for 2 hours, stirring occasionally.

5. In a large bowl, toss the chicory, romaine, radishes, cherry tomatoes and scallions. Add the lamb with its dressing and the currants in vinegar. Add the remaining dressing and toss well.
—Marcia Kiesel

COOKED BEEF SALAD WITH MUSTARD-ANCHOVY DRESSING

This recipe enables a small amount of leftover pot roast to be quickly transformed into a lively main-dish salad.
6 Servings

2 cups strips of pot roast (or other boiled or braised beef), each about 2 inches long and ½ inch wide
1 small red onion, thinly sliced
1 cup canned white kidney beans (cannellini), drained and rinsed
1 rib celery, trimmed and thinly sliced on the bias
16 oil-cured black olives, halved and pitted
1 large tomato—halved, seeded and diced
2 tablespoons parsley leaves
2 tablespoons Dijon mustard
1 large garlic clove
3 anchovy fillets, drained
2 tablespoons red wine vinegar
1 tablespoon drained capers
¾ cup olive or vegetable oil
Parsley sprigs, for garnish

1. Place the strips of beef, sliced onion, beans, celery, olives, diced tomato and parsley leaves in a mixing bowl.

2. Prepare the dressing: With a mortar and pestle or in a small bowl using the back of a wooden spoon, mash together the mustard, garlic, and anchovies to make a paste. Stir in the vinegar and capers. Add the oil gradually, whisking with a fork until the dressing is smooth and thick.

3. Pour the dressing over the salad and toss well. Garnish with the parsley sprigs.
—F&W

MADRIGAL SALAD

A tasty combination of ingredients, this warm salad is named after the late Rosie Madrigal, a childhood neighbor who frequently prepared it for our family. Have all the ingredients ready before you begin to cook the salad, as it is finished in only a few minutes.
4 Servings

½ pound flank steak
2 tablespoons olive oil
1 garlic clove, minced
One 35-ounce can Italian peeled tomatoes—well drained, seeded and cut into ½-inch dice
1 cup thinly sliced scallions, including some of the green tops
½ cup mayonnaise
½ teaspoon salt
½ teaspoon freshly ground pepper
¼ cup chopped parsley, for garnish

1. Trim away and discard any fat from the steak. Thinly slice the steak across the grain and then cut the slices into 3-inch-long strips.

2. Combine the oil and the garlic in a large nonreactive skillet and place it over moderate heat. When the oil is hot, add the meat and stir-fry over high heat for 1 minute; remove the meat with a slotted spoon and set it aside.

3. Add the tomatoes and scallions to the pan and cook over high heat, stirring gently, for about 30 seconds, or until the tomatoes are hot. Remove the pan from the heat and return the meat to it. Add the mayonnaise, salt and pepper; toss until the meat and vegetables are well coated with the mayonnaise.

4. Arrange the salad on a serving plate and sprinkle the chopped parsley over the top; serve hot.
—Jim Fobel

ORIENTAL BEEF SALAD

▼ Alsace Gewürztraminer, such as Hugel
6 Servings

1¾- to 2-pound flank steak
¾ cup dry white wine
½ cup soy sauce
¼ cup honey
2 garlic cloves, crushed
4 tablespoons shredded fresh ginger
¾ pound snow peas
1½ pounds asparagus, cut diagonally into 1-inch pieces
2 tablespoons Oriental sesame oil
¾ pound large mushrooms, cut into ¼-inch slices
1 tablespoon toasted sesame seeds

1. Lightly score one side of the steak on the diagonal at 1-inch intervals. In a glass or ceramic baking dish large enough to hold the meat, blend together the wine, soy sauce, honey, garlic and 3 tablespoons of the ginger. Add the steak, cover and marinate in the refrigerator

for at least 2 hours, turning every 30 minutes.

2. Preheat the broiler—with the broiler pan set 3 to 4 inches from the heat—for 10 minutes before you plan to cook the steak. Remove the meat from the marinade and pat dry on paper towels; reserve the marinade. Place the steak on the preheated pan and broil for about 3 minutes on each side for medium rare. Let cool to room temperature, cover and refrigerate.

3. Strain the reserved marinade into a small heavy saucepan. Bring to a boil over moderate heat and cook until reduced by half (to about ½ cup). Transfer to a small bowl; let cool, cover and refrigerate.

4. In a large pan of boiling salted water, blanch the snow peas until bright green, about 15 seconds. With a slotted spoon, remove to a bowl of cold water to stop the cooking, then drain. Add the asparagus to the same boiling water and blanch until bright green, about 30 seconds. Cool in cold water, then drain. Pat the vegetables dry, then cover and refrigerate.

5. About 30 minutes before serving, remove the steak, reduced marinade, snow peas and asparagus from the refrigerator. Mince the remaining 1 tablespoon ginger and add to the marinade. Add the sesame oil and blend well. Pour this dressing over the sliced mushrooms, toss and set aside.

6. Cut the steak across the grain on the diagonal into thin slices; if the slices are long, cut in half.

7. Just before serving, remove the mushrooms from the dressing with a slotted spoon; reserve the dressing. Arrange the steak, mushrooms, snow peas and asparagus decoratively on a platter or individual plates. Drizzle the reserved dressing over the meat and vegetables and garnish with the sesame seeds.

—Rick Ellis

THAI-STYLE BEEF AND LETTUCE SALAD WITH CHILE-LIME DRESSING

This dish was inspired by a Thai restaurant in New York's Greenwich Village.
❣ California Gewürztraminer, such as Alexander Valley Vineyards
4 Servings

1 small head of romaine lettuce
1 small head of leaf lettuce
1 small head of curly endive
¾ cup peanut oil
4 tablespoons fresh lime juice
3 tablespoons soy sauce
1 pound fairly lean steak, such as London broil, cut about 1 inch thick
1 tablespoon minced fresh ginger
2 medium garlic cloves, chopped
1 fresh jalapeño or other small hot green chile pepper, coarsely chopped
¼ cup minced fresh coriander (cilantro)
2 tomatoes (about 1 pound total), cut into 6 wedges each
1 medium red onion, sliced into rings

1. Trim away the ends and any wilted outer leaves of the romaine, leaf lettuce and curly endive. Separate the leaves, wash them well and dry. Tear the leaves into bite-size pieces, wrap and refrigerate. *(The greens can be prepared up to 1 day ahead.)*

2. In a medium bowl, combine 2 tablespoons of the peanut oil, 2 tablespoons of the lime juice and 1 tablespoon of the soy sauce. Trim any excess fat from the steak, add the steak to the bowl and marinate at room temperature, turning occasionally, for 1 hour.

3. In a food processor, combine ½ cup of the peanut oil, the remaining 2 tablespoons lime juice and soy sauce, the ginger, garlic and jalapeño. Process for 1 minute, or until smooth. Transfer the dressing to a bowl and set aside at room temperature.

4. In a medium skillet, warm the remaining 2 tablespoons oil over high heat until it smokes slightly. Remove the steak from the marinade, reserving the marinade. Add the steak to the skillet and sear, turning once, until browned, about 2 minutes on each side. Reduce the heat to moderate and continue to cook, turning once or twice, until the steak is medium-rare, about 5 minutes longer on each side.

5. Meanwhile, stir the reserved marinade into the chile-lime dressing. Put the greens in a large bowl.

6. When the steak is done, transfer it to a cutting board. Remove the skillet from the heat and add the dressing, stirring to scrape up any browned bits from the bottom of the pan. Cover the skillet to keep the dressing warm.

7. With a very sharp knife, cut the steak across the grain into thin slices. Pour the warm dressing over the greens in the bowl and toss well. Add the coriander and toss again.

8. Divide the greens among salad plates. Arrange the steak, with the slices overlapping slightly, on top of the greens. Garnish each plate with 3 tomato wedges and several onion rings. Spoon any dressing remaining in the bowl over the salads and serve at once.

—Michael McLaughlin

SOUTH SEA BEEF
SALAD
The special ingredient in this salad is fish sauce, which is available at all Asian markets. Buy the brand with the lightest color and use the sauce sparingly, as it can overwhelm subtler flavors.
4 Servings

½ pound beef tenderloin
½ pound snow peas
2 tablespoons fish sauce
   (nuoc mam)*
2 tablespoons fresh lime juice
1 tablespoon sugar
½ teaspoon Chinese chili sauce*
1 garlic clove, minced
1 medium red onion, thinly sliced
2 medium red bell peppers,
   thinly sliced
½ pound jicama, cut into thin strips
2 tablespoons chopped roasted
   peanuts
1 tablespoon grated lime zest
½ teaspoon freshly grated nutmeg
¼ cup minced fresh coriander
   (cilantro)
*Available at Asian markets

1. Place the meat in the freezer until very firm, about 20 minutes. Then cut into paper-thin slices. Stack the beef slices and cut into matchstick-size pieces. Put the meat on a plate and press plastic wrap directly on it so the meat is covered airtight. Refrigerate until ready to use.

2. In a large saucepan of boiling water, cook the snow peas for 10 seconds. Drain and rinse immediately under cold running water until cool. Drain well and pat dry. Cut the snow peas lengthwise into thin strips.

3. In a small bowl, combine the fish sauce, lime juice, sugar, chili sauce, garlic and 2½ tablespoons of water. Stir to dissolve the sugar.

4. In a large bowl, combine the red onion, red peppers, jicama, snow peas and half of the beef strips. Toss with the dressing and turn out onto a serving platter. Decorate with the remaining beef. Sprinkle the peanuts, lime zest, nutmeg and coriander over the top.
—Hugh Carpenter

THAI BEEF SALAD
I first tasted *yam neua* in the dining car of a train rolling through the jungle of south Thailand. It is usually eaten when people get together late in the afternoon to have a glass of Singha beer or Mekong whiskey.
6 Servings

1½ pounds beef tenderloin, at room
   temperature
¼ cup coarsely chopped fresh
   coriander (cilantro)
2 tablespoons coarsely chopped
   fresh mint
1 large jalapeño pepper, minced
3 large garlic cloves, crushed
   through a press
2 tablespoons nam pla*
2 tablespoons palm sugar* or light
   brown sugar
¼ cup fresh lime juice
1 head of leaf lettuce, leaves
   separated
2 tomatoes, thinly sliced
1 small cucumber, thinly sliced
1 small red onion, thinly sliced
Freshly ground black pepper
1 lime, cut into thin wedges
*Available at Asian markets

1. Preheat the oven to 500°. Set the beef tenderloin in a small baking dish and roast for 20 minutes, until rare. Let cool for 30 minutes, then refrigerate until cold, about 2 hours. Slice the tenderloin against the grain into ¼-inch-thick slices. Then cut into strips about ¼ inch wide.

2. In a large bowl, combine the meat strips, coriander, mint and jalapeño pepper and toss. In a small bowl, combine the garlic, *nam pla*, palm sugar and lime juice; mix well. Pour this dressing over the meat and toss to coat.

3. Cover a large plate or platter with lettuce leaves. Arrange the tomatoes around the outside, overlapping the slices as necessary. Arrange the cucumber slices inside the tomatoes and then the onion slices inside those.

4. Remove the meat from the dressing and mound it in the center of the vegetables. Pour any dressing that remains in the bowl over the meat. Cover the salad with a damp towel and refrigerate until chilled, at least 1 hour and up to 4 hours. Before serving, sprinkle with black pepper and garnish with the lime wedges.
—Jeffrey Alford

6
SIDE DISHES
VEGETABLES & GRAINS

ARTICHOKE HEARTS WITH MUSHROOMS
6 Servings

6 large artichokes
½ cup fresh lemon juice
2 tablespoons vegetable oil
⅔ cup extra-virgin olive oil
2 tablespoons dry vermouth
1 teaspoon oregano
½ teaspoon basil
¼ teaspoon thyme
½ teaspoon salt
6 ounces mushrooms, thinly sliced
3 tablespoons freshly grated Parmesan cheese

1. One at a time, using a stainless-steel knife, cut off the artichoke stems. Snap off all of the outer leaves by bending them back. With a knife, slice off the inner leaves about two-thirds of the way down, flush with the top of the choke. Trim away the dark, fibrous leaf ends to form a smooth, neat artichoke heart. To prevent discoloration, immediately place each heart in a large nonreactive saucepan filled with 1 quart of water, ¼ cup of the lemon juice and the vegetable oil. Cover with a piece of cheesecloth to keep the artichokes moist.

2. When all the artichokes are trimmed, place the saucepan over moderate heat, bring to a boil and cook the artichokes (still covered with cheesecloth) until tender when pierced with a knife, 30 to 40 minutes.

3. Meanwhile, in a medium bowl, whisk the remaining ¼ cup lemon juice, olive oil, vermouth, oregano, basil, thyme and salt until blended. Add the mushrooms and stir to coat well. Let marinate at room temperature, stirring occasionally, while the artichokes cook.

4. When the artichokes are tender, remove them with a slotted spoon and drain, inverted, on paper towels, until cool enough to handle. Scoop out the hairy chokes with a spoon.

5. Stir the Parmesan into the mushrooms and their marinade. Spoon half the mixture into a shallow dish. Arrange the artichoke hearts on top and fill them with the remaining mushrooms and marinade. Let marinate at room temperature for at least 1 and up to 6 hours.
—Diana Sturgis

BRAISED ARTICHOKES WITH SPINACH
6 Servings

6 large artichokes (1 pound each)
¼ cup fresh lemon juice
4 pounds spinach, large stems removed
2 tablespoons unsalted butter
6 medium garlic cloves, minced
¼ cup extra-virgin olive oil
¼ teaspoon freshly ground pepper
½ cup plus 2 tablespoons freshly grated Parmesan cheese (about 2 ounces)
1¾ cups chicken stock or canned low-sodium broth

1. Place the artichokes and lemon juice in a large bowl and add water to cover. Let soak for 30 minutes. With a small sharp knife, trim the artichokes: Cut off the stems and then peel off all the dark outer leaves. Cut off the leaves about 2 inches from the base (you should be able to see the rose-colored centers). Trim off the remaining leaves and any dark outer skin from the artichoke bottoms. Using a teaspoon, scrape out the hairy chokes. As each artichoke is trimmed, drop it back into the bowl of water and lemon juice.

2. Fill a large nonreactive flameproof casserole with water and bring to a boil over high heat. Reduce the heat to moderate, add the artichoke bottoms and simmer until fork-tender but still slightly firm, about 10 minutes. Drain and pat dry. Pour the water out of the casserole.

3. Rinse the spinach very well in several changes of water. Place the spinach (with some of the water still clinging to the leaves) in the casserole. Cover and steam over high heat, turning occasionally, until wilted, about 8 minutes. Drain the spinach and rinse with cold water to cool it down. Squeeze as much water as possible from the spinach, then coarsely chop.

4. Wipe the casserole dry, then add the butter and melt over moderate heat. Add the spinach, garlic and 2 tablespoons of the olive oil. Cook, stirring, until the garlic is fragrant, about 5 minutes. Remove from the heat. Stir in the pepper and ½ cup of the Parmesan cheese. Let cool.

5. Mound the spinach mixture in the artichoke bottoms. Set the casserole over moderately low heat. Add the remaining 2 tablespoons olive oil, the stuffed artichoke bottoms and the chicken stock. Cover and braise gently over low heat until the artichokes are just tender, about 30 minutes.

6. Sprinkle the remaining 2 tablespoons Parmesan on top of the artichokes and serve warm, with some of the broth spooned on top.
—Ann Chantal Altman

Vidalia Onion Tart (p. 90).

At left: Braised Pear, Celery and Endive with Parmesan and Basil (p. 49), Zucchini Ribbons with Arugula and Creamy Goat Cheese Sauce (p. 50), Caramelized Onion with Salsa (p. 49). Below, Potato Cases with Shiitake and Morel Filling (p. 52).

10-Ingredient Vegetable Fritters (p. 20).

ASPARAGUS WITH BUTTER VINAIGRETTE
6 Servings

2 pounds asparagus
3 tablespoons unsalted butter
1 tablespoon plus 1 teaspoon
    balsamic vinegar
¾ teaspoon salt
Freshly ground pepper

1. Steam the asparagus until just tender, about 3 minutes. Alternatively, cook the asparagus in a large skillet of boiling salted water for about 3 minutes.
2. Meanwhile, in a small nonreactive saucepan, melt the butter over moderate heat. Whisk in the vinegar and salt.
3. Transfer the asparagus to a large warm platter and pour the vinaigrette over it. Season with pepper to taste and serve warm.
—Lee Bailey

ASPARAGUS WITH HAZELNUT-ORANGE SAUCE
A coarse-textured, heady garnish for steaming hot asparagus.
4 Servings

6 tablespoons unsalted butter
½ cup hazelnuts or almonds,
    coarsely chopped
Grated zest of ½ orange
1 to 2 tablespoons orange juice
1 to 2 tablespoons orange liqueur
    (optional)
1½ pounds asparagus
Salt and freshly ground pepper

1. In a skillet, melt the butter over medium heat. Then add the nuts and toss in the butter until lightly golden, 5 to 6 minutes.
2. Add the orange zest and stir briefly. Add the orange juice, swirl to blend, then remove from the heat. Stir in the liqueur.
3. In a large skillet half filled with boiling salted water, cook the asparagus (the water should just barely cover the stalks) until crisp-tender, 2 to 3 minutes. Drain well. Sprinkle with salt and pepper to taste, then pour the hazelnut-orange sauce over them. Serve hot.
—F&W

GREEN BEANS AND ROASTED PEPPERS
6 Servings

1 pound green beans
2 large red bell peppers
⅔ cup olive oil
¼ cup fresh lemon juice
3 tablespoons chopped fresh dill
¾ teaspoon salt
½ teaspoon freshly ground pepper
¼ teaspoon sugar

1. Bring a large saucepan of water to a boil. Add the beans, let the water return to a boil and cook over moderate heat until the beans are crisp-tender, about 5 minutes. Drain into a colander and refresh under cold running water for 1 minute. Drain well and place in a medium bowl.
2. Roast the peppers over an open flame, or broil 4 inches from the heat, turning with tongs as the skin blisters and blackens, for about 5 minutes or until charred all over. Allow to cool for 1 minute; then seal in a plastic bag and let them "sweat" for 10 minutes. Rub off the charred skin and rinse briefly under cold running water; pat dry. Remove the stems, seeds and ribs. Slice the roasted flesh into ⅛-inch strips about 2½ inches long. Add the pepper strips to the bowl of beans.
3. In a small bowl, whisk the oil, lemon juice, dill, salt, pepper and sugar until blended. Pour the dressing over the vegetables, tossing to coat well. Allow to marinate at room temperature for at least 1 and up to 6 hours.
—Diana Sturgis

GREEN BEANS WITH COCONUT AND POPPED RICE
Browned, semi-popped rice is used here as a seasoning for the beans. As the rice cooks, it becomes an exotic spice with a deep-roasted taste. The other main flavoring here is coconut. This dish comes from South India, where vegetables are frequently parboiled before being stir-fried with seasonings.
6 Servings

¼ cup dried unsweetened shredded
    coconut* or 6 tablespoons
    freshly grated
⅓ cup boiling water
2 tablespoons coarse (kosher) salt
1½ pounds green beans, cut into
    1-inch lengths
¼ cup vegetable oil
½ teaspoon cumin seeds
1 teaspoon black mustard seeds*
1 tablespoon long-grain rice,
    preferably basmati*
3 dried red chiles
¼ teaspoon cayenne pepper
1 teaspoon sugar
¾ teaspoon table salt
2 tablespoons chopped fresh
    coriander (cilantro)
*Available at Indian markets

SIDE DISHES: VEGETABLES & GRAINS

1. If using dried coconut, place it in a small bowl, cover with the ⅓ cup boiling water and set aside for at least 15 minutes. Drain well.

2. In a large pot, bring 5 quarts of water to a rolling boil. Add the coarse salt and the beans and boil rapidly until the beans are just cooked slightly but still crisp, 2 to 3 minutes. Drain the beans and rinse under cold water; drain again.

3. In a large skillet, heat the oil over moderately high heat. Add the cumin and mustard seeds. When the mustard seeds start to pop, stir in the rice. When the rice browns (some grains might pop), after about 1 minute, stir in the dried chiles.

4. Add the green beans, cayenne, sugar and table salt; reduce the heat to moderate and stir-fry for 3 minutes. Stir in the fresh or drained coconut and fresh coriander, cover and reduce the heat to low. Cook for 1 minute. Transfer the beans to a bowl and serve hot.
—Madhur Jaffrey

SWEET-AND-SOUR BEETS

This recipe is easily prepared through Step 2 up to two days ahead. The finishing steps take about 10 minutes on the stovetop.
8 Servings

1½ pounds beets
2 tablespoons unsalted butter
2 tablespoons red wine vinegar
1 tablespoon sugar
½ teaspoon salt

1. Preheat the oven to 425°. Wrap the beets in aluminum foil and bake for 1 hour, or until tender when pricked with a knife. Remove from the oven and let cool.

2. Trim and peel the beets. Cut the beets into ⅛-inch-wide matchsticks. (The recipe may be prepared to this point up to 2 days ahead. Cover the beets and refrigerate.)

3. In a large skillet, melt the butter over moderately high heat. Add the beets and cook, tossing frequently, for about 2 minutes. Add the vinegar, sugar and salt and cook the beets until heated through, 3 to 5 minutes.
—W. Peter Prestcott

JADE FLOWER BROCCOLI

To the Chinese, broccoli represents jade, a symbol of health and youth.
6 Servings

2 nickel-size slices of fresh ginger, plus 2 teaspoons minced ginger
3 garlic cloves
1 teaspoon baking soda
1 pound broccoli florets, about 1½ inches in diameter
3 tablespoons peanut oil
¾ teaspoon salt

1. In a large saucepan, combine 8 cups of water, the slices of ginger, 2 of the garlic cloves and the baking soda. Bring to a boil over high heat.

2. Add the broccoli and cook until it turns bright emerald green, about 2 minutes. Drain and rinse under cold running water. Drain well and set aside.

3. Mince the remaining garlic clove. Heat a wok over high heat for 45 seconds. Add the peanut oil and swirl with a metal spatula to coat the sides of the wok. Add the salt and the minced ginger and garlic. Stir-fry until the garlic turns light brown, about 30 seconds. Add the broccoli and stir-fry until it is crisp-tender, 1½ to 2 minutes.

4. Transfer the broccoli to a serving dish. Let return to room temperature before serving. (This recipe can be prepared up to 1 day ahead. Let stand at room temperature for several hours, or cover and refrigerate.)
—Eileen Yin-Fei Lo

SPICY BRUSSELS SPROUTS

This recipe was created using leftover brussels sprouts to challenge anyone who wouldn't eat them in the first place. Expand this into a pilaf by folding in two cups of cooked rice.
6 Servings

1 tablespoon vegetable oil
3 tablespoons slivered almonds (1 ounce)
½ teaspoon yellow mustard seeds, crushed
¼ teaspoon cumin seeds
¼ teaspoon fennel seeds
3 cups cooked brussels sprouts, halved
⅛ teaspoon cayenne pepper
1 tablespoon finely chopped fresh ginger or Japanese pickled ginger
About 1 tablespoon fresh lime juice
½ teaspoon salt

1. In a large, heavy, nonreactive skillet, heat the oil over moderately high heat for about 30 seconds. Add the almonds and the mustard, cumin and fennel seeds. Cook, stirring, until the nuts and spices are fragrant and dark, about 15 seconds.

2. Reduce the heat to moderate and add the brussels sprouts, cayenne and ginger. Cook and stir until the vegetables are heated through and the spices are well distributed, about 4 minutes. Add a few tablespoons of water if the vegetables look dry.

3. Season to taste with up to 1 tablespoon of lime juice and the salt and serve hot. (The sprouts can be covered tightly and refrigerated for up to 1 day; reheat before serving.)
—Julie Sahni

RED CABBAGE WITH GRANNY SMITH APPLES

This piquant red cabbage with green apples is a classic combination that has a tart fruitiness.

12 Servings

3 tablespoons unsalted butter
1 onion, thinly sliced
1 garlic clove, crushed through a press
1 large head of red cabbage (about 2 pounds), finely shredded
2 medium Granny Smith or other tart green apples—quartered, cored and thinly sliced
2 tablespoons tarragon wine vinegar
2 teaspoons sugar
½ teaspoon mustard seeds
½ teaspoon salt
½ teaspoon freshly ground pepper

1. In a large nonreactive flameproof casserole, melt the butter over moderately high heat. Add the onion and sauté until softened but not browned, 3 to 4 minutes. Add the garlic and cook for 1 more minute.

2. Add the cabbage and toss. Cover and cook, stirring occasionally, until the cabbage wilts slightly, about 3 minutes.

3. Add the apples, vinegar, sugar, mustard seeds, salt and pepper. Toss well and continue to cook, stirring often, until the apples are tender, about 8 minutes. *(The recipe can be made up to 3 days in advance. Cover and refrigerate. Reheat before serving.)*
—Bob Chambers

CARROT AND CAULIFLOWER LOAF

Inspired by a vegetable pâté at a restaurant in Lyons, Léon de Lyon, this makes a lovely side vegetable with roasted meat, but an equally lovely appetizer served at room temperature.

6 to 8 Servings

1 head of cauliflower (about 2 pounds), broken into 1-inch florets
1½ pounds carrots, cut into ¼-inch slices
4 eggs
⅔ cup heavy cream, preferably not ultrapasteurized
½ teaspoon salt
Cayenne pepper
Freshly ground white pepper

1. Cook the cauliflower in a large pan of boiling salted water until very tender, 15 to 20 minutes. Remove with a slotted spoon and transfer to a sieve. Let drain for at least 10 minutes.

2. Meanwhile, return the water to a boil and add the carrots. Boil until tender, about 15 minutes. Drain them in a separate sieve.

3. Puree each vegetable separately in a food processor or food mill; then pass each through a fine-mesh sieve. Let the vegetables drain again in separate sieves for about 15 minutes.

4. With an electric mixer and in two separate bowls beat each of the vegetable purees with 2 of the eggs, ⅓ cup of the cream, ¼ teaspoon of the salt and a pinch each of cayenne and white pepper.

5. Preheat the oven to 400°. Butter a 4- to 6-cup loaf pan. First pour in the carrot puree, then the cauliflower.

6. Place the loaf pan in a larger pan filled with hot water that reaches about halfway up the sides of the pan. Bake in the middle of the oven for 1 hour, or until a cake tester comes out clean.

7. Remove the loaf pan from the water bath and let rest for about 1 hour, until lukewarm, before unmolding onto a platter. (Water may ooze out; just wipe dry with paper towels.)

8. Gently cut into ½-inch slices and serve lukewarm or at room temperature.
—Lydie Marshall

CARROTS GLAZED WITH LIME BUTTER

This simple recipe yields a refreshing vegetable side dish because the lime perfumes the sweetness of the carrots.

4 Servings

1 pound carrots (about 8 medium)
2 tablespoons sugar
¼ teaspoon salt
3 tablespoons unsalted butter, sliced
2 tablespoons fresh lime juice

1. Cut the carrots crosswise on the diagonal into ¼-inch slices. In a medium saucepan, bring the carrots and 4 cups of cold water to a boil over high heat and boil for 3 minutes; drain.

2. Return the carrot slices to the dry pan. Add 1 cup of water, the sugar and the salt. Bring to a boil over moderate heat and cook until the carrots are just tender and the liquid has reduced to 1 to 2 tablespoons, about 8 minutes. Stir in the butter and lime juice and serve hot.
—Jim Fobel

SAGE-GLAZED CARROTS

Sage has a natural affinity for root vegetables. Here, fresh sage leaves fried in butter are a crisp and pungent complement to glazed carrots.

12 Servings

2 pounds carrots, cut crosswise into 1½- to 2-inch lengths and quartered lengthwise
2 bunches of fresh sage (about 1 cup loosely packed leaves)
3 tablespoons unsalted butter
1 teaspoon sugar
Freshly ground white pepper

1. In a medium saucepan of salted water, bring the carrots to a boil over moderately high heat. Boil until they are crisp-tender, about 4 minutes. Drain the carrots into a colander immediately.

2. Chop enough of the sage leaves to measure 1 tablespoon; set aside.

3. In a large skillet, melt the butter over low heat. Arrange the remaining whole sage leaves flat in the pan and cook them until crisp and lightly browned, about 2 minutes on each side. Drain on paper towels.

4. Add the carrots and the reserved chopped sage to the skillet. Sprinkle the sugar on top and season with white pepper to taste; toss well. Cook over moderate heat for 3 to 4 minutes. Garnish with the sautéed sage leaves and serve. *(The recipe can be made up to 3 days in advance. Let the carrots cool, cover and refrigerate. Place the sage leaves in a separate bowl, cover tightly with plastic wrap and refrigerate. Before serving, reheat the carrots and garnish with the fried sage leaves.)*
—Bob Chambers

CHARRED CARROT PUREE

A couple of years ago, I read a recipe for a charred carrot soup (see page 80) by Anne Disrude, former *Food & Wine* associate test kitchen director, and was intrigued. So here are my pureed charred carrots, with thanks to Ms. Disrude for her inspiration.

6 Servings

6 tablespoons unsalted butter
1 tablespoon safflower or corn oil
3 pounds carrots, sliced ¼ inch thick
1 tablespoon lemon juice
½ teaspoon salt
¼ teaspoon freshly ground pepper
½ cup milk

1. In a large heavy skillet, melt 2 tablespoons of the butter in the oil over high heat.

2. Add the carrots and toss to coat evenly. Continue to cook, stirring them frequently, until they begin to char around the edges, 20 to 25 minutes.

3. Transfer the carrots to a food processor and add the remaining 4 tablespoons butter, the lemon juice, salt, pepper and milk. Puree until smooth. *(The puree can be made up to 1 day ahead; reheat over hot water in a double boiler.)*
—Lee Bailey

CURRIED CAULIFLOWER WITH PEAS

6 Servings

5 cups 1½-inch cauliflower florets (from a 2-pound head)
1 package (10 ounces) frozen peas or about 1½ pounds unshelled peas (to yield about 1¾ cups)
¼ cup dried currants
¼ cup vegetable oil
1 tablespoon curry powder
1 cup plain low-fat yogurt
¾ teaspoon salt

1. Bring a large saucepan of water to a boil. Add the cauliflower, let the water return to a boil and cook over moderate heat until the cauliflower is just tender, about 5 minutes. Drain into a colander and refresh under cold running water. Drain well and place in a large bowl.

2. In another saucepan of boiling water, cook the peas until just tender, about 2 minutes if using frozen, somewhat longer for fresh peas. Drain, refresh under cold running water and drain well. Add the peas and the currants to the cauliflower.

3. In a small skillet, cook the oil and curry powder over moderate heat, stirring, for 1 minute. In a small bowl, stir together the yogurt and salt. Scrape in the curried oil and stir until smooth.

4. Add the curried yogurt mixture to the vegetables and stir to coat well. Allow to marinate at room temperature for at least 1 and up to 6 hours.
—Diana Sturgis

MEXICAN CAULIFLOWER IN CHEESE SAUCE

Although especially complementary to Mexican dishes, this delicious vegetable side dish goes well with many standard main courses. In Mexico, heavy cream has a flavor and consistency close to crème fraîche. I substitute a combination of slightly reduced heavy cream and sour cream. You can prepare this dish a day ahead and serve it chilled or at room temperature. Garnish with the bacon and almonds just before serving.

6 Servings

- 1 head of cauliflower (about 2 pounds)
- 1 teaspoon salt
- 1 tablespoon vegetable oil
- 1 garlic clove, crushed through a press
- 1 cup heavy cream
- 1 cup shredded mild Cheddar cheese (4 ounces)
- ½ cup sour cream
- 6 slices hickory-smoked bacon, cut into ¼-inch squares
- 2 tablespoons sliced blanched almonds

1. Trim away about ½ inch from the bottom of the cauliflower stem; also remove any leaves. Place the whole head in a large saucepan and cover it with cold water. Add ½ teaspoon of the salt, cover the pan and bring the water to a boil over high heat. Reduce the heat slightly and boil until the cauliflower is just tender when pierced with a fork, about 10 minutes after the water comes to a boil. Let the cauliflower cool to room temperature in the water.

2. In a medium saucepan, heat the vegetable oil. Add the garlic and cook over low heat for 1 minute after the garlic begins to sizzle. Add the cream and bring to a boil over moderate heat. Boil gently for 5 minutes. Stir in the cheese and immediately remove from the heat. Stir in the sour cream and the remaining ½ teaspoon salt. Pour the sauce into a bowl, cover the surface with wax paper and let cool to room temperature.

3. In a medium skillet, cook the bacon over moderate heat until crisp and golden brown. Remove with a slotted spoon and drain on paper towels.

4. To serve at room temperature, simply place the cauliflower upright in a serving dish and pour the cheese sauce over it; sprinkle with the reserved bacon and the almonds. To serve chilled, pour the cheese sauce into a bowl just large enough to hold the head of cauliflower. Invert the head of cauliflower in the sauce, cover and refrigerate until ready to serve. Then transfer, right-side up, to a serving dish and pour the sauce over it. Garnish with the bacon and almonds.
—*Jim Fobel*

DEVILED CELERY ROOT

Shred the celery root with the shredding disk of a food processor or the second coarsest side of a four-sided grater.

6 Servings

- 5 tablespoons unsalted butter
- 1 pound celery root (celeriac), peeled and shredded
- 1 medium onion, chopped
- ½ of a medium red bell pepper, chopped
- 1 cup (loosely packed) fine fresh bread crumbs
- 2 hard-cooked eggs, chopped
- 2 tablespoons chopped flat-leaf parsley
- 1 tablespoon fresh lemon juice
- 2 tablespoons Dijon mustard
- ⅓ cup heavy cream
- ½ teaspoon hot pepper sauce
- ¼ teaspoon dry mustard
- ¼ teaspoon freshly ground black pepper

1. Preheat the oven to 375°. In a large heavy skillet, melt the butter over moderately high heat. Spoon 1 tablespoon of the melted butter into a small bowl and set aside.

2. Add the celery root, onion and red pepper to the skillet and sauté until the onion is golden, about 2 minutes. Reduce the heat to low, cover the skillet and simmer until the vegetables are tender, about 15 minutes.

3. Remove the skillet from the heat. Stir in ¾ cup of the bread crumbs, the eggs, chopped parsley, lemon juice, Dijon mustard, cream, hot sauce, dry mustard and black pepper.

4. Blend the remaining ¼ cup bread crumbs with the reserved 1 tablespoon melted butter. Mound the celery root mixture into 6 scallop shells. Sprinkle evenly with the buttered crumbs. Place the shells on a baking sheet and bake until lightly browned, about 25 minutes.
—*Jean Anderson*

SCALLOPED CELERY ROOT AND POTATOES

This dish tastes particularly good with lamb.

6 Servings

- 1¼ pounds boiling potatoes, peeled and thinly sliced
- 1 large celery root (celeriac), peeled and thinly sliced
- 2 tablespoons all-purpose flour
- 6 tablespoons unsalted butter, softened to room temperature
- 1 teaspoon Dijon mustard
- ¾ teaspoon salt
- 1 cup milk

1. Preheat the oven to 500°. In a large saucepan of lightly salted boiling water, cook the potatoes until just tender, about 5 minutes. With a slotted spoon, remove the potatoes to a colander and drain. Add the celery root to the boiling water and cook until tender, about 5 minutes. Drain well.

2. Meanwhile, in a medium bowl, whisk together the flour, butter and mustard until well blended.

3. In a large, buttered baking dish, arrange half the potatoes in a single layer. Spread one-third of the mustard butter on top. Season with ¼ teaspoon of the salt. Put all the celery root slices on top, spread with half of the remaining mustard butter and season with ¼ teaspoon of the salt. Cover with the remaining potatoes, mustard butter and ¼ teaspoon salt.

4. Pour in the milk and bake, uncovered, for 25 minutes. Reduce the oven temperature to 350° and bake for 5 to 10 minutes, or until golden brown.
—Lee Bailey

CELERY ROOT, GREEN BEAN AND POTATO PUREE

This soft, subtle dish pairs wonderfully with hearty roasts or steaks. Incidentally, if you have any of this puree left over, you can make patties out of it and fry them.
6 to 8 Servings

½ of a medium celery root (celeriac), peeled and cut into ½-inch dice
1 pound baking potatoes, peeled and cut into 1-inch cubes
1 pound green beans, cut into thirds
6 tablespoons cold unsalted butter
1 teaspoon salt

1. In a large saucepan, combine the celery root and potatoes. Cover with 2 inches of cold water and bring to a boil over high heat. Add the green beans and return to a boil. Reduce the heat to moderate and cook until the vegetables are tender, about 20 minutes; drain well.

2. Preheat the oven to 400°. Return the vegetables to the pan and cook over high heat, shaking the pan constantly to evaporate excess moisture, about 1 minute. Transfer the vegetables to a food processor or blender. Add 5 tablespoons of the butter and the salt. Puree, turning the machine quickly on and off, until smooth. (You may have to do this in two batches.)

3. Scrape the puree into a buttered 1½-quart casserole and smooth the top. Rub the remaining 1 tablespoon butter evenly over the top. *(The recipe can be prepared to this point up to 30 minutes ahead.)*

4. Bake the vegetable puree, uncovered, for 25 minutes.
—Lee Bailey

CORN CHILI

Fresh jalapeño, ground chiles and hot sauce give this warm, buttery side dish just the right kick.
6 Servings

4 cups fresh corn kernels (cut from 6 ears) or 2 packages (10 ounces each) frozen corn, thawed
4 tablespoons unsalted butter
1 medium red bell pepper, cut into ¼-inch dice
2 scallions, white and green portions thinly sliced separately
½ teaspoon ground mild chiles
½ teaspoon salt
½ teaspoon freshly ground black pepper
1 small jalapeño pepper—seeded and minced (about 1 tablespoon)
½ cup (loosely packed) fresh coriander (cilantro) leaves, finely chopped
Hot pepper sauce, to taste

1. In a medium saucepan, combine the corn with ½ cup of water. Cover and bring to a boil over moderately high heat. Cook the corn for 3 minutes, then drain in a colander and set aside.

2. Add the butter to the saucepan and melt over moderately low heat. Add the red bell pepper and the scallion whites and cook, stirring occasionally, until softened, about 3 minutes. Stir in the corn, ground chiles, salt and black pepper and cook, stirring occasionally, until heated through, about 3 minutes.

3. Stir in the jalapeño and the scallion greens. Remove from the heat and stir in the coriander. Season to taste with salt and hot sauce and serve warm.
—Tracey Seaman

CORN PUDDING WITH CHEESE AND CHILES

To get all of the sweet corn pulp off the cob, after cutting the kernels off with a sharp knife, scrape the cob again with the dull side of the knife blade.
6 Servings

4 fresh poblano or Anaheim chile peppers, preferably a combination of green and red chiles
2 cups corn kernels with pulp (from 5 ears of corn)
3 tablespoons cornstarch
6 ounces sharp Cheddar cheese
1 teaspoon unsalted butter
½ cup thinly sliced scallions
3 eggs
1 cup milk
½ cup chicken stock or canned broth
¾ teaspoon salt
1 teaspoon sugar

¼ teaspoon freshly ground
   black pepper
⅛ teaspoon freshly grated nutmeg
⅛ teaspoon cayenne pepper
4 ounces Monterey Jack cheese, cut
   into small dice (about ½ cup)

1. Roast the poblano or Anaheim chile peppers directly over a gas flame or under the broiler as close to the heat as possible, turning, until charred all over. Seal in a paper bag for 10 minutes; then rub off the blackened skin. Trim off the stems, ribs and seeds. Cut the peppers into ¼-inch dice.

2. Preheat the oven to 375°. In a food processor or blender, puree 1¼ cups of the corn kernels with the cornstarch.

3. Finely dice enough of the Cheddar cheese to measure ½ cup. Grate the remaining Cheddar.

4. In a small skillet, melt the butter over low heat. Add the scallions and cook until softened, 3 to 5 minutes. Remove from the heat and let cool slightly, about 10 minutes.

5. In a large bowl, combine the eggs, milk, chicken stock, salt, sugar, black pepper, nutmeg and cayenne. Add the corn puree and the remaining corn kernels with their pulp. Whisk to blend well. Stir in the sautéed scallions, roasted chiles, Monterey Jack cheese and diced Cheddar cheese.

6. Spoon into a well-buttered 8-inch square baking pan. Sprinkle the grated Cheddar cheese on top.

7. Place the pan in a deep roasting pan and pour in enough warm water to reach halfway up the sides of the baking pan. Bake for 50 to 55 minutes, until the pudding is golden around the edges and just set in the center. Serve hot.
—Richard Sax

CORN AND AVOCADO TORTA

Robert Del Grande recommends serving this as an accompaniment to grilled squab, but it is also delicious on its own.
4 Servings

1 ear of corn or ¾ cup corn kernels
2 tablespoons unsalted butter
1 red bell pepper, cut into ¼-inch dice
1 jalapeño pepper, seeded and minced
3 scallions, chopped
1 avocado, preferably Hass, cut into ¼-inch dice
¼ cup chopped fresh coriander (cilantro)
1 ounce goat cheese, crumbled
2 teaspoons walnut oil
1 teaspoon fresh lime juice
½ teaspoon salt
¼ teaspoon freshly ground black pepper
2 flour tortillas (7 inches in diameter)
½ cup sour cream
Sprigs of coriander, for garnish

1. Cut the kernels from the corn cob. In a large skillet, melt 1 tablespoon of the butter over moderate heat. Add the corn and cook, stirring, until golden brown, about 10 minutes.

2. Stir in the red pepper, jalapeño and scallions and cook until softened, about 5 minutes. Remove from the heat and let cool slightly.

3. Stir in the avocado, coriander, goat cheese, walnut oil and lime juice. Lightly mash the avocado to bind the ingredients. Season with the salt and black pepper.

4. Spread the avocado mixture over one of the tortillas. Cover with the second tortilla to make a sandwich, or *torta*.

5. In a large skillet, melt the remaining 1 tablespoon butter over moderate heat. Add the avocado *torta* and cook until golden brown on the bottom, about 2 minutes. Using a large spatula, carefully turn the *torta* and cook the other side until browned, about 2 minutes longer.

6. Using a sharp knife, cut the *torta* into quarters. Garnish each wedge with 2 tablespoons of the sour cream and a sprig of fresh coriander.
—Robert Del Grande

CALIFORNIA RANCHO-STYLE HOMINY

Hominy is a special variety of corn whose kernels have been soaked in a lye solution and washed to remove the hulls. For this recipe buy cooked hominy, which is sold canned in most grocery stores.
8 to 10 Servings

⅓ cup olive oil
2 large onions, coarsely chopped
4 garlic cloves, minced
1 tablespoon chili powder
1 tablespoon cumin
3 cans (16 ounces each) hominy, preferably yellow, drained and rinsed
½ teaspoon salt
1 teaspoon freshly ground pepper
½ pound Monterey Jack cheese, shredded (about 2 cups)
Fresh coriander (cilantro) leaves, for garnish

1. In a large skillet, heat the olive oil over moderately high heat. Add the onions and cook, stirring, until slightly softened, about 5 minutes. Add the garlic, chili powder and cumin; cook, stirring frequently, until the onions are softened and beginning to brown, about 10 minutes longer.

2. Add the hominy and cook, stirring, until heated through, about 5 minutes. Season with the salt and pepper. *(The recipe can be prepared to this point up to 1 day ahead. Cool, then cover and refrigerate. Reheat before proceeding.)*

3. To serve, fold the cheese into the warm hominy. Transfer to a serving dish, garnish with the coriander and serve warm.
—Susan Costner

GINGER-SPICED CUCUMBER

The method of salting the cut cucumber softens it just enough to remove the raw quality while leaving the fresh taste in. The effect of the hot ginger sauce on the cool cucumber is surprising and sublime. It goes well with almost any grilled meat.

4 to 6 Servings

1 European seedless cucumber (about 1 pound)
¾ teaspoon salt
1 teaspoon vegetable oil
1 teaspoon Oriental sesame oil
1 garlic clove, minced
1½-inch-long piece of fresh ginger, peeled and minced
⅓ cup sugar
⅓ cup distilled white vinegar

1. Rest the cucumber lengthwise against the handle of a wooden spoon or chopstick and cut on an angle into ¼-inch slices, stopping at the spoon so that you do not cut all the way through. Turn the cucumber over. Again resting it against the spoon, cut straight down into ¼-inch slices without cutting all the way through.

2. Sprinkle the cucumber all over and in between the slices with the salt. Wrap in a clean kitchen towel and refrigerate for at least 2 hours and up to 24 hours.

3. In a small saucepan, heat the vegetable oil and sesame oil over high heat. Add the garlic and ginger and cook, stirring, until aromatic, about 1 minute. Add the sugar and vinegar. Bring to a boil and cook, stirring, until the syrup is slightly thickened and reduced to about ⅓ cup, about 3 minutes. *(The sauce can be made up to 24 hours in advance; cover and refrigerate. Reheat before proceeding.)*

4. Arrange the cucumber on a serving platter, spreading the slices open slightly. Pour the ginger syrup over the cucumber.
—Marcia Kiesel

SAUTEED CUCUMBERS

This elegant dish makes a nice accompaniment to grilled or roasted meats, game, poultry or flavorful fish. Try it with salmon for a striking combination of green and pink.

6 Servings

2 large cucumbers, preferably European seedless
3 tablespoons unsalted butter
½ teaspoon salt
¼ teaspoon freshly ground white pepper

1. Peel the cucumbers and cut in half lengthwise. Using a teaspoon, scoop out and discard any seeds. Slice the cucumbers crosswise into ½-inch crescents.

2. In a large saucepan of boiling salted water, blanch the cucumbers for 1 minute after the water returns to the boil; then drain. Rinse under cold running water until cool. Drain well and pat dry. *(The recipe can be prepared to this point up to 1 hour ahead. Set aside at room temperature.)*

3. In a large skillet, melt the butter over moderately high heat until just foaming. Add the cucumbers, season with the salt and pepper and cook, tossing frequently, until heated through, 2 to 3 minutes. Serve hot.
—John Robert Massie

GRATIN OF EGGPLANT CAVIAR

In Provence, a poor man's caviar is a puree of eggplants, beaten with egg yolks and thickened with olive oil as for a mayonnaise. The "caviar" cooked with Gruyère and Parmesan cheeses makes a delicious gratin and goes well with poached fish or a broiled steak. This can be prepared a day ahead of time.

6 Servings

3½ to 4 pounds medium eggplants (about 3), unpeeled
2 large garlic cloves, cut in slivers
½ teaspoon thyme
3 egg yolks
⅓ cup plus 1 teaspoon light olive oil
⅓ cup freshly grated Gruyère cheese (1½ ounces)
½ teaspoon salt
Pinch of cayenne pepper
2 tablespoons freshly grated Parmesan cheese

1. Preheat the oven to 400°. Cut several incisions in each eggplant. Dip the garlic slivers in the thyme and put the garlic slivers in the eggplant cuts.

2. Place the eggplants on a jelly-roll pan and bake in the middle of the oven, turning 2 or 3 times, until the eggplants are soft to the touch, 30 to 60 minutes, depending on their thickness.

3. Peel the cooked eggplants, discarding the stems and peels. Puree the eggplant in a food processor for 1 minute or pass through the fine disk of a food mill.

4. Force the puree through a fine-mesh sieve to remove the seeds (if you

have a food mill with a very fine blade, this step may not be necessary).

5. Transfer the puree to a large mixing bowl. Beat in the egg yolks at medium speed. Slowly drizzle in ⅓ cup of the olive oil. Stir in the Gruyère, salt and cayenne.

6. Pour the eggplant mixture into a lightly buttered 1½-quart baking dish. Sprinkle ½ teaspoon oil over the surface to prevent it from drying out.

7. Bake for 20 minutes. Sprinkle the Parmesan cheese and remaining ½ teaspoon oil over the top. Continue to bake for 10 to 15 minutes, until lightly browned. *(The recipe can be made ahead of time and reheated in a 350° oven for 10 to 15 minutes.)*
—Lydie Marshall

EGGPLANT AND SPINACH TIMBALES
The Mediterranean flavor of these timbales makes them a perfect foil for lamb.
2 Servings

1 eggplant (about 1 pound)
1½ teaspoons salt
½ pound spinach, stemmed
3½ tablespoons extra-virgin olive oil
⅓ cup finely chopped onion
⅓ cup finely chopped red bell pepper
1 egg, beaten
1 tablespoon minced basil
1 tablespoon minced parsley
2 teaspoons Roasted Garlic Puree (p. 265)
¼ teaspoon freshly ground black pepper

1. Trim the eggplant and halve lengthwise. Using a sharp knife, score the flat surfaces with deep diagonal cuts. Sprinkle with 1 teaspoon of the salt and let drain on paper towels, cut-side down, for 1 hour.

2. Bring a large saucepan of water to a boil. Add the spinach leaves and blanch until wilted but still bright green, 5 to 10 seconds. Transfer to a colander and squeeze dry. Drain the leaves flat on paper towels.

3. Preheat the oven to 350°. Set the eggplant halves in a small roasting pan, cut-side up, and brush each half with 1 tablespoon of the olive oil. Bake until soft, about 1 hour. Let cool slightly, then use a spoon to scrape out the flesh; discard the skin. Finely chop the eggplant and place in a medium bowl.

4. In a small saucepan, heat 1 tablespoon of the oil. Add the onion and red pepper and cook over moderate heat, stirring frequently, until golden, about 5 minutes.

5. Add the onion to the chopped eggplant and blend in the egg, basil, parsley, Roasted Garlic Puree, black pepper and the remaining ½ teaspoon salt.

6. Brush two ⅔-cup ramekins with the remaining ½ tablespoon oil. Line the base and sides of the ramekins with spinach leaves, overlapping them on the sides and letting them extend over the rims. Fill the ramekins with the eggplant mixture and fold the overhanging spinach leaves over the surface. Fold any remaining spinach neatly over the tops. *(The recipe can be prepared to this point up to 1 day ahead. Cover and refrigerate.)*

7. Set the ramekins in a steamer over boiling water and steam until a knife inserted in the center comes out clean and hot to the touch, 20 to 25 minutes. Remove from the steamer, cover with foil and let rest in a warm place for about 10 minutes.

8. To serve, run a knife around the inside of the ramekins, drain off any excess liquid and invert each timbale onto a warmed dinner plate.
—Bob Chambers

BRAISED EGGPLANT AND TOMATO WITH SAGE
Serve this as a side dish or as a first course with grilled bread brushed with olive oil and rubbed with garlic.
2 Servings

2 small eggplants (about ½ pound each)
¾ teaspoon salt
¼ cup plus 2 tablespoons extra-virgin olive oil
½ pound plum tomatoes, halved
6 fresh sage leaves
3 garlic cloves, halved
Pinch of crushed red pepper

1. Trim the stem ends from the eggplants. Halve lengthwise. Sprinkle the eggplants with ½ teaspoon of the salt and let stand for 30 minutes. Gently squeeze excess liquid from the eggplants and blot dry with paper towels.

2. In a large skillet, heat ¼ cup of the oil. Add the eggplants in a single layer, cut-side down. Distribute the tomatoes, sage leaves and garlic on top. Sprinkle with the remaining ¼ teaspoon salt and the red pepper. Cover and cook over moderate heat for 20 minutes.

3. Add the remaining 2 tablespoons oil. Turn the eggplants, cover and cook until soft, 10 to 15 minutes longer.

4. Place 2 pieces of eggplant in each of 2 warmed shallow soup plates and top with the tomatoes and juices. (If you wish, slip off the tomato skins.)
—Anne Disrude

SKILLET RATATOUILLE

This ratatouille is a little unusual both in its composition and in how it is cooked. The vegetables are charred to give them a roasted taste, then they're perked up with soy sauce, sherry, vinegar and crushed red pepper. Both the texture and the flavor of this side dish improve as it sits.

4 Servings

1 medium red or yellow bell pepper, cut into ½-inch squares
2½ teaspoons salt
1 medium eggplant (about 1 pound), cut into 1-inch cubes
¼ cup tomato sauce, preferably homemade
2 tablespoons medium-dry sherry
1 tablespoon soy sauce
1 tablespoon red wine vinegar
1 teaspoon sugar
¼ teaspoon crushed red pepper
3 tablespoons peanut oil
1 small Spanish onion, diced
2 teaspoons minced garlic

1. In a small bowl, toss the bell pepper with ½ teaspoon of the salt. In a medium bowl, toss the eggplant cubes with the remaining 2 teaspoons salt. Set both vegetables aside for about 1 hour.

2. Rinse, drain and dry the bell pepper. Place in a bowl. Repeat the process with the cubed eggplant and place in a separate bowl.

3. In a small bowl, whisk together the tomato sauce, sherry, soy sauce, vinegar, sugar and crushed red pepper.

4. Heat a large cast-iron skillet over high heat just until it begins to smoke, about 1 minute. Add ½ tablespoon of the oil, wait for a few seconds and then add the bell pepper. Cook, stirring and pressing down lightly, until nicely and evenly charred, 2 to 3 minutes. Transfer to a bowl and set aside.

5. Reduce the heat to moderately high and add the remaining 2½ tablespoons oil to the skillet. Add the eggplant and cook, stirring and pressing down lightly, until beginning to char and soften, about 3 minutes.

6. Add the onion and cook, stirring until softened and beginning to brown, about 4 minutes. Add the garlic and cook, stirring, for 1 minute longer.

7. Stir the tomato sauce mixture and add it to the pan. Cook, stirring, until thickened, 1 to 2 minutes. Stir in the reserved bell pepper. Transfer the ratatouille to a serving bowl and let cool. *(The recipe can be made up to 5 days ahead and refrigerated, covered.)* Serve at room temperature.

—Karen Lee & Alaxandra Branyon

BAKED FENNEL AND BOSTON LETTUCE

From time to time I suppose it's usual for us to develop enthusiasms for particular foods. For me, at one point, it was fennel and cooked lettuce. I had always thought that fennel's flavor was too aggressive to suit me . . . until I got around to cooking it (up until then I had only had it raw, or not at all).

6 Servings

¼ pound pancetta or bacon, cut into 1-by-¼-inch strips
1 small carrot, coarsely chopped
1 small celery rib, coarsely chopped
1 small onion, coarsely chopped
2½ pounds fennel bulbs (about 3 large)
1 head of Boston lettuce (about ½ pound), leaves separated
¼ teaspoon freshly ground pepper
¾ cup chicken stock or canned broth
1 tablespoon fresh lemon juice
3 tablespoons unsalted butter
¼ teaspoon salt

1. Preheat the oven to 350°. In a small saucepan, blanch the pancetta in simmering water for 2 minutes; drain.

2. Sprinkle the pancetta and carrot, celery and onion over the bottom of a large shallow gratin or baking dish.

3. Bring a large saucepan of water to a boil over high heat. Add the fennel bulbs and cook for 2 minutes. Remove with a slotted spoon and let cool slightly. Add the Boston lettuce leaves to the boiling water and cook for 2 minutes; drain.

4. Cut the fennel bulbs lengthwise into thick slices. Lay the fennel on top of the chopped vegetables. Sprinkle with the pepper. Lay the wilted lettuce leaves on top to cover the fennel completely.

5. In a small saucepan, heat the chicken stock, lemon juice, butter and salt over moderately high heat until the butter melts; pour over the fennel. Cover the dish with a sheet of buttered aluminum foil and bake for 45 minutes, or until the fennel is tender.

6. Carefully lift the lettuce off the fennel and place the lettuce in a food processor. Arrange the fennel on a serving platter and cover with the foil to keep warm. Pour the pan juices along with the bacon, carrot, celery and onion into the processor with the lettuce. Puree to a sauce. Season with additional salt, pepper and lemon juice to taste.

7. Serve the fennel with a spoonful of the pureed vegetable sauce. Pass the remainder on the side.

—Lee Bailey

MIXED GREENS WITH CRACKLINS AND HOT PEPPER VINEGAR

Although I used broccoli rabe and escarole here, you can use mustard greens, collard greens or any combination of greens as an alternative. You must make the Hot Pepper Vinegar a week ahead, or you can just serve the greens with your favorite vinaigrette.

6 Servings

¼ pound double-smoked slab bacon, with rind, cut into ¼-inch cubes or 6 slices smoked bacon, sliced crosswise ⅛ inch thick
2 teaspoons olive oil
½ of a medium onion, chopped
½ cup chicken stock or canned low-sodium broth
½ pound broccoli rabe—trimmed, tough stems removed and cut into bite-size pieces
1 head of escarole (about 1 pound), torn into bite-size pieces
½ teaspoon salt
½ teaspoon freshly ground pepper
Hot Pepper Vinegar (see Note)

1. In a large saucepan, cook the bacon over moderately high heat until crisp and browned, about 6 minutes. Transfer to paper towels to drain.

2. Pour off the bacon fat. Add the olive oil and onion to the pan. Reduce the heat to moderate and cook until the onion is softened and lightly browned, about 5 minutes.

3. Add the chicken stock and scrape up any browned bits from the bottom of the pan. Add the broccoli rabe, cover and cook, stirring occasionally, until tender, about 10 minutes.

4. Add the escarole and cook, stirring, until wilted, about 5 minutes longer.

5. Season the greens with the salt and pepper and transfer to a large platter. Sprinkle the bits of bacon on top. Serve hot or warm, and pass the Hot Pepper Vinegar at the table.

NOTE: To make Hot Pepper Vinegar, add sliced fresh hot chile peppers (including seeds to taste, for a hotter flavor) to a bottle of red wine vinegar. Let sit at room temperature for 1 week before using.

—Lee Bailey

BRAISED MUSTARD GREENS

If you substitute fresh spinach for the mustard greens, the cooking time will be only 2 to 3 minutes, after the addition of the chicken stock.

6 Servings

2 tablespoons extra-virgin olive oil
1 medium onion, coarsely chopped
2 small shallots, minced
2 small garlic cloves, finely minced
2 tablespoons chicken stock or canned broth
2 pounds fresh mustard greens, rinsed, with large stems removed
2 teaspoons fresh lime juice
½ teaspoon salt
¼ teaspoon freshly ground pepper

1. In a large flameproof casserole, heat the oil over moderate heat. Add the onion, shallots and garlic and cook until golden, 8 to 10 minutes.

2. Add the chicken stock to the casserole. Arrange the mustard greens on top of the onion. Cover and cook, turning the greens several times, until tender, about 20 minutes. (The mustard greens will reduce in volume.) Season with the lime juice, salt and pepper. Toss and serve.

—Lee Bailey

STEWED KALE

This is a gratifyingly full-flavored vegetable dish; I can eat tons of it. It's important to clean the kale thoroughly as outlined here because its curly leaves do a great job of hiding dirt.

6 Servings

1 large bunch of kale (about 2 pounds), stems discarded
¼ cup olive oil
1 large onion, chopped
2 garlic cloves, minced
1¼ cups canned crushed tomatoes with their juices
¾ teaspoon salt
2 teaspoons red wine vinegar
Freshly ground pepper

1. Place the kale in a large bowl of warm water and rub the leaves with your hands to free any dirt. (Do this in batches if the leaves are cramped.)

2. Transfer the leaves directly to a large pot. Add ½ cup of water and bring to a boil over high heat. Cover, reduce the heat to moderate and steam the kale until tender, about 10 minutes. Drain the kale in a colander and set aside.

3. Meanwhile, in a large nonreactive skillet, heat the olive oil. Add the onion and cook over moderate heat until soft, about 10 minutes. Stir in the garlic and cook for 2 minutes. Add the tomatoes and salt and simmer until the sauce reduces slightly, about 5 minutes.

4. Coarsely chop the reserved kale and stir it into the sauce. *(The recipe can be made to this point up to 1 day ahead; cover and refrigerate. Bring to room temperature before proceeding.)* Cover and simmer until heated through, about 5 minutes. Stir in the vinegar and season to taste with pepper. Serve hot.

—Ken Haedrich

(RELATIVELY) QUICK GREENS IN POTLIKKER

Here's an updated version of the traditional boiled greens dish that is so popular throughout the South. Serve with chicken and corn bread or cornsticks for sopping up the potlikker or as a side dish with broiled or grilled pork chops.

6 Servings

1 tablespoon olive oil
1 pound smoked ham hocks or hog jowls
2 medium onions, chopped
6 large garlic cloves, slivered
12 whole black peppercorns
2 dried hot red peppers
1 teaspoon thyme
1 imported bay leaf
20 sprigs of parsley
3 pounds greens (collards, kale, mustard or turnip), trimmed of tough stems and coarsely chopped
Distilled white or cider vinegar, for serving

1. In a large saucepan, heat the oil over moderate heat. Add the ham hocks, onions and garlic and cook, stirring frequently, until the onions are golden, about 10 minutes.
2. Tie the peppercorns, hot peppers, thyme, bay leaf and parsley in a square of cheesecloth. Add to the saucepan along with 4 quarts of water. Bring to a boil, skim if necessary, reduce the heat and simmer, uncovered, for 1 hour.
3. Add the greens and simmer, uncovered, until tender, about 1 hour more.
4. Remove the greens to a bowl. Increase the heat to high and boil until the liquid is reduced to 2 cups. Return the greens to the pot and heat to warm through, about 3 to 5 minutes.
5. Ladle the greens and some potlikker into small bowls and serve as a side dish. Pass the vinegar in a cruet on the side to sprinkle on the greens.
—Sarah Belk

CEPES BORDELAISE
8 Servings

1 pound fresh cèpes (see Note)
4 tablespoons unsalted butter
3 garlic cloves, minced
1 teaspoon fresh lemon juice
¾ teaspoon salt
½ teaspoon freshly ground pepper

1. Trim the ends off the stems of the cèpes. Separate the stems from the caps. Mince the stems. Set the whole caps aside.
2. In a large skillet, melt the butter over moderate heat. Add the garlic and mushroom stems and sauté until softened, about 5 minutes.
3. Add the mushroom caps and 2 tablespoons of water to the pan. Cover, reduce the heat to low and cook gently (adding more water if the mushrooms begin to stick) until tender, about 40 minutes. *(The mushrooms can be made ahead to this point. Set aside, partially covered, at room temperature.)*
4. Before serving, reheat the cèpes. Season with the lemon juice, salt and pepper and serve hot.

NOTE: If fresh cèpes are not available, fresh shiitake mushrooms can be substituted.
—John Robert Massie

CIDER-BRAISED ONIONS

The easiest way to peel small white onions is to score the skin and surface layer from top to root end with a sharp paring knife. This gives you an incision to slip your thumbnail under, so that you can peel off the outer layer. Just try not to cut deeper than the outer layer of the onion.

6 Servings

1 pound small (1 inch) white onions
1 tablespoon unsalted butter
¾ cup fresh apple cider
¼ teaspoon soy sauce
1 tablespoon chopped parsley (optional)

1. Peel the onions, trimming off the root end as well as about ⅛ inch of the top. With a sharp paring knife, score a shallow cross in the root ends to ensure even cooking.
2. In a medium nonreactive skillet, melt the butter. Add the onions and stir to coat. Add the cider and soy sauce, increase the heat to high and bring to a boil. Reduce the heat to maintain a simmer, cover tightly and braise the onions until tender when pierced with a knife, about 10 minutes.
3. Uncover and increase the heat to high. Boil until the cider reduces to a light amber glaze, about 5 minutes. *(The recipe can be made to this point up to 1 day ahead; cover and refrigerate. Reheat before serving.)* Transfer the onions to a warm dish, scraping the glaze over them. Sprinkle with parsley if desired.
—Ken Haedrich

BAKED ONIONS WITH BALSAMIC VINAIGRETTE

Although you can use any good-quality onion in this recipe, sweet varieties like Vidalia or Walla Walla work best.

6 Servings

6 medium Vidalia onions, peeled
¼ cup plus 1 tablespoon olive oil
1½ teaspoons Dijon mustard
2 tablespoons balsamic vinegar
½ teaspoon salt
¼ teaspoon freshly ground pepper

1. Preheat the oven to 350°. Cut off the root and stem ends of the onions. Using a sharp knife, make a deep "X" on the top of each onion, cutting halfway down.

2. Place each onion on a 6-by-6-inch square of aluminum foil, with the "X" up. Top each with 1 teaspoon of the oil. Bring up the corners of each square of foil and twist them together tightly to completely enclose the onions. Place the onions, foil twists up, on a large baking sheet and bake for 40 minutes, or until very tender when pierced.

3. Meanwhile, in a small bowl, combine the mustard, vinegar, salt and pepper. Gradually whisk in the remaining 3 tablespoons olive oil until blended.

4. Unwrap the onions and arrange them on a serving platter. Spoon a scant tablespoon of the vinaigrette over each onion. Season with pepper to taste and serve hot, either halved or whole.
—Lee Bailey

TEMPURA ONION RINGS

4 to 6 Servings

4 cups vegetable oil, for frying
1½ cups cornstarch
3 large red onions (about 1¼ pounds), peeled and sliced crosswise ⅓ inch thick
1½ cups all-purpose flour
1 teaspoon salt, plus more for sprinkling
1½ cups coarsely crushed ice
Lemon wedges, ketchup or soy sauce, for serving

1. In a deep fryer or large saucepan, heat the oil to 350°. Meanwhile, put 1 cup of the cornstarch in a brown paper bag. Add the onion slices and shake well to coat. Transfer the coated onion rings to a large baking sheet.

2. In a large bowl, mix the remaining ½ cup cornstarch with the flour and 1 teaspoon salt. Add 1½ cups of cold water to the ice and dump the ice water all at once into the flour mixture. Stir vigorously with chopsticks or a fork to make a slightly thick but lumpy batter. Do not overmix. There will be pieces of ice and lumps the size of a nickel in the batter.

3. When the oil is hot, add as many coated onion rings to the batter as will fit comfortably in the fryer. (You will have to do this in about 5 batches.) Using chopsticks or a large fork, lift the rings out of the batter; make sure that there aren't any pieces of ice clinging to the onions. Add the rings to the hot oil and separate any that cling together. Fry the onion rings until golden brown all over, about 3 minutes. Transfer to paper towels or a paper bag to drain well. Repeat with the remaining onion rings and batter. Sprinkle the onion rings with salt if desired. Serve hot, warm or at room temperature with lemon, ketchup or soy sauce on the side.
—Marcia Kiesel

RICHLY GLAZED PEARL ONIONS

6 Servings

⅔ cup boiling water
⅓ cup dried porcini (or cèpes) mushrooms (about 1 ounce)
3 tablespoons unsalted butter
2 pints pearl onions, peeled
1 cinnamon stick, broken
1 cup dry red wine
1 cup chicken or beef stock or canned broth
½ cup chopped canned Italian plum tomatoes with some of their liquid
Salt and freshly ground pepper
Pinch of sugar (optional)

1. In a heatproof bowl, combine the boiling water and mushrooms. Set aside to soak until softened, about 20 minutes. Rinse the mushrooms in the soaking liquid and squeeze dry. Cut off any tough bits of stem. Coarsely chop the mushrooms and set aside. Set the soaking liquid aside to allow the grit to settle.

2. In a large nonreactive skillet, melt the butter over moderate heat until hot, about 2 minutes. Add the onions in an even layer. Add the cinnamon stick and cook, without stirring, until the onions are browned on one side.

3. Using tongs, turn each onion over and cook until dark brown on the other side, another 4 minutes. Pour in the red wine and boil for 2 minutes, then pour in the stock. Bring to a boil, cover and cook until the onions are tender and the liquid has reduced and thickened, about 20 minutes.

4. Uncover and stir in the reserved mushrooms and the tomatoes with their liquid. Pour in the porcini soaking liquid, stopping when you reach the grit on the bottom. Return to a boil, cover and simmer until the sauce is dark and syrupy, about 4 minutes. Season with salt and pepper to taste and the sugar if desired. *(The onions can be prepared up to 2 days ahead; cover and refrigerate. Reheat in a saucepan over moderately low heat, adding a little water if the sauce is too thick.)*
—Marcia Kiesel

SPICED TEMPURA PARSLEY
You can use curly parsley for this dish, but the flat-leaf variety holds more batter and is easier to handle. Serve this crunchy deep-fried parsley alongside grilled chicken or fish.
4 to 6 Servings

2 large bunches of flat-leaf parsley
1 tablespoon curry powder
1 teaspoon cumin
½ teaspoon cayenne pepper
1 teaspoon salt
½ teaspoon freshly ground black pepper
½ cup all-purpose flour
½ cup ice water
1 quart corn oil

1. Wash and thoroughly dry the parsley; separate it into large sprigs.
2. In a medium bowl, combine the curry, cumin, cayenne, salt, black pepper and flour. Pour in the ice water and stir briefly until just blended but still lumpy; do not overmix.
3. In a large heavy saucepan, heat the oil over moderately high heat until it reaches 375°. Reduce the heat to moderate. Dip 3 or 4 parsley sprigs at a time into the tempura batter, then place in the hot oil. Fry until golden brown, about 15 seconds. Drain on paper towels. Repeat with the remaining parsley. Serve hot.
—Marcia Kiesel

PLANTAIN GRATIN
Clara Lesueur, chef/owner of Chez Clara in Guadeloupe, sometimes sprinkles the plantains with grated Gruyère cheese before baking.
6 Servings

½ of a small onion, sliced
2 garlic cloves, lightly crushed
1 jalapeño pepper, halved and seeded
3 whole black peppercorns
3 cups milk
1½ sticks (6 ounces) unsalted butter
3 large yellow plantains, thinly sliced
¼ cup plus 1½ teaspoons all-purpose flour
½ teaspoon salt
1 teaspoon freshly ground pepper
¼ cup fresh bread crumbs

1. In a medium saucepan, combine the onion, garlic, jalapeño pepper and black peppercorns. Add the milk and bring to a simmer over moderately high heat. Cover and set aside to infuse for 25 minutes.
2. Preheat the oven to 475°. In a large skillet, melt 3 tablespoons of the butter over moderately high heat. Add half of the plantain slices and sauté, turning once, until browned, about 2 minutes per side. Repeat the process using 3 more tablespoons of butter and the remaining plantains.
3. In another medium saucepan, melt the remaining 6 tablespoons butter over moderate heat. Set aside 1½ tablespoons of the butter. Add the flour to the melted butter in the saucepan and cook, stirring, for about 2 minutes without allowing the flour to color to make a roux.
4. Strain the seasoned milk into the saucepan, increase the heat to moderately high and bring to a boil, whisking constantly, until thickened. Reduce the heat to moderate and cook, whisking frequently, until the sauce is smooth and no longer floury tasting, about 2 minutes. Season with ⅛ teaspoon of the salt and ¼ teaspoon of the pepper.
5. In a buttered medium gratin or baking dish, arrange the sautéed plantains in a single layer. Pour on one-third of the white sauce. Season with ⅛ teaspoon of the salt and ¼ teaspoon of pepper. Repeat the process twice more. Sprinkle the bread crumbs evenly over the top. Drizzle on the reserved 1½ tablespoons melted butter and bake for about 15 minutes, until bubbling and lightly browned on top.
—Clara Lesueur

CURRIED OVEN FRIES
4 Servings

3 Russet or other baking potatoes
⅓ cup clarified butter, melted
1½ teaspoons curry powder
½ to 1 teaspoon coarse (kosher) salt, to taste

1. Preheat the oven to 375°. Place a baking sheet in the oven to warm. Without peeling, cut the potatoes into 3-by-1-by-1-inch sticks. Rinse in cold water and pat dry.
2. In a large bowl, combine the butter with the curry powder. Add the potatoes and toss to coat all sides.
3. Remove the baking sheet from the oven. Arrange the potatoes on it in a single layer, return to the oven and bake for 20 minutes. Turn the potatoes and bake for 20 minutes longer. Turn again and bake for 15 minutes, until the potatoes are beginning to brown; then turn once more and bake for 10 minutes, or until browned all over.

4. Remove the potatoes from the oven and place in a large brown paper bag. Sprinkle in the salt, seal the bag and shake gently until the potatoes are evenly coated with salt. Serve the fries piping hot.
—Anne Disrude

TWICE-BAKED POTATOES WITH CREAM AND WILD MUSHROOMS
4 Servings

2 large Russet or other baking potatoes
1 ounce dried mushrooms, such as cèpes or morels
1 teaspoon salt
Pinch of freshly ground pepper
¾ cup heavy cream

1. Preheat the oven to 400°. Pierce the potatoes several times with a fork. Bake for about 50 minutes, or until tender when pierced with a fork.
2. Meanwhile, in a small bowl, soak the mushrooms in 1 cup of hot water until softened, about 30 minutes. Remove from the liquid, squeeze dry and coarsely chop.
3. Reduce the oven temperature to 350°. Scoop out the potatoes and place in a medium bowl (discard the skins or reserve for another use, such as the crisp skin croutons for the Potato Soup with Greens, p. 65). Coarsely break up the potatoes with a fork and mix in the chopped mushrooms, salt and pepper to taste.
4. Spoon the potatoes into a small gratin dish. Pour the cream on top. Bake, uncovered, for 30 minutes, or until bubbling and lightly browned.
—Anne Disrude

BAKED POTATO WITH AVOCADO PUREE AND CHIVES
2 Servings

1 large Russet baking potato
½ teaspoon vegetable oil
½ teaspoon salt
¼ teaspoon freshly ground pepper
½ of an avocado, mashed to a puree
2 teaspoons snipped fresh chives

1. Preheat the oven to 350°. Rub the potato with the oil and bake until soft, about 1 hour.
2. Split the potato lengthwise in half. Scoop out the pulp, leaving a thin shell of potato. Mash the pulp with a fork and season with the salt and pepper. Mix in the avocado puree until blended. Divide between the potato shells and sprinkle with the chives.
—Anne Disrude

POTATOES AND SUN-DRIED TOMATOES EN PAPILLOTE
Steaming the potatoes enclosed in paper facilitates the exchange of flavors between the potatoes and seasonings, resulting in a more intensely flavored potato.
4 Servings

1¼ pounds waxy (boiling) potatoes, cut into ⅜-inch dice
¼ cup finely chopped sun-dried tomatoes in oil, drained
1½ tablespoons minced parsley
3 tablespoons olive oil
½ teaspoon salt
¼ teaspoon freshly ground pepper

1. In a large pot of boiling salted water, cook the potatoes until fork-tender, about 10 minutes. Drain well. Place the potatoes in a large bowl.
2. Add the sun-dried tomatoes, parsley, oil, salt and pepper to the potatoes and toss until evenly mixed. *(The recipe may be prepared to this point 1 day in advance. Cover and refrigerate. Let the potatoes return to room temperature before proceeding.)*
3. Preheat the oven to 400°. Cut a sheet of parchment paper or aluminum foil into an 18-inch square. Mound the potatoes in the center. Fold up two sides over the potatoes and make a tight fold to seal. Fold the ends up and tuck under to form a neat but not tight package. Place the papillote on a baking sheet.
4. Bake in the center of the oven for 20 minutes. Serve the potatoes from the package or turn out into a serving dish.
—Anne Disrude

BASILICO POTATOES
These are the "bas'lacol' potatoes" that my friend Mick's mother used to make years ago. "She'd toss them with basil from our garden," he remembers, "for special summer suppers with barbecued steaks and sliced ripe tomatoes." They're wonderful, like crisp french fries flavored with the gutsy punch of fresh basil and plenty of fruity olive oil.
3 to 4 Servings

⅓ cup extra-virgin olive oil
3 to 4 medium baking potatoes, peeled and sliced ¼ inch thick
1 teaspoon salt
½ teaspoon freshly ground pepper
½ cup chopped fresh basil

1. Preheat the broiler. Line a baking sheet with heavy-duty aluminum foil. Drizzle the foil with 2 tablespoons of the olive oil. Arrange the potato slices in the pan, overlapping as little as possible. Drizzle the remaining olive oil over the potatoes and season with the salt and pepper. Turn the potatoes to coat generously all over.
2. Broil the potatoes 5 to 6 inches from the heat until golden brown, about 12 minutes. Carefully turn the potatoes

over to avoid tearing the foil. Broil about 12 minutes longer, until golden brown, checking during the last few minutes to prevent burning. Remove the pan from the oven and sprinkle the basil over the potatoes. Toss lightly and serve at once.
—Richard Sax

PAN-ROASTED POTATOES WITH LEMON AND MARJORAM

Next time you're in the mood for home fries, try this variation. They're irresistibly good.

4 to 6 Servings

Zest of 1 lemon, cut into thin julienne strips
½ pound thickly sliced smoked bacon, cut crosswise into 1-inch pieces
1½ pounds small red potatoes, cut into 1-inch cubes
1 large onion, chopped
1 teaspoon chopped fresh marjoram
½ teaspoon salt
½ teaspoon freshly ground pepper

1. Preheat the oven to 450°. In a small saucepan of boiling water, blanch the lemon zest for 1 minute. Remove the strips of zest with a slotted spoon and drain well on paper towels. *(The zest can be prepared 1 day ahead. Cover and refrigerate.)*

2. In the same water, blanch the bacon for 1 minute. Drain and dry the bacon well on paper towels.

3. In a large skillet, preferably cast iron, cook the bacon over moderate heat until crisp and lightly browned, about 3 minutes. Remove all but 3 tablespoons of the bacon fat. Increase the heat to high and when the fat is almost smoking, add the potatoes, onion and lemon zest. Stir to combine and cook for 1 minute.

4. Place the skillet on the bottom rack of the oven and cook for 15 minutes. Stir and then continue cooking until the potatoes and onions are lightly crisped and golden brown, another 10 minutes. Remove from the oven. Season the potatoes with the marjoram, salt and pepper and serve warm.
—Marcia Kiesel

POTATO AND ONION CROQUETTES WITH DEEP-FRIED PARSLEY

Deep-fried parsley is a classic accompaniment to savory croquettes. It has a wonderful nutty flavor and an unusual, delicate crunch. It is easy to prepare at the last minute, since the oil for frying the croquettes is already hot.

This croquette recipe is ideal for leftover mashed potatoes. Measure out 4½ cups and begin at Step 2. The croquette mixture should be chilled for at least 3 hours, so plan accordingly.

8 Servings

2 pounds all-purpose potatoes, peeled and quartered
3 tablespoons unsalted butter
1 medium onion, finely chopped
2 egg yolks
2 tablespoons heavy cream
1 teaspoon salt
¼ teaspoon freshly ground white pepper
⅛ teaspoon freshly grated nutmeg
1 cup all-purpose flour
2 whole eggs, beaten
2 cups fresh bread crumbs (made from 5 or 6 slices of firm-textured white bread, crusts removed)
1½ to 2 quarts peanut oil, for deep-frying
3 medium bunches of curly parsley—washed and thoroughly dried, large stems removed

1. Cook the potatoes in a large saucepan of boiling water until tender, about 15 minutes. Drain and rice the potatoes, or pass them through the medium disk of a food mill.

2. In a medium skillet, melt the butter over moderately high heat. Add the onion and cook until soft and beginning to color, about 3 minutes.

3. In a medium bowl, combine the potatoes, sautéed onion, egg yolks, cream, salt, pepper and nutmeg; mix until well blended. Cover the croquette and refrigerate until well chilled, about 3 hours or overnight.

4. Place the flour in a shallow dish, the whole beaten eggs in a second dish and the bread crumbs in a third shallow dish. To make each croquette, scoop out 1 tablespoon of the croquette mixture and quickly roll it into a small ball between your palms. Roll the ball in flour to coat all over; shake lightly on your fingers to remove any excess. Dip into the beaten eggs, letting excess drip back into the dish. Roll in the bread crumbs until completely coated. Roll lightly between your palms to remove excess crumbs.

5. In a deep-fat fryer or deep heavy saucepan, heat 2½ inches of oil to 375°. Fry the croquettes in batches without crowding for about 3 minutes, or until golden brown. Remove and drain on paper towels.

6. While the oil is still hot, place the parsley, a handful at a time, in the basket of the deep-fat fryer and, standing at arm's length, carefully lower into the hot oil (the oil will bubble and splatter because of the water content in the parsley). Cook for about 30 seconds, until dark green and crisp. Lift out and drain on paper towels.
—John Robert Massie

Saffroned Millet Pilaf with Roasted Red Peppers (p. 188).

Persian Rice (p. 197).

Cracked Wheat-Spinach Salad (p. 236).

CRISP POTATO-SCALLION ROAST
6 Servings

6 large baking potatoes (about 3 pounds), peeled and cut into 1-inch cubes
3 tablespoons unsalted butter
3 tablespoons vegetable oil
3 medium scallions, white and green parts coarsely chopped separately
¾ teaspoon salt
½ teaspoon freshly ground pepper

1. Bring a large saucepan of lightly salted water to a boil over high heat. Add the potatoes, return to a boil and boil for 5 minutes. Drain the potatoes in a colander and rinse under cold running water; drain well. *(The potatoes can be parboiled up to 4 hours ahead and refrigerated, covered.)* Pat the potatoes as dry as possible with paper towels.

2. Preheat the oven to 375°. In a 10-by-15-inch jelly-roll pan, combine the butter and the oil and place in the upper third of the oven until just melted, about 3 minutes. Add the potatoes, white part of the scallions, salt and pepper and toss to coat.

3. Bake the potatoes on a rack in the top third of the oven for 1 hour. Stir the potatoes well and bake them for 40 minutes longer, or until golden brown and crisp. Transfer to a bowl, sprinkle with the chopped scallion greens and serve hot.
—*Rick Rodgers*

POTATO CRISPS
When we made these fried potatoes, some preferred the crisp, thick-chip version, while others liked the chewy ones better. If you like yours chewy too, see the adjustments at the end of the recipe. The method for both types of crisps is the same; only the temperature of the oil and the cooking times vary.
4 to 6 Servings

1½ pounds Idaho potatoes, unpeeled
4 cups corn oil, for frying
Salt and freshly ground pepper
Malt vinegar, for serving

1. Using a sharp knife or a mandoline, slice the potatoes into ⅛-inch-thick rounds. Place in a large bowl of ice water and let soak for at least 45 minutes, adding more ice as necessary.

2. In a large skillet, heat the oil to 375° over high heat. Meanwhile, lay a clean kitchen towel on a work surface. Remove half of the potatoes from the ice bath and spread them out on a kitchen towel in a single layer. Cover with a second towel and pat thoroughly dry.

3. When the oil is hot, carefully add the drained potatoes and spread them out with tongs. Fry, turning occasionally, until well browned and crisp, about 10 minutes. As the potatoes are done, transfer them to paper towels to drain. Blot well. Repeat with the remaining potatoes. *(The potatoes can be fried up to 2 days ahead and stored in an airtight container.)* Serve the crisps warm or at room temperature, sprinkled with salt, pepper and malt vinegar.

CHEWY POTATO CRISPS
Heat the oil to 350°. Fry each batch of potatoes, turning, until evenly browned, 12 to 15 minutes. Drain well, sprinkle with salt and pepper and serve hot.
—*Tracey Seaman*

CHEESE AND ONION ROSTI
Note that the potatoes for these delicious pan-fried Swiss potato cakes must be cooked ahead of time and refrigerated. If you have leftover baked potatoes, you can use them here.
Makes 4 Pancakes

4 medium baking potatoes
2 medium onions, thinly sliced
½ cup lard
Salt and freshly ground pepper
About 2 tablespoons unsalted butter
2 tablespoons shredded Gruyère or Swiss cheese

1. In a large saucepan of boiling salted water, cook the whole, unpeeled potatoes over moderate heat, until a skewer or knife point easily pierces the outer ½ inch of potato but meets resistance in the center (the potatoes will finish cooking later), about 15 minutes. Drain the potatoes and rinse under cold running water. Refrigerate, uncovered, until well chilled, 3 to 4 hours or overnight.

2. Peel the potatoes. In a food processor fitted with a shredding blade or on the large holes of a hand grater, shred the potatoes lengthwise to make long, even shreds. Gently toss the potatoe shreds with the sliced onions.

3. In a heavy 8-inch skillet with sloping sides or in an omelet pan, melt 2 tablespoons of the lard over high heat until the lard begins to smoke.

4. Sprinkle one-fourth of the potato-onion mixture into the skillet. Season lightly with salt and pepper. Quickly push any stray shreds in from the outer rim of the pan to form an even round pancake. Lightly tamp down the top of the *rösti*.

SIDE DISHES: VEGETABLES & GRAINS

5. Reduce the heat to moderately high, and cook the *rösti*, shaking and rotating the pan occasionally to loosen the potatoes, until well browned on the bottom, about 3 minutes. (If the potatoes stick, remove the pan from the flame and rap it sharply on a hard surface, such as a cutting board, to loosen them.)

6. Add ½ tablespoon butter to the edge of the pan. Rotate quickly to melt and distribute the butter. Flip the *rösti* like a flapjack or turn over with a wide spatula.

7. Season lightly again with salt and pepper and sprinkle 1½ teaspoons of the Gruyère on top. Continue cooking over moderately high heat, adding additional butter if the pan becomes dry, until browned on the second side, about 3 minutes longer. Keep warm in a low oven while you make the other three *rösti* with the remaining ingredients.
—John Robert Massie

PARSLIED POTATO CAKES

These parslied potato cakes can be made well in advance. Once they are baked, let them cool, then cover tightly and freeze. The day you plan to serve them, thaw them, then reheat gently in a covered skillet, with additional oil and butter if needed.
Makes 12 to 14 Small Cakes

3 pounds boiling potatoes, peeled
⅓ cup chopped flat-leaf parsley
1½ teaspoons freshly ground pepper
1 teaspoon salt
1 garlic clove, crushed through a press
2 whole eggs plus 1 egg yolk
About 2 tablespoons unsalted butter
About ¼ cup olive oil

1. Grate the potatoes in a food processor fitted with the grating disk. In a strainer placed over a mixing bowl, squeeze the grated potatoes between your hands to remove any water. Reserve the liquid in the bowl.

2. Place the potatoes in a separate mixing bowl. Stir in the parsley, pepper, salt, garlic, whole eggs and egg yolk. Carefully pour off and discard the water from the reserved potato liquid, saving the potato starch that has settled at the bottom of the bowl. Scrape this starch back into the potato mixture and stir well to combine.

3. In a large skillet, melt 1 tablespoon of the butter in 2 tablespoons of the oil over low heat. When the butter has melted, remove the pan from the heat.

4. Preheat the oven to 400°. Stir the potato mixture to distribute the eggs evenly. Firmly pack the mixture into a round ¼-cup measure, squeezing out any excess liquid. Invert the measuring cup over the skillet and tap the bottom gently until the potato cake slides out. Repeat with the remaining potato mixture, placing the cakes 1½ inches apart in the pan. (Depending on the size of your skillet, it may be necessary to use 2 pans, or do this in 2 batches.)

5. Return the skillet to the stovetop and cook the potato cakes over moderately high heat until the bottoms are browned, about 2 minutes. Carefully turn over the potato cakes, adding an additional tablespoon butter and 2 tablespoons oil to the pan if needed. Cook the other side until browned, about 2 minutes longer.

6. Transfer the cakes to a baking sheet. Bake for 20 minutes in the middle of the oven.
—Bob Chambers

SPICED POTATO PANCAKES

These pancakes are held together by the starch from the potatoes. When the potatoes are rinsed in cold water, the starch will sink to the bottom of the bowl, where it looks like barely diluted cornstarch. To return the starch to the potatoes, pour off the water, leaving the starch at the bottom, and add the grated potatoes. A choice of flavorings is given. Pick your favorite, or divide the pancake mixture in thirds and sample all three.
4 to 6 Servings

3 medium all-purpose potatoes
3 eggs, beaten
¾ teaspoon salt
¼ teaspoon freshly ground pepper
1 tablespoon grated fresh ginger or 1 tablespoon minced jalapeño pepper or 1½ teaspoons grated orange zest
Safflower oil, for frying

1. Peel and coarsely grate the potatoes. Place in a large bowl, cover with cold water and stir with your hands to rinse. With a slotted spoon, remove the potatoes and squeeze dry in a kitchen towel. Let the water stand until the cloudy potato starch settles to the bottom of the bowl, about 5 minutes. Pour off the water, leaving the white starch at the bottom of the bowl.

2. Return the potatoes to the bowl. Add the eggs, salt, pepper and ginger (or other flavoring). Stir to combine. Preheat the oven to low. Line a serving platter with paper towels and place in the oven to warm.

3. In a large heavy skillet, warm ¼ inch of safflower oil over moderately high heat until a spoonful of the potato mixture immediately sizzles when dropped in. Drop the potatoes in heaping tablespoons and flatten with the back of the spoon to form pancakes 3 to

4 inches in diameter. Fry until well browned on one side, about 6 minutes. Carefully turn the pancakes and cook until browned on the other side, about 4 minutes.

4. Remove the pancakes to the serving platter and keep warm in the oven. Fry the remaining potatoes in the same manner. Serve as soon as possible.
—Anne Disrude

LYONNAISE POTATO CAKE

This earthy potato cake—a French country version of our American hash-brown potatoes—has onion in it, as any Lyonnaise dish should. The potatoes can be cooked and mashed, combined with the onion mixture and seasonings and placed back in the skillet ahead of time. However, for best results, the final cooking on top of the stove and under the broiler should be done just before serving.

6 to 8 Servings

2½ pounds red potatoes, peeled
¼ cup rendered chicken fat or butter
1 pound onions (about 3 medium), thinly sliced
¾ cup grated imported Swiss cheese
½ teaspoon salt
¼ teaspoon freshly ground pepper
2 tablespoons unsalted butter

1. Put the potatoes in a large saucepan and add cold water to cover. Bring to a boil and cook over moderate heat until tender, 25 to 30 minutes; drain. Set the potatoes aside until they are cool enough to handle, 10 to 15 minutes. Using a large spoon and fork, mash and crush the potatoes into pieces not larger than ½ inch.

2. In a large broilerproof skillet, preferably nonstick, heat the chicken fat over moderate heat. Add the onions and cook until lightly browned, 8 to 10 minutes. Add the onions, Swiss cheese, salt and pepper to the potatoes; blend well.

3. In the same nonstick skillet, melt the butter. Add the potato mixture and pack it down tightly. *(The recipe can be prepared ahead to this point. Set aside at room temperature for up to 3 hours.)*

4. About 40 minutes before serving, preheat the broiler. Heat the skillet with the potato mixture over moderate heat. Cover and cook until the bottom of the potato cake is browned, about 15 minutes. Transfer to the broiler, setting the pan about 6 inches from the heat. Broil until the top is nicely browned, 10 to 12 minutes. Remove the skillet from the oven and let stand for about 5 minutes. To serve, invert onto a serving plate and cut into wedges.
—Jacques Pépin

POTATO-MUSHROOM CAKES WITH CUMIN

Since I've always had a weakness for potatoes and mushrooms, I've combined them in these crisp fried cakes made of grated potatoes and dried porcini mushrooms. Hot out of the skillet, the cakes are sprinkled with chopped roasted cumin at the last minute.

Makes About 14 Cakes

2½ teaspoons cumin seeds
⅓ cup dried porcini mushrooms (½ ounce)
2 Idaho potatoes (1 pound total)
2 tablespoons unsalted butter, melted
2 tablespoons grated onion
¾ teaspoon salt
½ teaspoon freshly ground pepper
About ¼ cup vegetable oil, for frying

1. In a small dry skillet, toast the cumin seeds, shaking the pan occasionally, over high heat until fragrant, about 30 seconds. Coarsely chop with a large knife.

2. In a small bowl, cover the porcini with ½ cup of hot water. Set aside to soften, about 12 minutes. Rinse the mushrooms, squeeze dry, trim off any tough ends and coarsely chop.

3. Peel the potatoes and grate them on the coarse side of a box grater. Squeeze the potatoes to remove as much liquid as possible. Place the grated potatoes in a medium bowl and add the melted butter; toss well. Stir in the porcini, onion, salt, pepper and 2 teaspoons of the cumin.

4. In a large skillet, heat 2 tablespoons of the oil over moderately high heat until it begins to smoke, about 2 minutes. For each cake, spoon about 3 tablespoons of the potato mixture into the skillet and flatten to about 3 inches in diameter. Cook until the undersides are dark brown, about 2½ minutes, then flip and cook until the second side is brown, 2 minutes longer. Transfer to paper towels to drain, then place on a baking sheet and keep warm in a low oven. Repeat with the remaining oil and potato mixture. *(The cakes can be made up to 5 hours ahead, stored at room temperature and then reheated in a hot oven.)* Sprinkle the remaining cumin over the cakes before serving.
—Marcia Kiesel

SIDE DISHES: VEGETABLES & GRAINS

GRANDMA'S POTATO, RED PEPPER AND ZUCCHINI GRATIN

This simple vegetable gratin comes from Auberge de la Madone, a small family inn hidden high in the mountains above Nice. One summer's evening, sitting on the terrace beneath an ancient olive tree laden with olives, I was served this gratin as an accompaniment to a deliciously moist sautéed rabbit.

4 Servings

1 garlic clove, halved
¼ cup extra-virgin olive oil
2 pounds all-purpose or red potatoes, peeled and very thinly sliced
Salt
2 teaspoons minced fresh thyme or ½ teaspoon dried
2 red bell peppers, cut into thin rings
4 small zucchini (about 4 ounces each)

1. Preheat the oven to 350°. Rub the bottom of a medium gratin dish with the garlic. Grease lightly with some olive oil.
2. Arrange a layer of half the potatoes in the bottom of the dish, overlapping the slices as necessary. Season with salt to taste and ½ teaspoon of the thyme, and drizzle on 1 tablespoon of the olive oil. Add a layer of half the red peppers and then half the zucchini. Season again with salt and thyme and drizzle 1 tablespoon of the oil over the vegetables. Make a last layer of potatoes and end with a layer of peppers and zucchini, repeating the seasoning and oil for each layer.
3. Cover tightly with aluminum foil and bake until the vegetables are very soft and tender, about 1 hour.
—Patricia Wells

POTATO AND WILD MUSHROOM GRATIN
6 Servings

¼ pound pancetta or lean bacon, sliced ⅛ inch thick and cut into matchsticks
1 pound fresh wild mushrooms, such as chanterelles, morels and shiitakes
4 tablespoons unsalted butter
2 medium leeks, white and tender green, finely chopped
¾ teaspoon salt
¾ teaspoon freshly ground pepper
2 pounds red bliss or other waxy potatoes
1 cup heavy cream
1 cup milk
¼ teaspoon thyme
2 tablespoons freshly grated Parmesan cheese

1. In a large heavy skillet, cook the pancetta over moderate heat, stirring occasionally, until browned, about 10 minutes. Using a slotted spoon, transfer to paper towels to drain. Wipe out the pan.
2. Clean the mushrooms. Remove and discard any tough stems. Slice the mushrooms ¼ inch thick. Melt 2 tablespoons of the butter in the skillet and add the mushrooms. Cook over moderate heat, stirring, until softened and beginning to brown, about 15 minutes. Scrape the mushrooms into a medium bowl and add the pancetta.
3. Add the remaining 2 tablespoons butter to the skillet and melt over moderately low heat. Add the leeks and cook, stirring frequently, until thoroughly softened, about 10 minutes. Add the leeks to the mushrooms and pancetta. Stir in ¼ teaspoon each of the salt and pepper.
4. Preheat the oven to 350°. Peel the potatoes and slice them as thin as possible. Lightly butter a 2½- to 3-quart shallow gratin dish and evenly layer half of the potatoes in the bottom. Sprinkle with ¼ teaspoon each of the salt and pepper and spread the mushroom mixture on top. Cover with the remaining potatoes.
5. In a bowl, combine the cream and milk and pour over the potatoes. Sprinkle with the thyme, the remaining ¼ teaspoon each of salt and pepper and the Parmesan cheese. Cover loosely with foil and bake on the upper rack of the oven for 45 minutes. Uncover and bake for about 15 minutes longer, until the gratin is bubbling, the potatoes are very tender and the top is brown. Serve hot.
—Nancy Harmon Jenkins

POTATO AND JERUSALEM ARTICHOKE GRATIN

Rich with the nutty flavor of Jerusalem artichokes, this is no ordinary potato gratin. The French tired of Jerusalem artichokes during World War II, but they are now back in fashion. This recipe is from Paris chef Claude Udron, who makes it with Jerusalem artichokes from his garden in Normandy.

6 to 8 Servings

2 medium baking potatoes, peeled and sliced ⅛ inch thick
6 medium Jerusalem artichokes
1 tablespoon fresh lemon juice
2¼ cups heavy cream
1 teaspoon salt
½ teaspoon freshly ground white pepper
⅛ teaspoon freshly grated nutmeg

1. Preheat the oven to 375°. Rinse the potato slices in a colander to rid them of starch. Pat thoroughly dry, cover and set aside.

2. Peel the Jerusalem artichokes and drop them into a medium bowl of cold water mixed with the lemon juice. Slice them ⅛ inch thick and rinse in a colander under cold running water. Pat thoroughly dry.

3. Place a layer of the potatoes in a 9-by-13-by-2-inch baking dish and cover with a layer of Jerusalem artichokes; repeat with the remaining potatoes and Jerusalem artichokes. In a small bowl, combine the cream, salt, pepper and nutmeg; pour over the vegetables. Bake the gratin in the middle of the oven for about 1 hour, until the potatoes and Jerusalem artichokes are tender and the top is golden brown. Serve hot.
—Patricia Wells

SCALLOPED POTATOES WITH SWEET MARJORAM AND PARMESAN CHEESE
6 to 8 Servings

4 large baking potatoes, peeled and thinly sliced
1 teaspoon salt
¼ teaspoon freshly ground pepper
¼ teaspoon freshly grated nutmeg
2 medium garlic cloves, minced
¼ cup chopped fresh sweet marjoram
¼ cup freshly grated Parmesan cheese
2 cups heavy cream

1. Preheat the oven to 350°. Butter a 6 cup gratin dish or shallow casserole.
2. Layer one-fifth of the potato slices in the gratin dish and season with one-fourth each of the salt, pepper, nutmeg, garlic, marjoram and Parmesan. Repeat the layering 3 times. Top with a final layer of potato slices, overlapping them attractively.

3. Combine the cream with ½ cup of water and pour evenly over the potatoes. Cover the dish snugly with aluminum foil. Bake the potatoes for 1½ hours, then uncover and bake for 30 minutes longer, or until lightly browned. Remove from the oven and let stand for 10 minutes before serving.
—Jean Anderson

CALIFORNIA SCALLOPED POTATOES
Mild green chiles and Monterey Jack cheese give this casserole its easygoing character.
8 Servings

5 tablespoons unsalted butter, softened to room temperature
4 medium baking potatoes
1 cup shredded Monterey Jack cheese
1 cup shredded sharp Cheddar cheese
¼ cup all-purpose flour
1 teaspoon salt
½ teaspoon freshly ground black pepper
2 cans (4 ounces each) whole mild green chiles—rinsed, drained and cut into 1-inch squares
2½ cups milk

1. Preheat the oven to 400°. Grease the bottom and sides of a 15-by-9-by-2-inch baking pan with 2 tablespoons of the butter.
2. Peel the potatoes and slice them about 1/16 inch thick, placing the slices in cold water as you work. Toss the two cheeses together with the flour, salt and black pepper.
3. Drain the potatoes well and pat them dry. Separating the slices, arrange one-third of them in a layer in the buttered pan. Arrange half the chiles over the potatoes, topping them with one-third of the cheese mixture and dotting

them with 1 tablespoon of the remaining butter. Repeat the sequence and end with a final layer of potatoes, cheese and the remaining tablespoon of butter.
4. Pour the milk over the top and bake for 30 minutes. Reduce the heat to 350° and continue baking for an additional 20 to 30 minutes, or until the potatoes are tender when pierced with a fork and the top is golden brown.
5. Remove the casserole from the oven and allow to rest for 20 minutes before serving.
—F&W

COLD SCALLOPED POTATOES
Though traditionally served hot, this dish is absolutely delicious either cold or at room temperature so that it is perfect for picnics. Make it a day ahead.
4 Servings

2 tablespoons unsalted butter
1 large shallot, finely chopped
1 garlic clove, minced
¼ teaspoon crushed red pepper
2 pounds red potatoes, peeled and cut into ⅛-inch slices
1¾ cups heavy cream
1½ cups milk
½ teaspoon salt
¼ teaspoon freshly ground black pepper
½ cup grated Gruyère cheese
2 tablespoons freshly grated Parmesan cheese

1. Preheat the oven to 375°. In a large heavy saucepan, melt the butter over moderately low heat. Add the shallot and cook for 1 minute. Add the garlic and red pepper and cook until the shallot is softened, about 3 minutes longer. Add the sliced potatoes and toss

lightly. Pour in the cream, milk, salt and black pepper and heat to boiling, stirring occasionally; remove from the heat.

2. Transfer the potato mixture to a buttered large, shallow baking dish. Sprinkle the Gruyère and Parmesan cheese over the top. Bake for 15 minutes. Reduce the oven temperature to 350° and bake for 45 minutes longer, until bubbly and golden brown. Let cool, then cover and refrigerate.
—Phillip Stephen Schulz

MASHED POTATOES WITH ROMANO CHEESE
4 Servings

3 large baking potatoes (1½ pounds), peeled and cut into ¾-inch chunks
¾ cup milk
2 tablespoons unsalted butter
½ teaspoon salt
¼ teaspoon freshly ground black pepper
⅛ teaspoon cayenne pepper
¼ cup freshly grated Romano cheese
Pinch of paprika (optional)

1. In a large pot of boiling salted water, cook the potatoes, partially covered, until tender, 15 to 18 minutes. Drain the potatoes and return them to the pot.

2. Reduce the heat to moderate and cook for a few seconds to dry out the potatoes. Add the milk and butter and bring to a boil. Remove from the heat and add the salt, black pepper and cayenne.

3. Mash with a potato masher. Stir in the cheese until just incorporated. Mound the potatoes in a large serving bowl and dust with a pinch of paprika if desired. Serve hot.
—Jim Fobel

BASIL MASHED POTATOES
6 Servings

8 medium baking potatoes, unpeeled
1¾ teaspoons salt
2 cups (packed) basil leaves
½ cup freshly grated hard sheep's milk cheese, Parmesan or aged Gouda cheese (about 2 ounces)
3 tablespoons extra-virgin olive oil
¼ teaspoon freshly ground pepper
1½ to 2 cups milk, heated

1. Place the potatoes in a large heavy saucepan. Add 1 teaspoon of the salt and enough cold water to cover and bring to a boil over moderately high heat. Reduce the heat to moderate and simmer until tender, 35 to 40 minutes.

2. Meanwhile, in a food processor or blender, combine the basil, cheese, 2 tablespoons of the olive oil and ⅛ teaspoon of the pepper and puree, scraping as necessary.

3. Drain the potatoes and peel them while they are still warm. Pass them through a ricer or a food mill. Return the potatoes to the saucepan and add the remaining ⅛ teaspoon pepper. Using a wooden spoon, stir in 1½ cups milk, then add up to ½ cup more milk until the potatoes are as soft as you like them. Season with the remaining ¾ teaspoon salt and stir over low heat until hot, about 5 minutes. Stir in the reserved basil puree. Transfer the potatoes to a dish and drizzle the remaining 1 tablespoon oil on top.
—Mireille Johnston

RUTABAGA AND POTATO PUREE
6 Servings

6 slices of bacon
1 large rutabaga (2 pounds), peeled and cut into ½-inch chunks
2 large baking potatoes (1 pound), peeled and cut into 1-inch chunks
1 teaspoon salt
1 tablespoon unsalted butter
1 teaspoon sugar
½ teaspoon freshly ground pepper

1. In a heavy medium saucepan, fry the bacon over moderately high heat until it begins to brown and the fat is rendered, about 6 minutes.

2. Add the rutabaga and 4 cups of water to the saucepan and bring to a boil over moderately high heat. Cook until very tender, about 40 minutes. Drain the rutabaga and discard the bacon.

2. Twenty minutes before the rutabaga is done, in a medium saucepan, combine the potato chunks, ½ teaspoon of the salt and enough water to cover. Bring to a boil over high heat. Reduce the heat to moderately high and cook until tender, about 15 minutes. Drain well.

3. Place the rutabaga in a food mill fitted with a medium disk and puree into a medium bowl. Puree the potatoes through the food mill into the same bowl. Stir to combine. Alternatively, use a potato masher to thoroughly mash the rutabaga and potatoes.

4. Beat in the butter, sugar, pepper and the remaining ½ teaspoon salt. Serve warm. *(The puree can be transferred to a heatproof serving dish and kept in a low oven for up to 30 minutes.)*
—Jessica B. Harris

SCALLOPED SALSIFY WITH PARMESAN AND PECANS

Salsify, a particular delicacy in the Colonial South, has fallen so far from favor that few people know it. It's also known as oyster plant because its flavor has been likened to that of oysters; but I think it tastes more like artichokes. This dish is wonderful with roast chicken, turkey or pork.

4 Servings

1 pound salsify
2 tablespoons fresh lemon juice
4 tablespoons unsalted butter
3 tablespoons all-purpose flour
¼ teaspoon freshly ground pepper
⅛ teaspoon mace
¾ cup milk
¾ cup half-and-half or light cream
¼ cup freshly grated Parmesan cheese
½ teaspoon salt
⅔ cup fine fresh bread crumbs
⅓ cup finely ground pecans

1. Scrub the salsify well. Trim the stem and root ends. Peel the salsify and cut into 2-inch pieces; drop the pieces into 2 quarts of water mixed with the lemon juice as they are cut. Let soak for 20 minutes.

2. Preheat the oven to 375°. In a medium nonreactive saucepan, bring 1½ quarts of salted water to a boil. Drain the salsify and add it to the pan. Cover and cook over moderate heat until firm-tender, 15 to 20 minutes. Drain well and slice ⅛ inch thick.

3. Meanwhile, in a heavy medium saucepan, melt the butter over moderate heat; set aside 1 tablespoon of the melted butter in a small bowl. Whisk the flour, pepper and mace into the butter in the pan and cook, whisking, without browning, for 2 to 3 minutes. Stir in the milk and half-and-half and cook, whisking constantly, until thickened and smooth, 3 to 4 minutes. Remove from the heat. Blend in 3 tablespoons of the Parmesan cheese and the salt. Cover loosely and keep warm, stirring frequently to prevent a skin from forming on the surface.

4. Add the salsify to the sauce and toss lightly to mix. Transfer to a buttered, shallow 1-quart casserole or gratin dish.

5. In a small bowl, toss the bread crumbs and pecans with the remaining 1 tablespoon Parmesan cheese and reserved melted butter. Sprinkle evenly over the salsify. Bake for about 30 minutes, or until tipped with brown.

—Jean Anderson

BUTTERNUT SQUASH GRATIN

This is an adaptation of the rich, creamy pumpkin gratin served at Le Lafayette in Fort-de-France, Martinique's capital.

6 to 8 Servings

1 butternut squash (about 3 pounds)
4 cups milk
1¼ teaspoons salt
½ teaspoon freshly ground pepper
¼ teaspoon freshly grated nutmeg
1 garlic clove, crushed through a press
2 cups heavy cream

1. Preheat the oven to 375°. Halve the squash lengthwise. Scoop out the seeds and fibrous center and peel off the skin with a sharp knife. Cut into ¼-inch slices.

2. In a large saucepan, combine the squash with the milk and 2 cups of water. Add ½ teaspoon of the salt, ¼ teaspoon of the pepper, the nutmeg and the garlic. Bring to a simmer over moderately high heat. Reduce the heat to low and simmer until just tender, about 4 minutes; drain well.

3. Transfer the squash to a shallow 2-quart buttered gratin or baking dish and season with the remaining ¾ teaspoon salt and ¼ teaspoon pepper. Pour the cream on top and bake for about 20 minutes, until browned. Serve hot, directly from the gratin dish.

—Henri Charvet

ROASTED SHALLOTS

As they roast, these shallots caramelize on the outside and become tender and sweet. They make a fine accompaniment to any grilled meat, fish or fowl.

6 Servings

24 large shallots (about 1½ pounds), peeled but with the root ends left on
¾ teaspoon salt
¼ teaspoon freshly ground pepper
2 tablespoons extra-virgin olive oil
2 bay leaves
3 fresh thyme sprigs or ½ teaspoon dried

1. Preheat the oven to 350°. In a medium cast-iron skillet, toss the shallots with the salt, pepper and olive oil. Add the bay leaves and thyme, cover with foil and roast for 30 minutes.

2. Uncover and roast for 30 minutes longer, until the shallots are tender when pierced.

3. Increase the temperature to 450°. Roast the shallots for about 10 minutes, shaking the pan occasionally, until browned. Serve hot.

—Karen Lee & Alaxandra Branyon

BAKED BUTTERY SWEET POTATO CHIPS

Very thin slices of sweet potato are tossed with lemon juice, butter and a touch of sugar and then oven-baked in a skillet.

8 Servings

4 medium sweet potatoes (about 2 pounds total), peeled
½ teaspoon salt
½ teaspoon freshly ground pepper
2 teaspoons sugar
1 tablespoon fresh lemon juice
4 tablespoons unsalted butter, melted
Thin strips of lemon zest and a sprig of fresh mint, for garnish

1. Preheat the oven to 250°. In a food processor, or with a large sharp knife or mandoline, slice the sweet potatoes very thin, 1/16 to ⅛ inch; rinse in cold water. Drain and dry well with paper towels. Transfer to a large bowl and add the salt, pepper, sugar, lemon juice and butter; toss to coat well.

2. Layer the potatoes in a large ovenproof skillet, mounding them slightly in the center. Pour in any liquid from the bowl. Cover tightly with foil and bake for 50 minutes, or until tender. Serve straight from the skillet, garnished with strips of lemon zest and a sprig of mint.

—Lee Bailey

BAKED CHERRY TOMATOES WITH HERBS

Simplicity itself, these tomatoes are a fine last-minute accompaniment for broiled meats or fish.

4 Servings

3 tablespoons unsalted butter
2 tablespoons olive oil
2 or 3 garlic cloves, finely minced
1 pint cherry tomatoes
⅓ cup chopped fresh herbs, such as tarragon, basil, chervil, chives, parsley, or a combination
Salt and freshly ground pepper

1. Preheat the oven to 400°. Line a baking dish with foil, and place the butter, oil, and garlic in the dish. Place in the oven until the butter has melted. Spread the butter mixture over the bottom of the dish.

2. Add the tomatoes and sprinkle with the herbs. Shake the pan, coating the tomatoes evenly with butter, oil and herbs.

3. Return the pan to the oven and bake for 8 to 10 minutes, shaking the pan every couple of minutes, until the tomatoes are heated through and just beginning to soften. Do not overcook or the skins will split. Sprinkle with salt and pepper to taste and serve hot.

—F&W

SAVORY TOMATO CHARLOTTES

Served on a bed of Bibb lettuce and watercress, these fragrant tomato charlottes make a delicious accompaniment to an omelet at brunch or to cold meat at lunch or dinner.

4 Servings

1 loaf (1 pound) thinly sliced, firm-textured white bread, slightly stale, crusts removed
¾ cup olive oil
2 medium garlic cloves, bruised
2 large red bell peppers
2 medium shallots, minced
2½ pounds tomatoes—peeled, seeded and coarsely chopped, juices reserved
2 tablespoons chopped fresh basil
1½ teaspoons minced fresh thyme or ½ teaspoon dried
½ teaspoon salt

1. Line the bottom of four 10-ounce custard cups (about 4½ by 1¾ inches each) with a round of bread. Cut out 4 larger rounds of bread to fit the tops. Cut out 16 rectangles of bread, each 3½ by 1¾ inches. Fit 4 rectangles against the sides of each custard cup, trimming with scissors for an even fit.

2. In a large heavy skillet, warm ½ cup of the oil and the garlic over moderately low heat for 3 to 5 minutes to flavor the oil; do not let the garlic brown. Discard the garlic and pour the oil into a small bowl.

3. Remove the bread from the custard cups and lightly brush on both sides with the flavored oil. Working in batches, fry the bread in the same skillet over moderately high heat, turning once, until golden brown on both sides, about 1 minute on each side. Drain well on paper towels. Add more of the flavored oil as necessary, heating well before frying more bread.

4. While the bread is still warm and malleable, re-line the bottoms and sides of the molds with the fried bread pieces. Reserve the larger rounds to cover the molds.

5. Roast the peppers directly over a gas flame or under a broiler, turning occasionally, until blackened all over, 10 to 15 minutes. Seal in a paper or plastic bag and let steam for about 10 minutes. Rinse and peel. Remove the stems, seeds and ribs. Puree the peppers in a food processor or blender.

6. In a large heavy skillet, heat 1 tablespoon of the oil. Add the shallots and sauté over moderate heat until soft and translucent, about 2 minutes. Add the tomatoes and their juices and sauté over moderately high heat until the juices evaporate, about 10 minutes. Reduce the heat to low and stir in the red pepper puree, basil, thyme and salt; remove from the heat and let cool slightly.

7. Divide the filling evenly among the lined molds. Cover with the reserved bread circles. Place a square of plastic wrap or wax paper on top of each charlotte and weigh down with pie weights or an 8-ounce can. Refrigerate overnight or for up to 2 days.

8. To serve, remove the weights and let the charlottes return to room temperature. Preheat the broiler. Unmold the charlottes onto a cookie sheet. Lightly brush with the remaining 3 tablespoons oil. Broil about 4 inches from the heat until crisp on the top and warmed through, 3 to 4 minutes. Serve warm or at room temperature.
—Diana Sturgis

BAKED STUFFED TOMATOES WITH EGGPLANT AND ZUCCHINI

This dish is a play on the French ratatouille. American ingenuity turns it inside out; instead of adding chopped tomato to the vegetable mélange, the whole tomato becomes a container that holds the other vegetables. Serve at room temperature or slightly chilled.
4 Servings

4 firm, medium tomatoes
¾ teaspoon salt
3 tablespoons olive oil
1 small onion, finely chopped
2 tablespoons finely chopped green bell pepper
½ cup finely chopped zucchini
⅓ cup finely chopped eggplant
¼ teaspoon sugar
1 large garlic clove, minced
1 tablespoon fresh bread crumbs
2 tablespoons chopped fresh basil
1 teaspoon red wine vinegar
3 tablespoons minced parsley
⅛ teaspoon freshly ground black pepper

1. Preheat the oven to 350°. Slice the top quarter from the stem end of each tomato. Trim, chop and reserve the tops. Scoop out the insides and reserve for another use. Sprinkle the shells with ¼ teaspoon of the salt and invert on paper towels.

2. In a heavy medium skillet, heat 2 tablespoons of the oil over moderate heat. Add the onion, green pepper, zucchini, eggplant, sugar and reserved chopped tomato. Cook until the vegetables are softened, about 5 minutes. Add the garlic and cook for 1 minute longer. Transfer to a medium bowl.

3. Wipe out the skillet and add the remaining 1 tablespoon oil. Add the bread crumbs and sauté over moderate heat until golden, about 2 minutes. Add the toasted bread crumbs to the vegetables. Add the basil, vinegar, 2 tablespoons of the parsley, the black pepper and the remaining ½ teaspoon salt.

4. Using a paper towel, soak up any liquid in the tomato shells and fill them with the vegetable-crumb mixture. Set the stuffed tomatoes in a small baking dish and bake for 25 to 30 minutes, until the tomatoes are hot and softened but still firm enough to hold their shape. Sprinkle the tomatoes with the remaining 1 tablespoon parsley and serve at room temperature.
—Phillip Stephen Schulz

COUSCOUS-STUFFED TOMATOES

Quick-cooking, instant or precooked couscous are now widely available.
6 Servings

1 cup instant couscous
1½ cups boiling water
¾ teaspoon salt
1 small green bell pepper, chopped
1 medium red onion, chopped
⅓ cup dried currants
¼ cup white wine vinegar
2 teaspoons Dijon mustard
1 small garlic clove, minced
½ cup olive oil
½ cup chopped parsley
½ cup chopped fresh mint, plus mint sprigs for garnish
¼ teaspoon freshly ground black pepper
6 large tomatoes (about ½ pound each)

1. In a large bowl, stir the couscous into the boiling water with a fork; add ½ teaspoon of the salt and stir briefly. Cover and let stand until the water is absorbed, about 5 minutes.

2. Add the green pepper, onion and currants to the couscous. Stir to mix.

3. In a small bowl, whisk together the vinegar, mustard and garlic. Gradually whisk in the olive oil. Add the parsley and chopped mint and season with the remaining ¼ teaspoon salt and the black pepper. Pour the dressing over the couscous and toss well. Cover and refrigerate for at least 1 hour. (*The recipe can be prepared to this point up to 1 day ahead.*)

4. Cut the tomatoes in half horizontally and scoop out the core, seeds and most of the pulp. Place the tomato halves upside down on paper towels to drain for 10 minutes. Lightly season the tomato cavities with additional salt and black pepper.

5. Spoon the couscous filling into each tomato shell. Arrange the tomatoes on a large platter and garnish with mint sprigs.

—Melanie Barnard & Brooke Dojny

CALAMATA-STUFFED TOMATOES

These Mediterranean-flavored stuffed tomatoes take only five minutes to bake.

6 Servings

1½ cups coarse fresh bread crumbs
½ cup coarsely chopped Calamata olives
2 tablespoons freshly grated Parmesan cheese
1 garlic clove, minced
½ teaspoon freshly ground pepper
3 tablespoons extra-virgin olive oil
6 medium tomatoes

1. Preheat the oven to 500°. In a bowl, combine the bread crumbs, olives, Parmesan, garlic and pepper. Stir in the olive oil and set aside.

2. Shave a thin slice from the bottom of each tomato to level them. Cut the top third off the tomatoes and create a hollow inside. Reserve the tops for garnish.

3. Spoon ⅓ cup of the olive-bread crumb filling into each tomato and place on a baking sheet. Arrange the tops of the tomatoes, cut-side up, on the baking sheet. Bake on the top rack of the oven until the stuffing is brown and crusty on top, about 5 minutes. Garnish the stuffed tomatoes with the tomato tops.

—Marcia Kiesel

ORANGE-GLAZED TURNIPS

4 Servings

4 medium white turnips (1 to 1½ pounds)
2 cups orange juice
4 tablespoons unsalted butter
1 seedless orange, separated into segments and coarsely chopped
½ teaspoon salt
¼ teaspoon freshly ground pepper
¼ cup minced parsley

1. Halve the turnips lengthwise and then crosswise into ¼-inch half-rounds.

2. In a medium saucepan, combine the turnip slices and orange juice and bring to a boil over moderate heat. Reduce the heat and simmer the turnips, covered, until they are tender, 15 to 20 minutes. Remove the turnips with a slotted spoon, setting them aside on a plate.

3. Cook the orange juice remaining in the pan over moderate heat until reduced to ¼ cup. Stir in the butter, chopped orange, salt, pepper and parsley. Add the turnips and heat gently over low heat until the turnips are hot. Place in a dish and serve immediately.

—F&W

TURNIPS ANNA

Potatoes Anna, that chunky "pancake" made by overlapping paper-thin slices of raw potato in a pie plate or skillet, has always been a favorite of mine. Was it possible to prepare turnips the same way? Maybe, although I knew there were problems—potatoes are starchy enough to stick together, but turnips are not. The solution is simply to scatter an herbed mixture of flour and grated Parmesan cheese between the layers of sliced turnips. This dish is splendid with baked ham, roast pork, chicken, turkey, goose or duck.

In order for the turnips to cook through before the bottom layer overbrowns, they must be sliced as thin as tissue. The food processor does the job zip-quick, but you must use the extra-thin slicing disk.

6 to 8 Servings

5 tablespoons unsalted butter, melted
¼ cup bacon drippings (from about 6 strips of cooked bacon)
6 tablespoons all-purpose flour
6 tablespoons freshly grated Parmesan cheese
¾ teaspoon salt
½ teaspoon ground ginger
¼ teaspoon dry mustard
¼ teaspoon freshly ground pepper
¼ teaspoon finely crumbled thyme
¼ teaspoon finely crumbled rosemary
Pinch of freshly grated nutmeg
1½ pounds medium turnips, peeled and sliced paper thin

1. Preheat the oven to 450°. Brush the bottom and sides of a 9-inch glass pie plate generously with some of the melted butter; set aside.

2. In a small bowl, combine the bacon drippings and the rest of the melted butter. In another bowl, combine the flour, Parmesan cheese, salt, ginger, mustard, pepper, thyme, rosemary and nutmeg; spoon about ¼ cup of this mixture onto a small plate.

3. Pick out the prettiest and most uniform turnip slices. Dip one into the bacon fat and butter, then press one side into the seasoned flour on the plate. Lay the slice, floured-side up, in the center of the pie plate. Repeat dipping and placing the most uniform slices, overlapping them slightly and working out from the center in concentric rings, to cover the bottom and sides of the pie plate.

4. Continue layering the remaining turnip slices, in slightly overlapping concentric rings, brushing each layer

with the bacon fat and butter and scattering a little of the seasoned flour evenly on top, until all the turnips have been used.

5. Brush bacon fat and butter on the shiny side of a 9-inch square of aluminum foil and place, buttered-side down, on top of the turnips. Place a heavy 8- or 9-inch ovenproof skillet on top and press firmly; fill the skillet with pie weights or dried beans.

6. Bake in the middle of the oven for 30 to 35 minutes, until the bottom and sides are richly browned. (Set a pan on the rack below to catch any juices.)

7. Remove from the oven, lift off the skillet and peel off the foil. With a thin spatula, carefully loosen the turnip cake around the edges. Let cool for 5 minutes, then invert onto a heated serving plate. Cut into 6 or 8 wedges and serve at once.
—*Jean Anderson*

TURNIPS AND TURNIP GREENS WITH GINGER AND GARLIC

Southern greens are traditionally cooked with pieces of salted or smoked pork. Here soy sauce supplies the saltiness, and the meat is omitted entirely.
4 Servings

½ pound small white turnips, peeled and cut into ½-inch dice (or substitute whole, unpeeled, trimmed baby turnips)
2 tablespoons cold-pressed sesame oil or safflower oil
3 medium-large garlic cloves, minced
1 tablespoon minced fresh ginger
1½ pounds trimmed turnip greens (about 2½ pounds untrimmed), chopped into 1-inch pieces
2½ teaspoons sugar
2 tablespoons soy sauce or tamari

1. In a vegetable steamer over boiling water, steam the turnips, covered, until crisp-tender, 2 to 3 minutes. Do not overcook. Set aside.

2. In a large flameproof casserole, heat the oil over low heat. Add the garlic and ginger and cook, stirring frequently, until the garlic is fragrant but not browned, about 1 minute.

3. Add ¼ cup of water and increase the heat until the mixture simmers. Add the turnip greens by handfuls, letting them wilt slightly before adding the next batch. Cook, uncovered, stirring frequently, until the greens are just wilted and crisp-tender, about 5 minutes.

4. In a small bowl, combine the sugar and soy sauce. Add to the greens and toss to mix. Stir in the turnips and simmer, uncovered, to combine the flavors, about 3 minutes. With a slotted spoon, transfer the greens and turnips to a serving bowl and cover. Boil the liquid over high heat until reduced to a few tablespoons of syrupy glaze, about 5 minutes. Drizzle the glaze over the turnips and greens.
—*Sarah Belk*

LA MERE POULARD'S ZUCCHINI CREPES

These delightful crêpes from La Mère Poulard, the famous omelet restaurant at Mont-Saint-Michel in Normandy, make an excellent light side dish to accompany chicken or duck. They're also fine on their own with a tossed green salad.
Makes 8 to 10 Small Crêpes

2 medium zucchini (about 1 pound)
2 teaspoons salt
1 egg
2 tablespoons heavy cream
1 tablespoon all-purpose flour
2 garlic cloves, minced
About 2 tablespoons unsalted butter
About 2 tablespoons corn or peanut oil

1. Shred the zucchini in a food processor or with a hand grater. Sprinkle with the salt and drain in a stainless steel strainer for 30 minutes, tossing occasionally. Squeeze out excess water.

2. In a medium bowl, whisk the egg briefly. Add the heavy cream, flour and minced garlic and whisk to blend. Add the zucchini and, with a fork, toss into the egg mixture until just coated.

3. In a large skillet or on a griddle, melt 1 tablespoon of the butter in 1 tablespoon of the oil over moderately high heat. Spoon 1 heaping tablespoon of the zucchini batter into the pan. Spread with the back of a spoon to form an even 3-inch circle. Repeat to fill the pan. Cook until the underside is deep golden brown, 3 to 4 minutes. Using a wide, flat spatula, turn the crêpes. Press down on them and cook until the other side is browned, 2 to 3 minutes longer, adjusting the heat as necessary. Repeat with the remaining batter, adding additional butter and oil as necessary.
—*Patricia Wells*

ROASTED VEGETABLES WITH VINAIGRETTE

Roasting is one of the easiest, quickest and most delicious ways of cooking vegetables. The same method can be applied to asparagus, eggplant, broccoli or tomatoes.
6 Servings

4 medium zucchini (about 1 pound total), cut into 1½- to 2-inch chunks
4 medium yellow summer squash (about 1 pound total), cut into 1½- to 2-inch chunks

3 large red bell peppers (about 1½ pounds total), cut into 1½- to 2-inch pieces
½ cup olive oil
1½ tablespoons fresh lemon juice
1½ tablespoons white wine vinegar
1 teaspoon Dijon mustard
½ teaspoon grated lemon zest
¼ teaspoon salt
¼ teaspoon freshly ground black pepper

1. Preheat the oven to 500°. Place the zucchini, summer squash and red peppers in a large bowl. Drizzle with ¼ cup of the olive oil and toss well. Spread all the vegetables in a single layer on a large rimmed baking sheet or jelly-roll pan. Roast on the upper rack of the oven until crisp-tender and some of the edges begin to char, about 15 minutes.

2. Meanwhile, whisk the remaining ¼ cup olive oil with the lemon juice, vinegar, mustard, lemon zest, salt and black pepper until well blended.

3. Remove the vegetables from the oven and place in a shallow serving dish or on individual plates. Drizzle the vinaigrette over the vegetables while still hot. Serve at room temperature.
—*Melanie Barnard & Brooke Dojny*

MINIATURE VEGETABLES WITH LIME AND BASIL
6 to 8 Servings

16 miniature zucchini, 2 to 3 inches long
16 miniature carrots, 2 to 3 inches long, with 1 inch of green tops
¼ cup clarified butter
1 tablespoon olive oil
1 tablespoon walnut oil
16 cherry tomatoes
¼ cup fresh lime juice
1 cup fresh basil leaves, coarsely chopped
Salt and freshly ground pepper
½ cup freshly grated Parmesan cheese

1. Steam the zucchini and carrots until crisp-tender, 5 to 7 minutes. Arrange on a heated serving platter and cover loosely with foil to keep warm.

2. In a large nonreactive skillet, combine the butter, olive oil and walnut oil over moderately high heat. Add the tomatoes and sauté, shaking the pan, until slightly softened, about 2 minutes. With a slotted spoon, transfer the tomatoes to the platter.

3. Reduce the heat to moderately low. Add the lime juice and basil and cook, stirring, for 1 minute. Season with salt and pepper to taste.

4. To serve, pour the sauce over the vegetables and sprinkle the Parmesan cheese on top.
—*W. Peter Prestcott*

LIMA BEANS WITH BROWN BUTTER
Lima beans are one of the few vegetables that freeze well, so you can use frozen baby limas—I'll never tell.
6 Servings

4 cups shelled fresh baby limas or 2 packages (10 ounces each) frozen
4 tablespoons unsalted butter
1 tablespoon minced fresh mint (optional)
Salt

1. Steam the fresh lima beans for a few minutes until tender (the exact amount of time depends on their age), or cook the frozen according to the package directions.

2. In a medium skillet, melt the butter over moderate heat until browned, being careful not to let it burn, 3 to 5 minutes. Add the lima beans and toss to coat with the butter. Season with the mint and salt.
—*Lee Bailey*

WHITE LIMA BEANS WITH SORREL AND PARSLEY
This recipe is for a delicious side dish. The sorrel provides a tart bite that is just right with the creaminess of the beans. Big white limas, Great Northern and cannellini beans are all wonderful here, but the limas are especially good because they're so plump and moist and have a strong shape. Large limas require some special care in cooking because they can easily be overcooked. Before draining the soaked beans, run your hands through them and remove any skins that have loosened.

This side dish can be turned into a soup by simply adding two more cups of the reserved bean broth along with the beans in Step 4. Often there's a starchy residue at the bottom of the broth; if you're making soup, be sure to include it for body.
4 Servings

2 cups dried white lima beans (14 ounces), picked over
1 teaspoon salt
1 tablespoon unsalted butter
½ of a medium red or yellow onion, finely chopped
1 cup chopped parsley
6 ounces fresh sorrel leaves, large stems removed, leaves finely chopped
¼ cup heavy cream, or more to taste
1 teaspoon freshly ground pepper

1. In a large saucepan, soak the beans in plenty of cold water for at least

6 hours or overnight. Pour off the water, re-cover the beans with fresh water and bring to a boil. Boil the beans vigorously for 5 minutes. Drain the beans in a colander and rinse well to remove any scum.

2. Return the beans to the saucepan. Add 4 cups of water and bring to a boil over high heat. Reduce the heat to moderately low and simmer the beans very gently until tender, about 1 hour. Keep an eye on them to make sure they don't overcook. Toward the end of the cooking time, add 1 teaspoon of the salt. Drain the beans in a colander set over a bowl. Reserve the broth.

3. In a large skillet, melt the butter over moderate heat. Add the onion and ¾ cup of the parsley and cook, stirring occasionally, until the onion is softened, about 3 minutes. Add the sorrel and cook, stirring, until wilted, about 2 minutes. Stir in 1 cup of the reserved bean broth. Bring to a simmer and cook until the onion is soft and most of the broth has evaporated, about 12 minutes.

4. Stir in the cream, pepper and the beans. Cook gently, stirring, until the beans are heated through, about 2 minutes. Stir in the remaining ¼ cup parsley and season with salt to taste. Serve hot.
—Deborah Madison

BABY LIMA BEANS, CHERRY TOMATOES AND PEARS

Lima beans are the one bean I refuse to shell because they are extremely difficult to do, so here I've called for the frozen variety.

6 Servings

1½ cups cherry tomatoes
1 package (10 ounces) frozen baby lima beans, thawed
1 lemon
1 firm-ripe pear, preferably Anjou or Bosc—peeled, quartered and cored
2 tablespoons unsalted butter
¾ teaspoon salt
⅛ teaspoon freshly ground pepper

1. In a small pot of boiling water, blanch the cherry tomatoes for 1 minute; drain. Plunge the tomatoes into a bowl of cold water until the skins split. Peel off and discard the skins. Place the tomatoes in a small bowl.

2. Steam the lima beans until tender, 12 to 15 minutes.

3. Meanwhile, squeeze the juice from the lemon into a small bowl. Using a grater with large holes, shred the pear into the lemon juice, tossing to coat.

4. In a medium nonreactive saucepan, melt the butter over moderate heat. Add the pear and any juices, increase the heat to moderately high and cook, stirring constantly, until the pear is softened and the liquid is slightly reduced, about 1 minute.

5. Drain and discard any tomato juice that may have accumulated in the bowl. Add the tomatoes to the pears and toss over moderate heat for 1 minute. Add the lima beans, salt and pepper and cook, tossing constantly, until heated through, about 1 minute. Serve hot.
—Lee Bailey

BAKED BEANS WITH ONIONS AND ORANGE MARMALADE

The caramelized onions, marmalade and thyme give ordinary canned baked beans a sweetly delicious and unusual flavor. Serve these beans warm alongside sautéed ham or steaks.

4 Servings

2 tablespoons unsalted butter
2 medium onions, thinly sliced
¼ cup orange marmalade
1 teaspoon thyme
1 can (18 ounces) brick-oven baked beans
1 tablespoon Dijon mustard
⅛ teaspoon freshly ground pepper
4 slices of crisply cooked bacon, coarsely chopped

1. In a large heavy skillet, melt the butter over moderately low heat. Add the onions and cook, stirring occasionally, until very soft and lightly browned, 25 to 30 minutes.

2. Scrape the onions to one side of the pan, add the marmalade and cook until melted, 1 to 2 minutes. Stir together the onions and marmalade.

3. Increase the heat to moderate. Add the thyme, beans, mustard, pepper and bacon and cook until heated through, 1 to 2 minutes.
—Diana Sturgis

BAKED BEANS WITH SALSA

Fresh tomato salsa is the perfect lightener for baked beans. It introduces a clean acidity that balances the beans' sweet unctuous character. Cow peas hold their shape after cooking and so lighten the texture of the dish.

6 to 8 Servings

1 pound dried cow peas or black-eyed peas, rinsed and picked over
½ pound slab bacon, cut into 2-inch cubes
6 garlic cloves, unpeeled
1 medium onion, halved
3 imported bay leaves
1 dried hot red pepper
½ cup unsulphured molasses
10 plum tomatoes—peeled, seeded and cut into ½-inch chunks
1 cup thinly sliced scallions
¼ cup minced fresh coriander (cilantro)
2 jalapeño peppers, seeded and minced
1 tablespoon olive oil
1 tablespoon red wine vinegar

1 tablespoon fresh lime juice
1 teaspoon oregano
½ teaspoon cumin
½ teaspoon salt
½ teaspoon freshly ground black pepper
Cayenne pepper

1. In a large saucepan, cover the cow peas with 6 inches of water and soak overnight; or bring to a boil, cover, remove from the heat and let stand for 1 hour. Drain and rinse the beans.

2. Preheat the oven to 325°. In a large heavy flameproof casserole, combine the beans, bacon, garlic, onion, bay leaves, hot pepper and molasses. Add 3 to 4 cups of water to cover and bring to a boil over high heat. Cover, place in the oven and bake for 4 to 5 hours, or until the beans are tender. Check every hour and add additional water if necessary.

3. When the beans are done, if there is more than enough liquid to generously coat them, boil over moderately high heat to evaporate the excess. Remove and discard the bacon, onion, garlic and bay leaves. *(The beans can be made up to 5 days ahead. Cover and refrigerate, but let return to room temperature before proceeding.)*

4. Make the salsa 1 to 3 hours in advance. In a large bowl, combine the tomatoes, scallions, coriander, jalapeños, olive oil, vinegar, lime juice, oregano, cumin, salt, ¼ teaspoon of the black pepper and cayenne to taste.

5. Combine the beans (warm or at room temperature) with the salsa. Add the remaining ¼ teaspoon black pepper and more salt and cayenne to taste. Serve at room temperature.
—Anne Disrude

FRIJOLES NEGROS

Mexican cooks flavor this satisfying side dish with fresh *epazote* leaves, a mild herb. If you can't find it here, use a California bay laurel leaf instead; the flavor won't be the same, but it will be quite delicious.
8 Servings

1 pound dried black beans, rinsed and picked over
8 slices of hickory-smoked bacon, cut into ½-inch squares
1 large onion, chopped
4 garlic cloves, crushed through a press
12 fresh epazote leaves or 1 California bay laurel leaf
1 teaspoon salt, or more to taste

1. Place 6 cups of cold water in a large flameproof casserole. Add the beans and bring to a boil over high heat. Boil for 2 minutes, turn off the heat and let the beans soak, covered, for 1 hour.

2. Bring to a boil over high heat; reduce the heat to a simmer.

3. Meanwhile, in a medium skillet, cook the bacon over moderate heat, stirring frequently, until the fat is rendered and the bacon is crisp and golden brown. Remove the bacon with a slotted spoon and add it to the beans. Discard all but 2 tablespoons of the bacon fat.

4. Add the onion to the drippings and sauté over low heat until softened and translucent, about 5 minutes. Add the garlic and cook for 1 minute longer. Scrape the onion, garlic and bacon fat into the beans. Add the epazote or bay leaf and simmer, stirring occasionally, until the beans are tender, about 2 hours. If the beans take longer to cook, add ½ cup water for each additional ½ hour of cooking time. Season with the salt after the beans are tender.
—Jim Fobel

FRENCH-STYLE WHITE KIDNEY BEANS

The flavor of garlic and olive oil and the crunch of baby string beans will take you straight to the Mediterranean. Serve this warm dish, French style, with broiled or sautéed lamb steaks.
2 to 3 Servings

¼ pound young string beans, preferably haricots verts
¼ cup extra-virgin olive oil
1 large garlic clove, crushed
¼ teaspoon thyme
1 can (19 ounces) white kidney beans, rinsed and drained
Salt and freshly ground pepper

1. Cook the string beans in a medium saucepan of boiling salted water, uncovered, until crunchy but still tender and bright green, 2 to 3 minutes. Drain and set aside.

2. In a heavy medium saucepan, combine the olive oil and garlic. Cook over low heat for 10 minutes, without browning the garlic, to infuse the oil with flavor. Discard the garlic.

3. Add the thyme and kidney beans to the oil in the saucepan. Cook over low heat for 5 minutes. Add the string beans and toss gently. Season with salt and pepper to taste. Serve warm or at room temperature.
—Diana Sturgis

LAFAYETTE'S SPICY BAKED RED BEANS

For a taste of Martinique, serve these beans with steamed white rice.
6 Servings

1 pound small red beans
1 medium onion, peeled and stuck with 4 whole cloves
1 carrot, finely chopped
4 sprigs of fresh thyme or ½ teaspoon dried

2 bay leaves
¼ cup minced flat-leaf parsley
2 tiny dried red chiles
¼ pound lean salt pork, cut into ¼-inch dice
4 to 6 tablespoons unsalted butter
2 shallots, finely chopped
½ cup finely chopped scallion greens
3 garlic cloves, crushed through a press

1. In a large flameproof casserole, cover the beans with at least 2 inches of water and bring to a boil over moderate heat. Remove from the heat, cover and set aside for 1 hour. Drain the beans and rinse under cold running water.

2. Return the beans to the casserole. Add the onion stuck with cloves, the carrot, thyme, bay leaves, 3 tablespoons of the parsley and the chiles. Add fresh cold water to cover by at least 2 inches. Bring to a boil, reduce the heat to moderate and simmer until the beans are just tender, about 1 hour.

3. Meanwhile, in a medium saucepan of boiling water, blanch the diced salt pork for 4 minutes. Drain and dry on paper towels. In a medium skillet, melt the butter over moderately high heat. Add the shallots, scallion greens and salt pork and cook until the salt pork browns, about 5 minutes.

4. Preheat the oven to 400°. Drain the beans, reserving 2 cups of the liquid. Add the salt pork mixture to the beans along with the garlic and remaining 1 tablespoon parsley; mix well. Pour in the reserved cooking liquid and bake, covered, for 40 minutes, or until the beans are tender. Remove the sprigs of thyme, bay leaves and chiles before serving.
—Henri Charvet

SPICY BLACK-EYED PEAS
These black-eyed peas are seasoned three times, producing three quite different effects. First, cumin seeds are sautéed in oil with onion and garlic and left in the sauce to stew. Then cumin and mustard seeds are popped in hot oil and stirred into the peas. At the very end, a kind of salsa of raw seasonings is added.
6 Servings

PEAS
¼ cup vegetable oil
1 teaspoon cumin seeds
1 medium onion, finely chopped
4 garlic cloves, finely chopped
5 plum tomatoes (fresh or drained canned), peeled and chopped
2 packages (10 ounces each) frozen black-eyed peas, partially thawed
¾ teaspoon salt
¼ teaspoon cayenne pepper
1 teaspoon ground coriander

RAW SEASONINGS
1 tablespoon fresh lemon juice
1 fresh green chile, finely chopped
1 tablespoon finely chopped onion
2 tablespoons finely chopped fresh coriander (cilantro)
⅛ teaspoon salt

COOKED SEASONINGS
3 tablespoons vegetable oil
½ teaspoon cumin seeds
½ teaspoon black mustard seeds
3 dried red chiles
1 garlic clove, finely chopped

1. To cook the peas: In a medium saucepan, heat the oil over moderately high heat. When hot, put in the cumin seeds and cook for 10 seconds. Add the chopped onion and garlic and fry, stirring frequently, until the onion turns brown at the edges, about 3 minutes. Stir in the tomatoes and cook for 1 minute.

2. Add the black-eyed peas, salt, cayenne, ground coriander and 2 cups of water. Stir well and bring to a boil. Cover partially, reduce the heat to low and cook until the peas are tender, about 40 minutes.

3. For the raw seasonings: Combine all the ingredients in a small bowl; set aside.

4. For the cooked seasonings: When the peas are tender, in a small skillet, heat the oil over moderately high heat. Add the cumin and mustard seeds. As soon as the mustard seeds begin to pop, after a few seconds, stir in the dried chiles. Add the garlic and when it begins to brown, scrape the oil and seasonings into the black-eyed peas and stir to blend. *(The recipe can be prepared to this point up to 1 day ahead. Let cool, then cover and refrigerate. Reheat before proceeding.)* Mix in the raw seasonings just before serving.
—Madhur Jaffrey

KASHA WITH WILD MUSHROOMS AND TOASTED WALNUTS
12 Servings

1 ounce dried cèpe or porcini mushrooms
4 to 5 cups hot chicken stock or canned broth
3 tablespoons unsalted butter
1½ cups chopped onions
½ pound small fresh shiitake mushrooms, stems removed, or button mushrooms, sliced
2 large eggs, lightly beaten
2 cups medium kasha
Salt
1 cup (about 4 ounces) coarsely chopped walnuts

2 tablespoons plus ¼ teaspoon
    walnut oil
3 tablespoons minced parsley

1. Soak the dried mushrooms in 2 cups of the hot chicken stock until soft, about 30 minutes. Remove the mushrooms and squeeze dry over the bowl of soaking liquid. Rinse the mushrooms quickly under cool water and pat dry. Chop coarsely and set aside. Strain the soaking liquid through a strainer lined with a dampened paper towel. Add enough chicken stock to make a total of 4 cups liquid; set aside in a medium saucepan.

2. In a large heavy skillet, melt 1 tablespoon of the butter. Add the onions and cook over moderately low heat, stirring occasionally, until soft but not brown, about 10 minutes. Push the onions to one side. Add the remaining 2 tablespoons butter and the fresh mushrooms to the skillet. Cook over moderate heat until the mushrooms are tender, about 5 minutes. Add the reserved chopped cèpes and cook with the onions and fresh mushrooms, stirring, for 2 minutes. Transfer to a bowl and set the skillet aside for cooking the kasha.

3. Bring the reserved stock to a boil over moderate heat; set aside over low heat. In a large bowl, toss the eggs with the kasha until the grains are thoroughly moistened and all the egg has been absorbed. Place the egg-coated kasha in the skillet and cook over moderately high heat, stirring constantly to break up any lumps, until the grains are hot, dry and mostly separate, about 4 minutes.

4. Add the hot stock to the kasha. Stir to moisten thoroughly, then cover tightly and steam over low heat for 10 minutes. Stir in the mushroom mixture and season with salt to taste. Cover and cook until all the liquid is absorbed and the grains are fluffy and separate, about 10 minutes. Remove the skillet from the heat and let stand, uncovered, stirring occasionally, until the kasha has cooled to room temperature. *(The recipe can be prepared ahead to this point. Transfer the kasha to a covered container and refrigerate for up to 3 days or freeze for up to 1 month. Let return to room temperature before proceeding.)*

5. A few hours before serving, preheat the oven to 300°. Toss the walnuts with ¼ teaspoon of the walnut oil. Place on a baking sheet and roast for 10 minutes to crisp them and bring out their full flavor.

6. About 30 minutes before serving, preheat the oven to 350°. Stir the walnuts into the kasha and transfer to an attractive shallow baking dish. Cover tightly and bake, stirring 2 to 3 times, for 20 minutes or until heated through. Stir in the remaining 2 tablespoons walnut oil and the parsley. Season with salt to taste. Fluff the kasha with a large fork and serve while still hot.

—*Leslie Newman*

SAFFRONED MILLET PILAF WITH ROASTED RED PEPPERS

Small, round and yellow, millet is an important grain in Indian and African cooking. It has a pleasant, nutty flavor, but unlike other grains, such as rice, the seeds do not end up completely separate, nor do they cook evenly. Toasting millet in butter, as in this recipe, enhances its flavor and keeps it moist.

4 Servings

2 medium red bell peppers
1 tablespoon olive oil
Pinch of saffron threads
3 tablespoons unsalted butter
1 small onion, finely chopped
½ teaspoon salt
1 small bay leaf
⅛ teaspoon turmeric
1 cup millet
2 tablespoons chopped fresh basil
1 tablespoon chopped parsley
2 teaspoons chopped fresh
    marjoram or ¾ teaspoon dried
Freshly ground black pepper
Finely chopped basil or parsley,
    for garnish

1. Roast the red peppers directly over a gas flame or under the broiler, turning frequently, until the skins are well charred all over. Place the peppers in a bowl, cover with plastic wrap and set aside to steam for 15 minutes.

2. Peel the peppers. Remove and discard the cores, seeds and veins. Cut the peppers into small squares and place in a medium bowl. Strain any pepper juices over the peppers.

3. In a small skillet, warm the oil with the saffron threads over low heat for 1½ minutes. Remove from the heat and let stand for 5 minutes, then stir the saffron oil into the peppers.

4. In a heavy medium saucepan, melt 1½ tablespoons of the butter. Add the onion, salt, bay leaf and turmeric and cook until the onion is softened, about 5 minutes.

5. Meanwhile, in a heavy medium skillet, melt the remaining 1½ tablespoons butter over moderate heat. Add the millet and toast, stirring frequently, until the grains begin to color and pop, 3 to 4 minutes.

6. Stir the millet into the onion mixture with 2 cups of water and bring to a boil. Cover and cook over very low heat, stirring occasionally, until the millet is slightly chewy but not crunchy, about 20 minutes. Stir in the basil, parsley, marjoram and the roasted peppers with their juices.

7. Transfer the millet to a serving bowl and fluff with a fork. Season with black pepper to taste and garnish with additional basil. Serve hot.

—*Deborah Madison*

Scallop, Mussel and Asparagus Salad with Orange-Saffron Dressing (p. 43).

Above, Pipérade Pie (p. 91).
Left, Lima Beans with Brown Butter (p. 184).

MIXED GRAIN TRIO

At his restaurant in Chicago, Charlie Trotter accents three harmonious, grain-based mixtures with a pungent pesto-like sauce of tarragon, parsley, spinach and basil.

4 Servings

¼ cup barley
¼ cup quinoa
½ cup instant couscous
2 tablespoons pine nuts
1 tablespoon plus 1 teaspoon olive oil
6 ounces chanterelle mushrooms, stemmed and coarsely chopped
4 ounces shiitake mushrooms, stemmed and finely chopped
1 shallot, minced
Salt and freshly ground black pepper
1 tablespoon finely chopped scallions
1 tablespoon finely chopped parsley
¼ cup fresh corn kernels
½ of a small red bell pepper, cut into ¼-inch dice
3-inch piece of cucumber—peeled, seeded and cut into ¼-inch dice
1 tablespoon chopped fresh mint
1 tablespoon chopped basil
Herb Coulis (recipe follows)

1. In a medium saucepan, combine the barley and 1¼ cups water. Cover and cook over low heat until tender, about 45 minutes. Drain and set aside.

2. In a small saucepan, combine the quinoa and ½ cup water. Cover and cook over low heat until all the water has been absorbed and the quinoa is cooked, about 15 minutes. Set aside.

3. In a small saucepan, bring ⅓ cup water to a boil over high heat. Stir in the couscous, cover and remove from the heat. Let stand for 5 minutes, then uncover and fluff lightly with a fork. Set aside.

4. Preheat the oven to 400°. Place the pine nuts in a pie plate and bake until browned, about 4 minutes. Set aside. Leave the oven on.

5. In a small skillet, heat 1 tablespoon of the olive oil over moderately high heat. Add the chanterelles and cook, stirring occasionally, until tender, about 4 minutes. Using a slotted spoon, transfer the chanterelles to a small bowl. Increase the heat to high and add the remaining 1 teaspoon olive oil to the skillet. When hot, add the shiitakes and the shallot and cook, stirring constantly, until wilted, 2 to 3 minutes. Season with a pinch of salt and pepper.

6. In a medium bowl, combine the shiitakes with the reserved quinoa and the chanterelles. Season with salt and pepper.

7. In another bowl, combine the barley with the scallions, parsley, corn kernels and red pepper. Season with salt and pepper.

8. In a third bowl, combine the couscous, cucumber, mint, basil and toasted pine nuts. Pack each of the three grain mixtures into four ½-cup ramekins, making a total of 12.

9. Arrange the ramekins in a large roasting pan. Pour enough hot water into the pan to reach halfway up the sides of the ramekins. Cover with foil and bake for about 20 minutes, until the grains are heated through.

10. To serve, spread one-fourth of the Herb Coulis on each serving plate and invert one of each molded grain mixture onto each plate. Serve warm.
—*Baba S. Khalsa*

HERB COULIS
Makes ⅔ Cup

½ cup (packed) fresh tarragon leaves
1 cup (packed) spinach leaves, coarsely chopped
½ cup (packed) parsley leaves
¼ cup (packed) basil leaves, coarsely chopped
2 tablespoons chopped fennel fronds
1 tablespoon extra-virgin olive oil
½ teaspoon fresh lemon juice
Salt and freshly ground pepper

In a food processor, combine the tarragon, spinach, parsley, basil and fennel. Puree until a coarse paste forms. With the machine running, slowly pour in the olive oil and ⅓ cup water. Add the lemon juice and season with salt and pepper to taste.
—*Baba S. Khalsa*

SPINACH-RICE TIMBALES WITH LEMON CREAM
6 Servings

2 cups heavy cream
Zest of 2 lemons, removed in large strips with a vegetable peeler
½ cup white rice, cooked (to yield about 1¾ cups)
½ cup wild rice, cooked (to yield about 1½ cups)
2 pounds spinach—stemmed, rinsed and finely chopped (reserve 6 small spinach leaves for garnish)
¾ teaspoon freshly ground pepper
½ teaspoon salt
1½ teaspoons fresh lemon juice
½ cup half-and-half

1. In a heavy medium saucepan, simmer the heavy cream and lemon zest until the cream has reduced by half, about 1 hour.

2. Strain the cream to remove the zest and let cool to room temperature. Cover and refrigerate for several hours or overnight, until thickened to the consistency of sour cream.

3. In a medium bowl, combine the cooked white and wild rice, chopped spinach, pepper, salt, 1 teaspoon of the lemon juice and ½ cup of the thickened lemon cream; mix well.

4. Firmly pack about ¾ cup of the spinach-rice mixture into each of six ½-cup timbale molds or paper cups or plastic glasses. Unmold onto small plates.

5. Make a sauce by combining the remaining lemon cream, the remaining ½ teaspoon lemon juice, the half-and-half and salt to taste.

6. Pour about 2 tablespoons of the sauce around each timbale, garnish with a small spinach leaf and serve at room temperature.
—Anne Disrude

PARSLEY-RICE SOUFFLE

This is not a tall, puffy soufflé, but rather a dish that comes out with a modest domed top. It has the advantage of being stable, unlike other soufflés, so it won't fall before you serve it.
4 to 6 Servings

3 tablespoons unsalted butter
1 small onion, minced
¾ cup rice
3 cups chicken stock or canned broth
¼ teaspoon salt
½ cup minced parsley
2 eggs
1 cup milk

1. Preheat the oven to 350°. Grease a 2-quart soufflé dish with 1 tablespoon of the butter.

2. In a medium saucepan, melt the remaining 2 tablespoons butter over moderate heat. Add the onion and cook until it is softened and translucent, about 5 minutes.

3. Add the rice and stir to coat. Add the stock and salt and bring to a boil. Cover, reduce the heat to low and cook until the rice is very tender, about 20 minutes.

4. Stir the parsley into the rice. Let cool for 5 minutes. Beat in the eggs one at a time, then stir in the milk and mix until well blended.

5. Pour the parslied rice into the buttered soufflé dish and bake for about 50 minutes, or until set and lightly browned on top.
—Lee Bailey

FRIED-RICE PATTIES

To add some flavor to the rice, cook it in chicken stock or canned low-sodium broth. The rice can be cooked one day ahead. Let return to room temperature before proceeding.
6 Servings

4 tablespoons unsalted butter
1 medium onion, finely chopped
4 cups cooked rice
¼ cup plus 1 tablespoon all-purpose flour
2 eggs, lightly beaten
¼ teaspoon salt
¼ teaspoon freshly ground pepper
¼ cup minced parsley
2 tablespoons olive oil

1. In a small saucepan, melt 2 tablespoons of the butter over moderate heat. Add the onion and cook, stirring, until softened, about 5 minutes.

2. In a medium bowl, combine the cooked rice with the flour, tossing with a fork until all the grains are coated. Add the beaten eggs, salt, pepper, parsley and cooked onion and mix well.

3. Using a 1-cup measure, form 12 patties about 3 inches in diameter. Press lightly to flatten. To firm the patties, place them between 2 sheets of wax paper and microwave at High power for 1 minute. Alternatively, place the patties on a tray or baking sheet lined with wax paper and refrigerate for 1 hour.

4. In a large heavy skillet, melt the remaining 2 tablespoons butter in the olive oil over moderately high heat. When the fat is foaming, carefully place half of the patties in the pan. Cook until golden, about 2 minutes per side. Drain on paper towels. Fry the remaining patties. Serve hot.
—Lee Bailey

WILD RICE AND APPLE CIDER PILAF

6 Servings

2 cups unsweetened apple cider
1 cup wild rice
6 tablespoons unsalted butter
½ cup sliced almonds
½ cup raisins
¾ teaspoon salt
½ teaspoon freshly ground pepper
½ teaspoon cinnamon
¼ teaspoon freshly grated nutmeg

1. In a large nonreactive saucepan, bring the cider and 1½ cups of water to a boil over high heat. Reduce the heat to moderately low and stir in the wild rice. Simmer, covered, until the rice is tender but still firm, 30 to 40 minutes. Strain, discarding the cooking liquid.

2. In a large skillet, melt the butter over moderate heat. Add the almonds and raisins and cook, stirring, until the almonds are lightly browned, about 5 minutes.

3. Add the salt, pepper, cinnamon and nutmeg. Stir in the wild rice and cook until warmed through, about 1 minute.
—F&W

WILD MUSHROOM AND SCALLION PILAF
6 to 8 Servings

2 ounces dried porcini mushrooms
1 cup boiling water
2 tablespoons unsalted butter
2 tablespoons peanut oil
2 cups converted rice
3 cups hot chicken stock or canned broth
1 cup thinly sliced scallions (white part only)
½ cup freshly grated Parmesan cheese
½ cup coarsely chopped pecans
Salt and freshly ground pepper

1. Place the dried mushrooms in a medium bowl. Cover with the boiling water and soak until softened, about 30 minutes.
2. In a medium skillet, melt the butter in the oil over moderately high heat. Add the rice and sauté, stirring until translucent, about 5 minutes.
3. Add the chicken stock, stir well and bring to a boil. Reduce the heat to low, cover tightly and simmer until almost all of the liquid has been absorbed, about 20 minutes.
4. Preheat the oven to 375°. Drain the mushrooms in a cheesecloth-lined sieve set over a bowl. Add the strained liquid to the rice. Cover and simmer for 5 minutes.
5. Rinse the mushrooms under warm water. Drain on paper towels and chop coarsely. Add the mushrooms, scallions, Parmesan and pecans to the rice. Toss gently to combine. Season with salt and pepper to taste. *(The recipe can be prepared several hours ahead to this point. Let the rice return to room temperature before proceeding.)*
6. Spoon the pilaf into an 8-cup ovenproof serving dish and bake for 20 minutes, until the top is golden.
—*W. Peter Prestcott*

SPICED POTATO PILAF
6 Servings

6 tablespoons vegetable oil
1 medium onion, chopped
1 medium boiling potato, peeled and cut into ¼-inch dice
½ teaspoon turmeric
½ teaspoon garam masala*
⅛ teaspoon cayenne pepper
2 cups rice
1 teaspoon salt
3 cups hot water
*Available at Indian markets and specialty food stores

1. In a large heavy saucepan, warm the oil over moderate heat. Add the onion and stir until it just begins to brown. Add the potato, turmeric, garam masala and cayenne. Cook, stirring, for 2 minutes.
2. Stir in the rice and salt; fry and stir another 3 minutes, reducing the heat if the rice seems to stick. Pour in the hot water and bring to a boil. Turn the heat to very low, cover tightly and cook for 25 minutes. Do not open the pot during this period.
3. Remove the rice from the heat and let rest, still covered, for another 10 minutes. Serve hot.
—*Madhur Jaffrey*

CHINESE ALMOND RICE
Almonds are widely eaten in China, where they are thought to bring good health. What I have created with them is a variation on a classic Hakka dish, in which rice and other vegetables, meats or shellfish are steamed in closed lotus leaves. I use foil instead and make the almonds the dominant ingredient.
6 Servings

2 cups rice, preferably short-grain, though long-grain can be used
6 dried Chinese black mushrooms* (about ½ ounce)
¼ cup dried shrimp*
½ cup whole blanched almonds
6 ounces Chinese sausages*
1 tablespoon peanut oil
2 garlic cloves, minced
4 water chestnuts, preferably fresh,* peeled and finely diced
2 scallions, cut into ¼-inch slices
1½ tablespoons oyster sauce*
1 tablespoon Oriental sesame oil
1½ teaspoons double-dark ("C") soy sauce*
1½ teaspoons light soy sauce
1½ teaspoons sugar
½ teaspoon salt
Pinch of freshly ground white pepper
*Available at Asian markets

1. Wash the rice 3 times in cold water, rubbing it between your hands. Drain well. Put the rice in a medium saucepan and add 1¾ cups of fresh cold water to cover. Soak for 2 hours.
2. Meanwhile, in a small bowl, cover the dried mushrooms with hot water and soak for 30 minutes. In a separate bowl, cover the dried shrimp with hot water and soak for 30 minutes. Rinse and drain the mushrooms and shrimp. Cut both into ¼-inch dice.
3. Meanwhile, place the almonds in a small saucepan. Add 2 cups of water.

SIDE DISHES: VEGETABLES & GRAINS

Bring to a boil, reduce the heat to low and simmer, partially covered, for 30 minutes. Drain and set aside.

4. Place the Chinese sausages on top of the rice and soaking water. Bring to a boil over high heat and continue to boil, stirring occasionally, uncovered, until most of the liquid has evaporated, about 7 minutes. Reduce the heat to low, cover and cook, stirring 2 or 3 times, until the rice is tender, 7 to 10 minutes. Remove the sausages from the rice. As soon as they are cool enough to handle, cut them into ¼-inch dice.

5. Place a wok over high heat for about 45 seconds. Add the peanut oil and swirl with a metal spatula to coat the sides of the wok. Add the minced garlic and cook, stirring, until light brown, about 10 seconds. Add the diced mushrooms and shrimp and stir-fry for 1 minute. Remove the mushrooms and shrimp from the wok and set aside.

6. Place the cooked rice in a large bowl. Add the sausages, mushroom and shrimp mixture, almonds, water chestnuts, scallions, oyster sauce, sesame oil, both soy sauces, the sugar, salt and white pepper. Mix well.

7. Mound the rice mixture in the center of a 24-inch length of heavy-duty aluminum foil. Form a bundle by folding the foil over to enclose the rice completely. *(The recipe can be made to this point 1 day in advance. Refrigerate and let return to room temperature before steaming.)*

8. Place the bundle of rice, folded-side down, on a steamer rack. Bring the water to a boil over high heat. Cover and steam until the rice is heated through, about 30 minutes. Transfer the bundle to a warmed serving dish and cut a round hole in the top. Scoop out the rice to serve.

—Eileen Yin-Fei Lo

YELLOW RICE

Yellow rice, flavored with coconut milk, turmeric and cumin, is the mainstay of the Indonesian buffet called a *rijsttafel*.
12 Servings

4 cups rice, rinsed and drained
½ teaspoon cumin
⅜ teaspoon turmeric
1 teaspoon salt
1½ cups homemade coconut milk, or ¾ cup canned unsweetened* mixed with ¾ cup water
*Available at Asian markets

1. In a large saucepan or flameproof casserole, combine the rice, cumin, turmeric, salt, coconut milk and 5¼ cups of water. Let soak for 30 minutes.

2. Bring to a boil over moderate heat. Reduce the heat to low and cook, covered, for 15 minutes. Remove the pan from the heat, stir the rice once and let stand, still covered, for 10 minutes.

—Copeland Marks

CURRIED BASMATI RICE

In the Indian kitchen, rice is served mostly as an accompaniment and rarely as a main course.
6 Servings

2 cups basmati rice*
2¼ cups plus 3 tablespoons beef stock or canned broth
2½ tablespoons curry powder (see Note)
¼ cup peanut or other vegetable oil
1 tablespoon minced fresh ginger
1 medium garlic clove, minced
2 large onions, finely chopped
¾ teaspoon salt
½ of a medium green bell pepper, cut into ¼-inch dice
½ of a medium red bell pepper, cut into ¼-inch dice
¼ cup minced fresh coriander (cilantro)
*Available at Indian and specialty food markets and some supermarkets

1. In a large saucepan, combine the rice and enough water to cover. Rub the rice between your hands a few times. Drain well in a colander. Repeat this procedure 2 more times. Return the rice to the saucepan, add 2 cups of the beef stock and set aside to soak for 1 hour.

2. In a small bowl, mix the curry powder with 2 tablespoons of the beef stock to form a thick paste. Set aside.

3. In a small saucepan, heat 1½ tablespoons of the peanut oil over high heat until a wisp of white smoke appears, about 1 minute. Add the ginger and garlic and cook, stirring constantly, until the garlic is golden, about 2 minutes. Stir in the reserved curry paste and the remaining 5 tablespoons beef stock. Reduce the heat to low, cover and simmer for 15 minutes, stirring occasionally. Remove from the heat and set aside, covered.

4. In a medium skillet, heat the remaining 2½ tablespoons peanut oil over moderate heat until a wisp of white smoke appears, about 1 minute. Add the onions and cook, stirring occasionally, until soft, about 7 minutes. Increase the heat to moderately high and stir-fry until lightly browned, about 8 minutes.

5. Add the onions, the reserved curry mixture and the salt to the rice and mix well. Bring to a boil over high heat and cook, stirring often, for 3 minutes. Reduce the heat to moderately low, cover and cook until the rice is tender but firm, 8 to 10 minutes.

6. Stir in the diced green and red bell peppers and the coriander. Transfer to a bowl and serve hot.

NOTE: Curry powder loses its potency quickly. If you have an old jar of it, purchase a new one from a good source.

—Eileen Yin-Fei Lo

FLAGSTAFF GREEN RICE

Madeira gives this unusual version of green rice an extra dimension.

8 to 10 Servings

12 medium scallions
2 tablespoons olive oil
2 serrano chiles or small jalapeño peppers, seeded and finely chopped
2 tablespoons Madeira or sherry
2 cups rice
1 teaspoon salt
1 teaspoon freshly ground black pepper
2 cups canned chicken broth diluted with 1½ cups water
½ cup minced fresh coriander (cilantro)
½ cup minced parsley

1. Thinly slice the scallions; reserve the white and green parts separately. In a large saucepan, heat the olive oil over moderately high heat. Add the scallion whites and the chiles and cook until softened but not browned, about 3 minutes.

2. Stir in the Madeira, rice, salt and black pepper. Add the diluted chicken broth and bring to a boil. Cover and reduce the heat to low and cook until the rice is tender and the liquid has been absorbed, about 20 minutes.

3. Fluff with a fork and stir in the coriander, parsley and reserved sliced scallion greens. Transfer to a serving dish and serve warm.

—Susan Costner

COFFEE-FLAVORED NUTTY PULAU

This excellent *pulau*—an Indian rice dish in which the rice, nuts, fruits and spices are all cooked together from the beginning—is flavored with coffee in addition to the usual spices. While coffee is not native to India, it is now grown and consumed there in abundance and accounts for an increasing amount of India's export earnings. Mysteriously, the good Indian coffees actually have overtones of cinnamon, cloves and cumin—three spices used regularly in Indian cooking.

6 Servings

¼ cup raisins
¼ cup walnuts
¼ cup pecans
3 tablespoons vegetable oil
3-inch cinnamon stick, halved
6 whole cloves
6 cardamom pods
¾ cup mixed dried fruit
1½ cups basmati rice* or converted long-grain white rice
1½ cups strongly brewed coffee
*Available at Indian and specialty food markets and some supermarkets

1. Mix together the raisins, walnuts and pecans. In a heavy medium saucepan, heat the oil over moderately high heat. Add the cinnamon, cloves, cardamom, ½ cup of the nuts and raisins and ½ cup of the dried fruits. Sauté until the nuts are lightly browned, about 15 seconds. Stir in the rice and add the coffee and 1½ cups of hot water.

2. Bring to a boil, reduce the heat to low, cover and simmer until the liquid is absorbed and the rice is tender, 25 to 30 minutes. Garnish with the remaining ¼ cup nut mixture and dried fruits.

—Gaylord India

PERSIAN RICE

This rice is an adaptation of an ancient dish once made for kings. A blend of spice, rice, fruit and nuts, it has complex flavors, interesting textures and a warm yellow glow.

6 to 8 Servings

1 tablespoon unsalted butter
2 tablespoons light olive oil
1 large onion, chopped
4 garlic cloves, minced
1 dried red chile, seeded and finely chopped
½ teaspoon cumin seeds, crushed
2 cardamom pods, seeds removed and crushed
½ teaspoon ground ginger
½ teaspoon (firmly packed) saffron threads
4½ cups warm chicken stock or canned low-sodium broth
2 cups rice, rinsed well
1 cinnamon stick
1 bay leaf
¼ cup dried currants
10 small, whole dried apricots, quartered
¾ teaspoon salt
½ cup sliced unblanched almonds
2 tablespoons pine nuts
¼ cup shelled unsalted pistachio nuts
Freshly ground black pepper
3 tablespoons fresh coriander (cilantro) leaves, torn

1. In a large flameproof casserole, melt the butter in the olive oil over low heat. Add the onion and garlic and cook, stirring, until softened but not browned, about 5 minutes.

2. Increase the heat to moderate and add the chile, cumin seeds, cardamom and ginger. Cook, stirring, until fragrant, about 2 minutes. Add the saffron and chicken stock, mix well and remove from the heat. Cover and let the saffron permeate the stock for 5 minutes.

SIDE DISHES: VEGETABLES & GRAINS

3. Preheat the oven to 400°. Bring the stock to a simmer over moderate heat. Stir in the rice, cinnamon stick, bay leaf, currants, apricots and the salt. Return to a simmer, then reduce the heat to low. Cover and cook for 17 minutes. If the rice is still too wet, cook longer.

4. Meanwhile, place the almonds and pine nuts on a baking sheet and toast in the oven for about 4 minutes, until golden. Add the pistachios and toast for 1 more minute. Transfer to a plate to cool.

5. Uncover the rice and stir once with a fork to fluff up. Season with additional salt and black pepper to taste. Add the nuts and toss. Sprinkle the coriander on top and serve.
—Marcia Kiesel

FRIED GINGER RICE
This is a simple dish of fried ginger, scallions, butter and rice. When sticks of fresh ginger are deep-fried, they turn crunchy with only a hint of spiciness.
4 to 6 Servings

3 ounces fresh ginger, peeled
½ cup vegetable or corn oil
1½ cups long-grain rice, preferably basmati,* rinsed
2 large scallions, thinly sliced
1 tablespoon unsalted butter
Salt and freshly ground pepper
*Available at Indian and specialty food markets and at some supermarkets

1. Thinly slice the ginger crosswise. Stack the slices and cut crosswise into small sticks.

2. In a medium saucepan, heat the oil over high heat until hot, about 4 minutes. Test by dropping in a piece of ginger; it should sizzle immediately. Add all the ginger and stir briefly to separate. Cook until light brown and starting to crisp, about 5 minutes. Using a slotted spoon, transfer the fried ginger to paper towels.

3. In another medium saucepan, combine the rice with 1¾ cups of water and bring to a boil over high heat. Reduce the heat to low, cover and cook for 16 minutes. Remove from the heat and keep covered for 5 minutes.

4. Fluff the rice and stir in the scallions, butter, salt and pepper to taste, and the reserved fried ginger and serve at once.
—Marcia Kiesel

ORANGE PISTACHIO COUSCOUS
4 Servings

¼ cups chicken stock or canned low-sodium broth
Finely grated zest of 1 medium orange
2 tablespoons unsalted butter
½ teaspoon salt
1½ cups instant couscous
2¼ teaspoons orange flower water*
½ cup shelled roasted pistachios,* coarsely chopped (about 2½ ounces)
*Available at Middle Eastern markets

1. In a medium saucepan, combine the stock, orange zest, butter and salt. Bring to a boil over high heat. Add the couscous; stir well. Cover, remove from the heat and let stand for 5 minutes.

2. Sprinkle the orange flower water over the couscous and add the pistachios. Fluff lightly with a fork.
—Linda Burum

7

SIDE DISHES
SALADS

SOUTHERN GREENS SALAD WITH PEPPERED PECANS

Barton Levenson devised this salad for Oakland's Gulf Coast Oyster Bar. The original recipe called for collard greens, kale, mustard and red chard, but other hardy greens such as spinach, escarole, beet or turnip greens, or Belgian or curly endive in combination with watercress can be substituted with great success. Toss in the pecans at the end to keep them crisp.

2 Servings

¼ pound bacon
1½ tablespoons olive oil, preferably extra-virgin
¼ cup coarsely chopped red onion
1 small garlic clove, minced
¼ teaspoon coarsely ground pepper
1 tablespoon balsamic vinegar
½ teaspoon grated lemon zest
½ teaspoon fresh lemon juice
2 teaspoons minced fresh parsley
½ teaspoon minced fresh thyme or ⅛ teaspoon dried
¼ teaspoon coarse (kosher) salt
1½ quarts mixed hearty salad greens (see headnote, above)
¼ cup coarsely chopped Peppered Pecans (recipe follows)

1. In a medium skillet, cook the bacon over moderate heat, turning, until browned and crisp, 7 to 10 minutes. Transfer to paper towels and pat dry; chop coarsely. Reserve 1½ tablespoons of the rendered bacon fat.

2. Heat a large stainless steel bowl or wok over moderate heat, grasping the rim with a pot mitt. Add the reserved bacon fat and chopped bacon, the olive oil, red onion, garlic and pepper. Sauté until the onion is slightly softened, about 1½ minutes.

3. Add the vinegar, lemon zest, lemon juice, 1 teaspoon of the parsley, the thyme and salt. Cook until the onion is almost translucent, about 30 seconds.

4. Add the greens all at once and toss quickly over heat until the greens are coated with dressing and barely warmed but not wilted, about 30 seconds longer.

5. Remove the bowl from the heat and season with additional salt and pepper to taste. Toss in the chopped Peppered Pecans, then quickly arrange the salad on two heated plates. Sprinkle the reserved 1 teaspoon parsley on top and serve.

—*Barbara Tropp*

PEPPERED PECANS

These are also delicious as an appetizer, with a glass of rich red wine.

Makes About 1 Cup

¼ cup sugar
1 tablespoon coarse (kosher) salt
1½ to 2 tablespoons coarsely ground black pepper
1 cup (4 ounces) pecan halves

1. In a small bowl, blend the sugar, salt and 1½ to 2 tablespoons pepper, to taste. Set the seasoning mixture aside.

2. Heat a large, very heavy skillet, preferably cast iron, over high heat until it is hot enough to vaporize a bead of water on contact. Add the pecans and cook, tossing constantly, for 1 minute to bring the nut oil to the surface.

3. Sprinkle the nuts with half the seasoning mixture. Shake the pan vigorously until the sugar melts, about 1 minute. Add the remaining seasoning mixture and continue shaking the pan until the sugar again melts and coats the pecans. Immediately turn the nuts onto a baking sheet, spreading them apart. Let cool; then seal in a plastic bag.

—*Barbara Tropp*

SALAD OF SPINACH CROWNS

Spinach crowns are the root ends of the plant with about 2 inches of the stem attached. The Chinese call them "parrot beaks," because of their colorful red tips. Save the spinach leaves for another recipe.

The crowns tend to toughen as they sit, so prepare this salad no more than an hour before you plan to serve.

2 Servings

16 to 20 spinach crowns (roots and 2 inches of the stem from about 2 pounds of fresh spinach)
¼ cup soy sauce
1 teaspoon Oriental sesame oil
1 teaspoon rice vinegar

1. Trim off any woody parts from the roots and remove any stems that look bruised or spoiled. Rinse well in at least two changes of clean, cold water (sand tends to collect between the stems).

2. In a medium saucepan of boiling salted water, blanch the crowns until tender but still bright green, about 4 minutes. Drain and rinse under cold running water until cool.

3. In a medium bowl, combine the soy sauce, oil and vinegar with 2 tablespoons of water. Add the crowns and toss to coat with the dressing. Divide between two plates and pour any dressing that remains in the bowl over the salads.

—*Anne Disrude*

MIXED GREEN SALAD WITH SOURED CREAM AND BLUE CHEESE DRESSING

The combination of creamy blue cheese and fresh, crisp salad greens is a welcome interlude between the main course and the dessert.

12 Servings

⅓ cup heavy cream
1½ tablespoons tarragon wine vinegar
6 ounces Saga blue cheese, rind removed, at room temperature
¼ cup milk
¼ cup mayonnaise
1 tablespoon grainy mustard
½ teaspoon tarragon
¼ teaspoon salt
Freshly ground pepper
1 medium head of romaine lettuce
1 medium head of curly leaf lettuce
2 bunches of arugula or watercress
2 pints of cherry tomatoes (yellow and red if available), halved

1. In a mixing bowl, combine the cream with the vinegar. Stir until the cream thickens, about 1 minute. Break up the blue cheese and blend it into the cream with a fork, leaving some lumps. Whisk in the milk, mayonnaise, mustard, tarragon, salt, and pepper to taste and blend well. *(The dressing can be made up to 2 days in advance. If it gets too thick, it can be easily thinned again with a small amount of milk.)*

2. Thoroughly rinse and drain the various greens. Tear into bite-size pieces. Place in a salad bowl and toss with the tomatoes. *(The vegetables can be prepared up to 1 day ahead. Store the washed greens in airtight plastic bags in the refrigerator. Store the cherry tomatoes uncut.)*

3. Divide the salad evenly among 12 salad plates. Drizzle 2 tablespoons of dressing over each portion and serve.
—*Bob Chambers*

SALADE BRESSANE

This salad has a mild dressing, which goes particularly well with the delicacy of Boston lettuce. Because it contains so little vinegar, you can enjoy your wine down to the last drop. Be sure to toss the lettuce and dressing just at the last moment, as Boston lettuce tends to wilt quickly.

6 to 8 Servings

2 heads of Boston lettuce (as white as possible), about 1 pound
1 tablespoon red wine vinegar
½ teaspoon salt
½ teaspoon freshly ground pepper
⅓ cup heavy cream

1. Remove any bruised leaves from the heads of lettuce and trim off the dark green tops of both heads. Cut out the cores of the lettuces and separate the leaves. If the central ribs of the outside leaves are tough or fibrous, remove them. Split the inner leaves in half through the ribs. Wash gently; dry well.

2. In a large salad bowl, combine the vinegar, salt and pepper. Stir in the cream until blended. The dressing will thicken.

3. At serving time, add the lettuce to the dressing and toss to coat.
—*Jacques Pépin*

FIELD SALAD WITH SHALLOTS AND CHIVES

This simple variation on a familiar theme comes from Paris chef Marcel Baudis. The combination of shallots and chives is often used in Baudis's native town of Montauban in the southwest of France. Marinating the shallots in oil softens their often harsh flavor.

4 to 6 Servings

¼ cup minced shallots
¼ cup extra-virgin olive oil
2 teaspoons red wine vinegar
2 teaspoons sherry vinegar
Salt and freshly ground pepper
6 cups bite-size pieces of mixed greens, such as chicory (curly endive), radicchio, watercress, mâche, dandelion greens and arugula
⅓ cup chopped chives

1. In a small bowl, stir 2 tablespoons of the shallots into the oil. Set aside at room temperature for at least 1 hour and up to 24.

2. In another small bowl, combine the red wine and sherry vinegars. Add the shallot-oil mixture and stir to blend thoroughly. Season with salt and pepper to taste.

3. In a salad bowl, toss the salad greens with the remaining 2 tablespoons shallots and the chives. Pour the vinaigrette over the greens and gently toss until thoroughly and evenly coated. Adjust the seasonings, if necessary, and serve immediately.
—*Patricia Wells*

GREEN GULCH LETTUCES WITH SONOMA GOAT CHEESE

Colorful greens, pecans, oranges and goat cheese make this a superb salad.
6 Servings

1 bunch of watercress
1 small head of red leaf lettuce
1 small head of Boston or butter lettuce
1 small head of romaine lettuce
¾ cup pecan halves (about 3 ounces)
3 large navel oranges
1 tablespoon balsamic vinegar
½ cup light olive oil
1 teaspoon minced shallot
Salt and freshly ground pepper
5 ounces of Sonoma goat cheese or other mild goat cheese

1. Preheat the oven to 350°. Trim, rinse and dry the watercress, red leaf, Boston and romaine lettuces. Tear into bite-size pieces.
2. Arrange the pecans on a baking sheet and bake until they begin to brown, 5 to 8 minutes.
3. Remove the orange zest from one of the oranges and mince enough to measure ½ teaspoon. Over a bowl to catch the juices, peel the remaining oranges. Remove the outer membranes from the orange sections and cut the individual segments apart. Reserve 3 tablespoons of the juice.
4. In a small bowl, combine the vinegar, oil, shallot, reserved orange juice, orange zest and salt and pepper to taste.
5. In a large bowl, toss together the greens, orange segments, pecans and vinaigrette. Arrange the salad on serving plates and crumble some of the goat cheese over the top of each serving.
—Annie Somerville

SPINACH SALAD WITH FETA, MINT AND OLIVES

Crumbled feta and Greek olives add wonderful flavor to this dish. For best results, toss the greens with the vinaigrette and hot oil just before serving.
4 Servings

3 large garlic cloves, minced
½ cup plus 2 tablespoons extra-virgin olive oil
4 slices of French bread, cut ¼-inch thick
1 tablespoon sherry vinegar
⅛ teaspoon salt
⅛ teaspoon freshly ground pepper
5 to 6 cups spinach, or 3 cups spinach and 2 cups curly endive or escarole, stemmed and washed
½ cup thinly sliced red onion
2 ounces crumbled feta cheese or freshly grated Parmesan (about ½ cup)
4 Calamata or other brine-cured black olives
½ teaspoon minced fresh mint

1. Preheat the oven to 350°. In a small bowl, combine the garlic and ½ cup of the olive oil.
2. Brush both sides of each bread slice with some of the garlic oil; reserve the remainder. Arrange the bread on a baking sheet and bake until lightly browned, 5 to 7 minutes.
3. In a medium bowl, combine the vinegar, remaining garlic-oil mixture, the salt and pepper.
4. In a salad bowl, toss together the spinach, onion, feta cheese and olives. Add the vinaigrette and toss again.
5. In a small skillet, heat the remaining 2 tablespoons oil. Pour the hot oil over the salad and toss. Add the mint and additional salt and pepper to taste. Serve with the toasted croutons.
—Annie Somerville

QUATORZE'S CHICORY AND BACON SALAD

This recipe for a classic French salad called *frisée aux lardons* makes two very generous portions.
2 Servings

2 shallots, minced
2 tablespoons Dijon mustard
2½ tablespoons red wine vinegar
⅛ teaspoon freshly ground pepper
½ cup extra-virgin olive oil
⅔ pound slab bacon, cut into cubes
½ loaf of French bread, crust removed, bread cut into ½-inch cubes
1 head of chicory (curly endive), torn into large pieces (about 6 cups)
Salt (optional)

1. In a small bowl, whisk together the shallots, mustard, vinegar and pepper. Gradually whisk in the olive oil in a thin stream.
2. In a medium skillet, cook the bacon over moderate heat until browned and crisp, about 10 minutes. Remove with a slotted spoon and set aside.

3. Pour off and discard all but ¼ cup of the fat from the pan. Add the bread cubes to the hot bacon fat and sauté over moderately high heat, tossing until lightly browned, about 5 minutes. Remove and set aside.

4. Return the bacon to the skillet and toss until heated through. Remove from the heat and stir in the dressing.

5. Place the chicory in a large bowl. Add the hot dressing, bacon and bread cubes to the bowl and toss well. Salt lightly if desired. Serve immediately.
—*Quatorze, New York City*

CHARDENOUX'S SALAD OF BLUE CHEESE, NUTS AND BELGIAN ENDIVE

This is one of my favorite wintertime salads: crunchy Belgian endive, freshly cracked walnuts and piquant blue cheese, all tossed in a lemony dressing of fragrant hazelnut oil.
4 to 6 Servings

2 tablespoons fresh lemon juice
¼ teaspoon salt
¼ cup hazelnut oil or extra-virgin olive oil
2 pounds Belgian endives (about 8 medium heads)
1 cup walnut pieces, freshly cracked
6 ounces imported French Roquefort or Fourme d'Ambert cheese, crumbled

1. In a small bowl, combine the lemon juice and salt; stir to dissolve. Gradually whisk in the oil in a thin stream until the dressing is blended.

2. Rinse the endives, pat dry and separate the leaves. Place the whole leaves in a large salad bowl. Sprinkle the walnuts and crumbled cheese over the endive. Whisk the dressing, drizzle over the salad and toss to coat.
—*Patricia Wells*

PORT AND STILTON SALAD
8 Servings

1½ tablespoons unsalted butter
2 teaspoons safflower oil
1 garlic clove, unpeeled but bruised
¾ cup walnuts
1 large head of red leaf lettuce, separated into leaves and tough ends removed
2 small heads of Bibb lettuce, separated into leaves
Port Dressing (p. 258)
2 Belgian endives, thinly julienned
½ pound Stilton cheese, coarsely chopped

1. In a medium skillet, melt the butter in the oil over moderate heat. When the foam subsides, add the garlic and cook for 2 minutes. Discard the garlic.

2. Add the walnuts and cook, shaking the pan frequently, until toasted to a dark golden color. Drain the nuts on paper towels.

3. In one bowl, toss the lettuce leaves with ⅔ cup of the Port Dressing. In another bowl, gently toss the julienned endives with the remaining dressing.

4. On each of 8 chilled salad plates, place 1 large leaf of red lettuce. Arrange 3 Bibb lettuce leaves in a cloverleaf shape on top. Neatly bunch the endive julienne in a fringe below the Bibb.

5. Place a heaping spoonful of the Stilton in the middle of each salad and top each with 1½ tablespoons of the toasted walnuts.
—*W. Peter Prestcott*

MIXED LEAF SALAD WITH ROASTED WALNUTS AND ROQUEFORT

Slow roasting is the key to bringing out the true character of the walnuts in this salad. Any mix of greens can be used as the base, so don't feel confined by those listed below.
2 Servings

½ cup coarsely chopped walnuts
1 tablespoon red wine vinegar
1 tablespoon Dijon mustard
⅛ teaspoon salt
3 tablespoons extra-virgin olive oil
2 teaspoons minced parsley
Freshly ground pepper
4 cups mixed salad greens, such as arugula, radicchio, chicory (curly endive) and red leaf lettuce
⅓ cup crumbled Roquefort cheese

1. Preheat the oven to 275°. Spread the walnuts on a baking sheet and roast in the oven for 45 minutes, until fragrant and golden. Let cool.

2. In a small bowl, combine the vinegar, mustard and salt. Gradually whisk in the olive oil. Stir in the parsley and season with pepper to taste.

3. In a large bowl, toss the salad greens with the roasted walnuts, Roquefort and the vinaigrette. Divide the salad between 2 plates and serve immediately.
—*Bob Chambers*

RUSSIAN SALAD WITH FRESH PEACHES
4 Servings

1 cup diced carrots
¾ cup cut green beans, ½-inch lengths
⅓ cup fresh peas
2 peaches
2 teaspoons fresh lemon juice
½ cup mayonnaise
1½ tablespoons sour cream
2 teaspoons Dijon mustard
2 tablespoons thinly sliced scallions
1 tablespoon chopped parsley
½ teaspoon salt
¼ teaspoon freshly ground black pepper

1. Cook the carrots, green beans and peas separately in a generous amount of boiling, salted water until just tender. Refresh under running cold water to stop the cooking; drain.
2. Plunge the peaches into boiling water for a few seconds, then cool them quickly in cold water. Peel off the skins, slice, then cut the peaches into ½-inch cubes. Toss with the lemon juice to prevent discoloration. Combine the vegetables and peaches.
3. In a small bowl, combine the mayonnaise, sour cream, mustard, scallions, parsley, salt and pepper and blend well. Pour the dressing over the salad, toss and chill for up to 2 hours before serving. Taste and adjust the seasoning, adding salt and pepper if needed.
—F&W

PEACH AND WATERCRESS SALAD
This delicious and innovative peach salad accompanied with a loaf of good bread makes an ideal light lunch for an unbearably hot day.
4 Servings

1 teaspoon minced fresh ginger
1 tablespoon fresh lemon juice
1½ tablespoons red wine vinegar
¼ teaspoon salt
¼ teaspoon freshly ground pepper
¼ cup light olive oil
4 thick slices of bacon (about ¼ pound)
4 cups Boston lettuce, torn into bite-size pieces
1 bunch of watercress, large stems removed
2 large peaches, peeled and sliced
2 tablespoons minced chives

1. In a small bowl, combine the ginger, lemon juice, vinegar, salt and pepper. Whisk in the olive oil and set aside.
2. In a medium skillet, fry the bacon over moderate heat until crisp, about 5 minutes. Drain on paper towels and crumble into small pieces.
3. Arrange the lettuce and watercress on each of 4 salad plates. Divide the peach slices and bacon among the plates. Pour the vinaigrette over the salads and garnish with the minced chives.
—Marcia Kiesel

TENDER GREENS WITH PROSCIUTTO, CROUTONS AND WALNUT OIL VINAIGRETTE
When making salads to be served with wine, avoid using bitter greens. Use a mild vinegar and dilute it with other liquids. Also, supply some textural bumps for the wine to grab on to, such as the croutons in the following recipe. Here's a perfect salad for wine.
2 Servings

½ cup walnut oil
3 medium garlic cloves, smashed and peeled
1 cup bread cubes (½-inch)
Salt and freshly ground pepper
1 tablespoon red wine vinegar with 5 to 6 percent acidity
1½ teaspoons rich chicken stock or canned low-sodium broth
1½ teaspoons dry white wine
4 paper-thin slices of prosciutto di Parma (about 2 ounces)
6 cups (loosely packed) mild greens, such as Bibb, Boston or red-leaf lettuce

1. In a heavy medium skillet, heat ¼ cup of the walnut oil over moderately low heat. Add the garlic and cook, stirring, until golden, about 10 minutes. Remove and discard the garlic. Increase the heat to moderately high. Add the bread cubes and sauté, turning, until golden brown and crisp on all sides, about 5 minutes. Season with a pinch each of salt and pepper and drain on paper towels; set aside.
2. In a small bowl, whisk the vinegar, chicken stock and wine. Add ⅛ teaspoon each of salt and pepper, then gradually whisk in the remaining ¼ cup walnut oil.

3. Cut each slice of prosciutto into 4 lengthwise strips. In a large bowl, lightly coat the greens with the dressing. Add the croutons and toss well. Place the salad on 2 large plates and arrange the prosciutto strips decoratively on top. Serve immediately.
—David Rosengarten

SALAD OF MIXED GREENS AND HERBS WITH ORANGES
8 Servings

2 navel oranges
1 tablespoon minced shallots
1 tablespoon honey mustard
2 teaspoons cider vinegar
1 teaspoon balsamic vinegar
¼ cup plus 1 tablespoon olive oil
Salt and freshly ground pepper
12 cups mixed baby lettuce leaves
1 cup (loosely packed) flat-leaf parsley leaves
1 cup (loosely packed) mint leaves
1 bunch of chives, cut into 2-inch lengths

1. Using a sharp knife, peel the oranges, making sure to remove all the bitter white pith. Working over a bowl, cut the oranges in between the membranes to release the sections.
2. In a small bowl, combine the shallots, mustard and vinegars. Gradually whisk in the olive oil. Season the dressing with salt and pepper to taste.
3. In a large salad bowl, combine the lettuce, parsley, mint and chives. Pour 3½ tablespoons of the dressing over the salad and toss well. Add the orange segments and toss lightly. Add more dressing if desired. Serve immediately.
—Elizabeth Woodson

GREEN SALAD WITH LEMON AND FENNEL
To get most of the work for this salad out of the way ahead of time, rinse, dry and trim the greens up to six hours before serving. Wrap in a dampened kitchen towel and refrigerate. They will be nicely crisped by the time the salad is assembled.
8 Servings

1 lemon
3 fennel bulbs
¼ cup safflower oil
¼ cup light olive oil
1 teaspoon crushed fennel seeds
½ teaspoon salt
¼ teaspoon coarsely cracked pepper
1 bunch of watercress, large stems removed
3 heads of Bibb lettuce, separated into leaves
1 head of Boston lettuce, torn into pieces

1. Remove the lemon zest in long strips with a vegetable peeler. Cut lengthwise into very thin strips. Wrap in plastic wrap to keep moist.
2. Remove and discard the green stalks of the fennel. Cut out the cores and remove any tough outer sections. Cut the bulbs lengthwise into thin strips. In a medium bowl, toss the fennel strips with the juice of half of the lemon. Cover with ice water and let stand until crisp, about 30 minutes. Drain and spin or pat dry.
3. In a small bowl, whisk together the safflower and olive oils, fennel seeds, salt, pepper and the strained juice of the remaining ½ lemon.
4. To assemble the salad, toss the fennel with 2 tablespoons of the dressing. In a separate bowl, combine the watercress and Bibb and Boston lettuce. Toss the greens with the remaining dressing.
5. Divide the greens among 8 large chilled plates. Place a small mound of fennel in the center. Top the fennel with lemon strips and an additional sprinkling of cracked black pepper.
—Anne Disrude

GREEN BEANS MIMOSA
6 to 8 Servings

2 pounds green beans, trimmed
5 hard-cooked eggs
1 tablespoon white wine vinegar
3 tablespoons vegetable oil
½ teaspoon salt
⅛ teaspoon freshly ground black pepper
Pinch of freshly grated nutmeg
3 tablespoons Dijon mustard
6 tablespoons heavy cream
1 tablespoon tarragon vinegar
⅛ teaspoon freshly ground white pepper
½ pound bacon, crisp-cooked and crumbled

1. Slice the green beans French-style, cutting them in half lengthwise between the seams. In a vegetable steamer, steam the beans until almost tender, 4 to 6 minutes. Refresh the beans under cold running water and drain. Set aside.
2. Cut the eggs in half lengthwise. Remove the yolks and rub them through a sieve over a bowl; reserve. Using the fine side of a grater, grate the egg whites into a separate bowl; reserve.
3. In a small bowl, whisk the white wine vinegar with the vegetable oil, ¼ teaspoon of the salt, the black pepper and nutmeg. Set the vinaigrette aside.
4. In another small bowl, whisk the mustard and cream together. Stir in the tarragon vinegar, the remaining ¼ teaspoon salt and the white pepper.

5. In a large bowl, mix the beans with the grated egg white. Add the vinaigrette and toss well. Arrange the salad in the center of a large serving platter and scatter the bacon around the edge. Place 8 small mounds of the egg yolks around the edge of the dish. Pour the mustard-cream sauce over the salad, or serve it separately. Serve the salad slightly chilled or at room temperature.
—F&W

GREEN BEANS IN AQUAVIT

This salad tastes best when allowed to chill overnight.
8 Servings

2 tablespoons caraway seeds
¼ cup olive oil
½ teaspoon anchovy paste
2 tablespoons sherry vinegar or white wine vinegar
1½ pounds green beans
1 pound small red potatoes
5 tablespoons aquavit
8 scallions, thinly sliced (including some of the green tops)
Freshly ground pepper

1. Grind the caraway seeds to a fine powder with a mortar and pestle or a spice grinder and transfer to a large bowl. Stir in the olive oil, anchovy paste and vinegar; mix well and set aside.
2. Cook the beans in a large pot of boiling salted water until barely tender, about 7 minutes. With a slotted spoon, transfer the beans to a large bowl and pour the dressing over them; reserve the cooking water.
3. Return the cooking water to a boil, add the potatoes and cook until they are just tender, 10 to 15 minutes. Transfer to a bowl of cold water to stop the cooking. Quickly peel the potatoes, cut them in half lengthwise and then crosswise into ⅛-inch slices, placing them in a bowl as you work. While they are still warm, toss the slices with 3 tablespoons of the aquavit. Cool to room temperature.
4. Add the potatoes to the beans, sprinkle them with the scallions and the remaining 2 tablespoons aquavit and gently toss. Add salt and pepper to taste. Cover and refrigerate overnight, allowing the salad to come to room temperature before serving.
—F&W

GARDEN GREEN BEAN AND TOMATO SALAD WITH WARM BACON AND PINE NUT DRESSING

This green bean and tomato salad, with its hot sweet-and-sour dressing, also makes a wonderful first course.
4 Servings

¼ cup pine nuts
1¼ pounds green beans
2 medium beefsteak tomatoes, cut into wedges
6 ounces smoky slab bacon, trimmed of excess fat and cut into ¼-inch dice
3 tablespoons extra-virgin olive oil
3 tablespoons red wine vinegar
½ teaspoon sugar
½ teaspoon salt
⅛ teaspoon freshly ground pepper

1. Preheat the oven to 375°. Spread the pine nuts in a shallow baking pan and toast in the oven until golden brown, about 10 minutes.
2. In a large saucepan of boiling salted water, cook the beans, stirring once, until just tender, about 5 minutes. Drain immediately and rinse the beans under cold water. Drain the beans thoroughly and arrange on a platter. Place the tomato wedges around the beans.
3. Meanwhile, in a small skillet, cook the bacon over low heat until crisp and golden, about 10 minutes. Pour off all but 2 tablespoons of the bacon fat. Add the olive oil to the skillet. Stir in the vinegar, sugar, salt and pepper. Bring to a boil and simmer, stirring, until the sugar is dissolved, about 1 minute. Add the pine nuts and pour the dressing over the beans and tomatoes. Serve at once.
—Michael McLaughlin

GREEN BEAN AND BENNE SALAD

In the South's Low Country, sesame (benne) seeds turn up in biscuits, soups, candies, and in this delightful green salad.
8 Servings

¼ cup sesame seeds
2 pounds green beans, stemmed, but with the tender young green tip left intact
1 teaspoon crushed red pepper
¼ cup extra-virgin olive oil
2 tablespoons fresh lemon juice
2 garlic cloves, minced
Salt and freshly ground black pepper

1. Place the sesame seeds in a small dry skillet and roast over moderate heat, stirring frequently, until golden brown, about 5 minutes. Set aside.
2. In a large pot of rapidly boiling water, cook the green beans just until they lose their raw flavor, 3 to 5 minutes. Immediately drain in a colander and rinse under cold running water. Drain well.
3. Place the beans in a bowl and add the roasted sesame seeds, red pepper, olive oil, lemon juice and garlic. Toss well. Season with salt and black pepper to taste. Serve at room temperature.
—John Martin Taylor

BEET AND CHICORY SALAD

Beets baked in this manner are sweetly mellow, tender and tasty. The method works equally well for beets to be served hot as a vegetable course.
6 Servings

3 pounds fresh beets
¾ teaspoon salt
½ teaspoon freshly ground pepper
Finely grated zest of 1 lemon
Juice of 2 lemons
About 3 tablespoons olive oil
1 head of chicory (curly endive), torn into bite-size pieces
¼ cup chopped parsley
Red Wine Vinaigrette (p. 258), as needed
½ cup coarsely chopped walnuts, toasted (see Note)

1. Preheat the oven to 450°. Trim off the leafy tops of the beets, if there are any, leaving 1- to 2-inch stems. Rinse the beets, shake off most of the water and wrap the beets tightly in aluminum foil. Place the foil package on a baking pan and bake until the beets are just tender, about 45 minutes. Open the foil and cool the beets to room temperature.
2. When the beets are cool enough to handle, peel them. Halve each beet lengthwise, then cut crosswise into ¼-inch half-rounds. Place the slices in a large bowl and add the salt, pepper, lemon zest and juice, and enough olive oil to coat the ingredients lightly. Toss, then allow to marinate for at least 20 minutes, or up to several hours, at room temperature.
3. Toss the chicory with the parsley in enough of the Red Wine Vinaigrette so that the leaves are lightly and evenly coated. Arrange the chicory in a bowl or on a platter and arrange the sliced beets over the greens. Scatter the walnuts over the salad, then drizzle any marinade remaining in the bowl of beets over all. Serve immediately.

NOTE: To toast the walnuts, bake in a 300° oven for a few minutes, turning frequently, until the walnuts are just hot through; be careful not to overheat them. Cool to room temperature.
—F&W

ASPARAGUS AND BEET SALAD WITH TARRAGON
4 Servings

2 medium beets, with root ends and 1 inch of the tops left on
8 large asparagus spears
2 teaspoons finely chopped fresh tarragon or ½ teaspoon dried
2 tablespoons olive oil
2 teaspoons fresh lemon juice
Salt and freshly ground pepper

1. In a small heavy saucepan, cook the beets in boiling water to cover until tender, 30 to 45 minutes. Drain, let cool to room temperature, then peel and cut into 2-inch julienne strips.
2. In a large pot of boiling salted water, cook the asparagus until tender but still bright green, 3 to 4 minutes. Drain and rinse under cold running water, drain well. Cut the spears into 2-inch lengths, then quarter lengthwise.
3. Toss the beets and asparagus with the tarragon, olive oil and lemon juice. Season with salt and pepper to taste. Marinate for at least 30 minutes at room temperature to let the flavors develop before serving.
—Anne Disrude

HEARTS OF PALM, BEET AND ENDIVE SALAD
12 to 16 Servings

3½ pounds beets
1 pound radishes, sliced into ¼-inch rounds
1 can (16 ounces) hearts of palm, drained and cut into ¼-inch rounds
1½ pounds celery (about 2 bunches), cut crosswise into ¼-inch slices
2 pounds firm, ripe tomatoes— halved, seeded and cut into ½-inch dice
7 scallions, minced
2½ cups minced fresh dill
½ cup safflower oil
¼ cup walnut oil
¼ cup extra-virgin olive oil
¼ cup sherry vinegar
3 tablespoons fresh lemon juice
2 teaspoons salt
1½ teaspoons coarsely cracked pepper
2 pounds Belgian endives, leaves separated

1. Preheat the oven to 400°. Wrap the beets in aluminum foil, about 3 to a package. Bake the beets in the middle of the oven for 1¼ to 1½ hours, or until tender when pierced with a skewer or knife tip. Remove from the oven and open the packets of foil. When the beets are cool enough to handle, peel them under cold running water and pat dry. Slice into ¼-inch rounds and let cool completely.
2. In a large bowl, toss the radishes, hearts of palm, celery, tomatoes and scallions with 2 cups of the dill.

3. In a medium bowl, whisk together the safflower, walnut and olive oils, the vinegar, lemon juice, salt and pepper. Pour ½ cup of this dressing over the radish mixture and toss well.

4. In a medium bowl, toss the beets with half of the remaining dressing. In another bowl, toss the endive spears with the remaining dressing.

5. On a large round or oval platter, make a bed of the radish mixture. Arrange alternating rows of overlapping beets and endive spears. Sprinkle with the remaining ½ cup dill. Serve at room temperature.
—*W. Peter Prestcott*

PINEAPPLE AND BEET SALAD
Beets retain more of their flavor and color when they are baked rather than boiled.
8 Servings

2½ pounds medium beets
1 large fresh pineapple (3 to 3½ pounds)
¼ cup plus 2 tablespoons fresh lemon juice
½ teaspoon salt
Freshly ground pepper

1. Preheat the oven to 400°. Wrap the beets in aluminum foil. Bake in the middle of the oven for 1¼ to 1½ hours, or until tender when pierced with a skewer. Remove from the oven and open the foil packet. When the beets are cool enough to handle, peel them under cold water and pat dry. Cut into 1-inch chunks.

2. Cut the rind off the pineapple, removing the "eyes." Quarter the pineapple lengthwise. Cut the core portion from each section and discard. Cut the remaining pineapple into 1-inch chunks. Put the beets and pineapple in separate bowls; cover and refrigerate.

3. Shortly before serving, in a medium bowl, toss the pineapple with the beets, lemon juice, salt and pepper to taste.
—*Bob Chambers*

BAKED BEET AND ENDIVE SALAD
6 Servings

2 bunches of beets, with ½ inch of tops
2 teaspoons green peppercorn mustard
¼ cup balsamic vinegar
¾ teaspoon salt
1 teaspoon freshly ground pepper
¼ cup plus 2 tablespoons safflower oil
¼ cup light olive oil
4 Belgian endives, separated into leaves
2 tablespoons chopped chives

1. Preheat the oven to 425°. Line a large roasting pan with foil. Place the beets in the pan in a single layer. Cover the pan with foil and bake until the beets are tender, about 1 hour.

2. When the beets are cool enough to handle, remove the tops and slip off the skins. Cut the beets into ½-inch wedges.

3. In a small bowl, combine the mustard, vinegar, salt and pepper. Slowly whisk in the safflower and olive oils.

4. In a medium bowl, toss the beets with three-fourths of the vinaigrette. Arrange the endive leaves on a large plate, mound the beets on top, pour on the remaining dressing and sprinkle with the chopped chives.
—*Lee Bailey*

BEET AND AVOCADO SALAD WITH TARRAGON VINAIGRETTE
This beet and avocado salad is simple but takes on a new dimension when drizzled with a mustard vinaigrette flavored with fresh tarragon.
4 Servings

1½ pounds small beets, tops removed
1 small garlic clove, minced
1 teaspoon Dijon mustard
2 tablespoons chopped fresh tarragon
1 tablespoon raspberry or red wine vinegar
½ teaspoon salt
¼ teaspoon freshly ground pepper
⅓ cup olive oil
2 cups chicory (curly endive), torn into bite-size pieces
2 cups Boston lettuce, torn into bite-size pieces
1 avocado, halved and thinly sliced

1. In a medium saucepan, place the beets and enough water to cover by 1 inch. Bring to a boil over high heat. Reduce the heat to low and boil the beets until tender, 20 to 30 minutes.

2. Meanwhile, in a small bowl, combine the garlic, mustard, tarragon, vinegar, salt and pepper. Whisk in the olive oil.

3. Drain the beets, then peel and trim them. Slice the beets lengthwise and then cut them into ¼-inch julienne strips.

4. Divide the curly endive and Boston lettuce among 4 salad plates. Arrange the beet and avocado slices over the lettuce. Just before serving, drizzle the vinaigrette over the salads.
—*Marcia Kiesel*

Thai Rice Noodle Salad (p. 116) and Broiled Thai Chicken Salad (p. 137).

Soft-Shell Crab and Hazelnut Salad (p. 123).

SHREDDED RAW BEET AND CELERY SALAD

6 Servings

4 medium beets, peeled and shredded
3 cups shredded celery (7 to 8 ribs)
¾ cup White Wine Vinaigrette (p. 258)
1 teaspoon Dijon mustard
Salt and freshly ground pepper

1. Combine the shredded beets and celery in a medium bowl. Pour the White Wine Vinaigrette over the vegetables and toss until thoroughly coated. Add the mustard and stir to blend. Stir in salt and pepper to taste.

2. Cover the bowl tightly and refrigerate for at least 6 hours (to "relax" the texture and flavors). You may also refrigerate this salad up to two or three days before serving.

—F&W

WARM STRING BEAN SALAD

6 Servings

1¼ pounds string beans
2 tablespoons balsamic vinegar
1½ teaspoons fresh lemon juice
¾ teaspoon dry mustard
¾ teaspoon salt
Freshly ground pepper
¼ cup plus 1 tablespoon olive oil
1 garlic clove, minced
¼ cup finely chopped red onion (optional)

1. In a large pot of boiling salted water, cook the string beans until just tender but still slightly crunchy, about 3 minutes; drain, refresh and drain again thoroughly.

2. Meanwhile, in a small bowl, combine the vinegar and lemon juice. Whisk in the mustard, salt and pepper to taste; then whisk in the oil in a thin stream. Add the garlic.

3. To serve, place the beans on a serving platter. Pour the vinaigrette over the beans. Sprinkle the onion on the top and season with additional salt and pepper to taste. Serve warm.

—Lee Bailey

WARM BROCCOLI SALAD WITH TOASTED PINE NUTS

You can also serve this warm salad right along with the main course, as a piquant side vegetable.

6 to 8 Servings

2 bunches of broccoli
2 tablespoons fresh lemon juice
½ teaspoon dry mustard
¾ teaspoon salt
¼ teaspoon freshly ground white pepper
4 dashes of hot pepper sauce
2 tablespoons safflower or corn oil
2 tablespoons extra-virgin olive oil
½ of a small red onion, minced
¼ cup toasted pine nuts

1. Using a small sharp knife, cut the florets from the broccoli and separate into 1- to 1½-inch pieces. Trim off the ends of the stalks. Peel the remaining stalks and cut crosswise into ¼-inch-thick slices. Steam the broccoli until just tender, 4 to 5 minutes.

2. Meanwhile, in a small bowl, whisk the lemon juice, mustard, salt, white pepper and hot sauce. Whisk in the safflower oil and olive oil in a thin stream.

3. Transfer the warm broccoli to a medium bowl and toss with the vinaigrette. Add the onion and pine nuts and toss to mix. Serve warm.

—Lee Bailey

CABBAGE SALAD WITH GARLIC

This salad offers a striking visual presentation by contrasting white against red cabbage.

6 Servings

4½ cups finely shredded green cabbage (about ¾ pound)
4½ cups finely shredded red cabbage (about ¾ pound)
4 or 5 garlic cloves, crushed through a press
1 can (2 ounces) flat anchovy fillets, drained
1 tablespoon red or white wine vinegar
½ cup olive oil or vegetable oil
½ teaspoon freshly ground pepper
¼ teaspoon salt
¼ cup small sprigs of parsley, for garnish

1. Place the green and red cabbage in two separate bowls and set aside.

2. On a work surface, using a sharp heavy knife, chop and mash together the crushed garlic and anchovy fillets into a puree.

3. In a medium bowl, combine the vinegar, oil, pepper and salt. Stir in the garlic-anchovy puree. (Do not make this sauce in a food processor, because it will become too thick.) Divide the sauce between the two bowls of cabbage and toss well to combine.

4. Place the red cabbage in a pretty glass or crystal bowl. Make a well in the center to form a nest. Mound the green cabbage in the center. Garnish with the parsley.

—Jacques Pépin

BABY CARROT AND TURNIP SALAD

Sweet, young carrots and turnips join forces in this salad that's dressed with a grainy mustard vinaigrette.

6 to 8 Servings

1½ pounds baby turnips, peeled
1½ pounds baby carrots, peeled
1 teaspoon salt
½ teaspoon sugar
2 tablespoons grainy mustard
¼ cup red wine vinegar
¾ cup extra-virgin olive oil
Freshly ground pepper
2 tablespoons minced fresh chives

1. Put the turnips and carrots in a large skillet or flameproof casserole. Add the salt, sugar and cold water to cover. Bring to a boil. Reduce the heat to moderately low and simmer until the vegetables are tender when pierced with a knife, 8 to 10 minutes. Drain well.

2. Meanwhile, in a small bowl, whisk the mustard with the vinegar and oil. Season the dressing with salt and pepper to taste.

3. Toss the warm vegetables with the dressing. Let cool and refrigerate. Before serving, sprinkle with the chives and toss well.

—Bob Chambers

CARROT AND ORANGE SALAD WITH BLACK MUSTARD SEEDS

6 Servings

1 navel orange
1 pound carrots, coarsely grated
2 tablespoons fresh lemon juice
¼ teaspoon cayenne pepper
½ teaspoon sugar
1 teaspoon salt
2 tablespoons vegetable oil
1 teaspoon black mustard seeds

1. Using a sharp knife, peel the orange; make sure to remove all of the bitter white pith. Slice the orange crosswise ⅛ inch thick. Stack the slices and cut them into 6 wedges.

2. In a large serving bowl, combine the orange, carrots, lemon juice, cayenne, sugar and salt. Mix well.

3. In a small skillet, heat the oil over moderately high heat. When hot, add the mustard seeds. When the seeds begin to pop, scrape the oil and seeds over the salad. *(The recipe can be prepared to this point up to 4 hours ahead. Cover and refrigerate.)* Toss the salad well before serving.

—Madhur Jaffrey

CUCUMBER AND CHIVE SALAD

4 Servings

4 medium cucumbers, peeled and sliced as thin as possible
4 cups ice water
2 tablespoons salt
3 tablespoons cider vinegar
1 tablespoon minced fresh dill
¼ cup sliced chives or scallion greens
2 teaspoons sugar

1. Place the cucumber slices in a large, shallow dish; cover with the ice water and sprinkle with the salt. Refrigerate for 1 hour.

2. Drain the cucumbers in a colander and rinse thoroughly under cold running water to remove the salt. Drain completely.

3. Place the cucumbers in a shallow dish. Add the vinegar, dill, ¼ cup cold water, the chives and sugar. Gently toss to coat the cucumbers. Cover the dish and refrigerate at least 2 hours or as long as 6 hours before serving.

—F&W

COLD CUCUMBER "FETTUCCINE" SALAD

This dish is a culinary pun; the "fettuccine" is really thin ribbons of cucumber, smoked ham and scallion, tossed together to resemble a pasta salad.

4 Servings

2 European seedless cucumbers, unpeeled
2 teaspoons coarse (kosher) salt
2 teaspoons Oriental sesame oil
2 eggs, beaten
3 tablespoons light olive oil
1 tablespoon rice vinegar
¼ teaspoon Dijon mustard
¼ teaspoon coarsely cracked pepper
1 small bunch of scallions, cut into thin 2-inch strips
¼ pound thinly sliced smoked ham, such as Black Forest, cut into ¼-inch-wide strips as long as possible

1. Using the ¼-inch julienne blade of a mandoline, cut the cucumbers lengthwise into long strands, rotating the cucumbers to avoid cutting the seed core in the center; discard the seed core. (Alternatively, use a swivel-bladed vegetable peeler to remove long, thin ribbons of cucumber, rotating to avoid the seed core. Roll up the ribbons and thinly slice the cylinders crosswise to make long, thin strips.)

2. Place the cucumber in a colander set over a bowl or in the sink. Sprinkle with the salt and toss to mix. Let drain for 30 minutes. Rinse under cold running water, then drain on paper towels.

3. Brush a heavy medium skillet, preferably nonstick, with some of the sesame oil and heat until shimmering. Pour in 2 to 3 tablespoons of the egg; swirl to cover the bottom of the pan and cook over moderate heat until set but not browned, about 2 minutes. Turn and cook the second side until dry,

about 15 seconds. Slide this egg pancake onto a work surface and let cool. Repeat with the remaining egg, brushing the skillet with sesame oil before making each pancake. You will be able to make about 3 pancakes.

4. When cooled, roll up the egg pancakes and slice crosswise at ¼-inch intervals to make thin strips.

5. In a small bowl, whisk together the remaining sesame oil (about 1 teaspoon), the olive oil, vinegar, mustard and pepper until blended.

6. In a serving bowl, toss together the strips of cucumber, egg, scallion and ham. Toss with the dressing and serve at once at room temperature.
—Anne Disrude

MARINATED CUCUMBER SALAD
6 Servings

- 2 medium cucumbers, very thinly sliced crosswise
- 2 small onions, halved and thinly sliced crosswise
- ½ cup distilled white vinegar
- 2 tablespoons (packed) dark brown sugar
- Pinch of salt
- 6 coriander seeds
- 5 allspice berries, crushed
- 3 whole cloves
- 1 cinnamon stick

1. In a medium nonreactive bowl, layer the cucumbers and onions. Set aside.

2. In a small nonreactive saucepan, whisk the vinegar with the brown sugar and salt. Stir in the coriander seeds, allspice berries, cloves, cinnamon stick and ¼ cup of water. Bring to a boil over moderate heat. Reduce the heat to moderately low and simmer for 5 minutes.

3. Pour the liquid over the cucumbers and onions. Place a plate on top to weigh down the salad. Refrigerate for at least 6 hours or preferably overnight. Serve chilled or at room temperature.
—Jessica B. Harris

FENNEL AND EGGPLANT SALAD
Thin strips of sun-dried tomatoes would make a nice addition to this salad, as would a garnish of chopped pistachios.
8 Servings

- 3 eggplants (about 1½ pounds each), cut into ½-inch cubes
- ¾ cup extra-virgin olive oil
- ½ teaspoon freshly ground black pepper
- 3 pounds fennel bulbs, trimmed, greens reserved
- 3 yellow bell peppers, cut into thin strips
- ¼ cup fresh lemon juice
- 2 tablespoons white wine vinegar
- 1 teaspoon dry mustard
- ½ teaspoon salt
- 3 tablespoons minced reserved fennel greens or parsley
- 1 small garlic clove, minced

1. Preheat the oven to 450°. In a large bowl, toss the eggplant with ½ cup of the olive oil and the black pepper.

2. Spread the eggplant cubes out in a single layer on 2 baking sheets. Roast for 45 minutes, stirring every 5 to 7 minutes, until tender and browned.

3. Meanwhile, halve the fennel bulbs lengthwise and cut out the cores. Slice crosswise ⅛ inch thick.

4. In a large skillet, combine the fennel, bell pepper strips and ¼ cup of water. Cook over high heat, stirring constantly, until the water has evaporated and the vegetables are crisp-tender, 3 to 4 minutes. Transfer the vegetables to a large bowl.

5. In a bowl, combine the lemon juice, vinegar, dry mustard, salt, fennel greens, garlic and remaining ¼ cup olive oil.

6. Add the eggplant to the fennel-pepper mixture. Pour on the dressing and toss well. Cover and refrigerate. Serve at room temperature.
—Bob Chambers

FENNEL SALAD
6 Servings

- 6 medium fennel bulbs (about 5 pounds total), trimmed, feathery leaves reserved
- 2 navel oranges
- 2 small lemons
- ¼ cup plus 2 tablespoons extra-virgin olive oil
- 1 teaspoon salt
- 1 teaspoon coriander seeds
- ¼ teaspoon freshly ground pepper
- 1 tomato—peeled, seeded and diced
- ¼ cup black Niçoise olives in brine, pitted (and halved if large)

1. Halve the fennel bulbs lengthwise and remove the tough cores. In a food processor fitted with a slicing disk, thinly slice the fennel crosswise. Spread in a flat serving dish.

2. Using a small sharp knife, peel the oranges, removing all the bitter white pith. Working over the bowl of a food processor, cut between the membranes to release the sections. Discard the membranes. Repeat with the lemons. Add the olive oil, salt, coriander seeds and pepper to the citrus sections in the bowl and puree.

3. In a medium nonreactive saucepan, combine the citrus dressing with the tomato and cook over low heat until lukewarm, 2 to 3 minutes. Pour over the fennel and toss. Season with additional salt to taste. Sprinkle the olives and the reserved feathery fennel leaves on top.
—Mireille Johnston

SUGAR SNAP PEA AND SUMMER SQUASH SALAD WITH RYE CROUTONS

This light and simple salad full of fresh seasonal vegetables makes a lovely summer dish.

3 to 4 Servings

¼ cup chopped fresh dill
¼ teaspoon crushed dill seed
1 teaspoon Dijon mustard
1 tablespoon Champagne vinegar or white wine vinegar
¼ teaspoon salt
¼ teaspoon freshly ground pepper
3 tablespoons vegetable oil
2 tablespoons light olive oil
3 slices of rye bread, crusts removed, cut into ½-inch dice
2 tablespoons unsalted butter, melted
½ pound sugar snap peas, trimmed
½ pound summer squash, cut into ⅜-inch-thick slices

1. Preheat the oven to 350°. In a small bowl, combine the fresh dill, dill seed, mustard, vinegar, salt and pepper. Whisk in the vegetable and olive oils.

2. Place the bread cubes on a baking sheet and drizzle with the melted butter. Toss to coat evenly. Bake until crisp and golden brown, about 12 minutes. *(The croutons can be made up to 1 day ahead. Store them in a covered container at room temperature.)*

3. In a medium saucepan of boiling salted water, cook the sugar snap peas until bright green and crisp-tender, about 2 minutes. With a slotted spoon, remove to a colander and rinse under cold running water until cool. Add the squash to the boiling water and cook until crisp-tender, about 2 minutes. Remove and rinse as above. Drain and dry the sugar snap peas and summer squash on paper towels. Cut the sugar snap peas into 1-inch pieces. *(This recipe can be prepared to this point up to 4 hours ahead. Cover and refrigerate the vegetables until ready to use.)*

4. Place the sugar snap peas and summer squash in a large salad bowl. Add the dressing and croutons and toss. Serve immediately.

—Marcia Kiesel

FRESH PEA SALAD WITH PANCETTA

The salty, cured flavor of the pancetta is a fitting complement to the sweet and tender peas. If pancetta is unavailable, use lightly smoked, thick-sliced bacon.

6 Servings

2 teaspoons salt
4 pounds of fresh peas, shelled (4 cups) or 2 packages (10 ounces each) frozen baby peas
½ pound pancetta, sliced ¼ inch thick
2 large shallots, minced
3 tablespoons red wine vinegar
½ cup heavy cream
2 tablespoons fresh lemon juice
¼ teaspoon freshly ground pepper

1. In a heavy medium saucepan, bring 6 cups of water to a boil with 1½ teaspoons of the salt. Add the peas and cook until tender, 8 to 10 minutes for fresh or 1 minute for frozen. Drain and rinse under cold water to stop the cooking process; drain well. Place the peas in a bowl.

2. Stack the pancetta slices and cut them in thirds. Then slice crosswise ¼ inch wide to make small strips. In the saucepan used for the peas, cover the pancetta with cold water and bring to a boil. Immediately remove from the heat and drain well. Return the pancetta to the pan and cook over moderate heat until some of the fat has been rendered, about 5 minutes.

3. Add the shallots and cook, stirring frequently, until the shallots are translucent and the pancetta begins to brown. Drain off the fat. Add the pancetta and shallots to the peas and toss.

4. Add the vinegar to the same saucepan, and cook over moderate heat for 30 seconds, scraping the bottom of the pan to release the flavorful bits. Pour the vinegar into a small bowl. Whisk in the cream, lemon juice, pepper and remaining ½ teaspoon salt. Add to the peas and pancetta and toss well. Serve at room temperature.

—Bob Chambers

SUGAR SNAP PEAS IN TOASTED SESAME SEED VINAIGRETTE

A crisp salad should not marinate for too long, but the snap peas can be blanched and the dressing made ahead and refrigerated. Let the dressing return to room temperature and toss the salad just before serving.

4 Servings

1 pound sugar snap peas, stem ends and strings removed
¼ cup sesame seeds
1 small garlic clove, crushed through a press
1 teaspoon Dijon mustard
1 tablespoon fresh lemon juice
¼ teaspoon salt
¼ teaspoon freshly ground pepper
1 tablespoon red wine vinegar
¼ cup olive oil
1½ teaspoons Oriental sesame oil

1. In a large saucepan of boiling salted water, cook the sugar snap peas until crisp-tender, about 30 seconds. Drain, rinse under cold running water and then drain well.

2. In a heavy skillet, toast the sesame seeds over moderate heat, stirring, until nut brown, about 3 minutes. Pour onto a plate to cool.

3. In a medium bowl, combine the garlic, mustard, lemon juice, salt and pepper. Blend in the vinegar. Whisk in the olive oil and sesame oil. Add the sugar snap peas and toasted sesame seeds and toss. Serve at room temperature or slightly chilled.
—Phillip Stephen Schulz

SARDINIAN ROASTED PEPPER SALAD

Charring peppers on an outdoor grill gives them great flavor. This savory yet fruity salad goes well with grilled lamb or chicken. It keeps for about a week in the refrigerator.
6 Servings

6 medium bell peppers, preferably 2 red, 2 green and 2 yellow
2 tablespoons balsamic or red wine vinegar
½ cup raisins
1 small fresh hot red pepper, seeded and minced (about 1 teaspoon) or ½ teaspoon crushed red pepper
1 small garlic clove, minced
6 plum tomatoes (about 1 pound)— peeled, seeded and coarsely chopped—or 1 can (20 ounces) Italian peeled tomatoes, drained and coarsely chopped
6 tablespoons olive oil, preferably extra-virgin
¼ teaspoon oregano
½ teaspoon salt
¼ teaspoon freshly ground black pepper

1. Light the charcoal or preheat the broiler. Grill or broil the bell peppers 3 to 4 inches from the heat, turning, until completely charred, 10 to 20 minutes. Let cool for 1 minute, then seal in a brown paper bag and let steam for 10 minutes.

2. Remove the peppers from the bag and scrape off the charred skin. Halve the peppers and remove the stems, seeds and ribs. Cut the peppers into ¼-inch strips. *(The recipe may be prepared 1 day ahead to this point. Wrap and refrigerate. Let return to room temperature before continuing.)*

3. In a small nonreactive saucepan, bring the vinegar to a boil. Reduce the heat to moderately low, add the raisins and simmer, covered, for 15 minutes.

4. In a large bowl, place the roasted bell peppers, the hot pepper, garlic, tomatoes, oil, oregano, salt and black pepper. Add the raisins, toss and let marinate at room temperature for at least 30 minutes before serving.
—Beverly Cox

URAB URAB

This light vegetable and greens salad from Indonesia is tossed with a sweet-and-hot dressing and sprinkled with unsweetened dried coconut.
4 to 6 Servings

½ pound bean sprouts (about 3 cups)
¼ pound green beans, cut into 2-inch pieces
2 teaspoons instant tamarind or tamarind paste (see Note)*
1 garlic clove, crushed through a press
1 fresh hot red chile, seeded and minced, or 1 teaspoon crushed red pepper
½ teaspoon sugar
½ teaspoon salt
½ bunch of watercress, tough stems removed (about 2 cups)
¼ cup dried unsweetened coconut*
*Available at Asian markets and health food stores

1. Place the bean sprouts in a large heatproof bowl. Cover with boiling water and let stand for 2 minutes. Drain and rinse in cold water. Drain; let dry on paper towels. Wrap and refrigerate.

2. In a medium saucepan of boiling salted water, blanch the green beans until just tender, about 3 minutes. Drain and rinse under cold water; drain well.

3. In a small bowl, mix the tamarind, garlic, chile, sugar and salt to make the dressing.

4. Toss the bean sprouts, green beans and watercress together. Add the dressing and coconut and toss again to mix. Serve at room temperature.

NOTE: If using tamarind paste, measure 2 teaspoons and soak in 2 tablespoons of water for 30 minutes. Strain through a sieve; discard the pulp.
—Copeland Marks

FRESH TOMATO SALAD

Waverley Root once wrote of how he ordered a simple salad of fresh tomatoes in every great restaurant in Paris, assuming that only a truly great establishment would bother to prepare such a simple item perfectly (he was often disappointed). The simplicity belies the sheer magnificence of this dish when prepared with perfectly ripe, meaty tomatoes.
1 Serving

1 tomato
Coarse (kosher) salt
Freshly ground pepper
1 to 2 tablespoons olive oil
1 to 2 tablespoons wine vinegar
1 to 2 tablespoons chopped fresh basil (or other fresh herb)
1 tablespoon chopped parsley (optional)

1. Cut the tomato into thick slices. Arrange a layer in the bottom of a small glass or porcelain serving dish (a soufflé dish works well).

2. Sprinkle the tomato slices lightly with salt and pepper. Then douse them with some of the oil and vinegar and top with the basil and the parsley.

3. Continue to layer the tomato slices in the dish, sprinkling each layer as directed above. Sprinkle the top layer with extra herbs and allow the tomatoes to marinate at room temperature for at least an hour before serving. Spoon the marinating juices over the slices as you serve the salad.
—F&W

ORIENTAL-STYLE TOMATO SALAD
8 Servings

1 ounce (about 14 medium) dried shiitake mushrooms
2 cups boiling water
3 tablespoons Oriental sesame oil
¼ cup soy sauce
¼ cup rice vinegar
1 cup sliced scallions (about 10)
2 cans (8 ounces each) water chestnuts, drained and cut into 2-by-⅛-inch julienne strips
1 tablespoon shredded fresh ginger
3 pounds (about 6 large) firm-ripe tomatoes, seeded and cut into ½-inch cubes

1. Place the mushrooms in a medium bowl. Pour on the boiling water and let the mushrooms soak for 30 minutes, until softened. Drain and squeeze out the excess water. Trim off any woody stems and cut the caps into ⅛-inch slivers.

2. In a medium bowl, combine the mushroom slivers, sesame oil, soy sauce and vinegar; mix well.

3. Stir in the scallions, water chestnuts and ginger. Add the tomatoes and toss gently. Cover and refrigerate until chilled, at least 1 hour, before serving.
—Jim Fobel

TOMATO, CUCUMBER AND PINEAPPLE SALAD
4 Servings

3 medium cucumbers—peeled, seeded and cut into ½-inch cubes
1 tablespoon plus ½ teaspoon salt
1 large pineapple
3 large tomatoes, seeded and cut into ½-inch pieces
½ cup finely chopped mint
1½ teaspoons sugar
2 tablespoons white wine vinegar
2 tablespoons fresh lemon juice
1 tablespoon olive oil
3 tablespoons thinly sliced chives or scallion greens
Mint sprigs, for garnish

1. Place the cucumbers in a colander and sprinkle with 1 tablespoon of the salt. Toss and set over a bowl to drain for about 30 minutes. Rinse, drain and pat dry with paper towels.

2. Cut the pineapple lengthwise (through the leafy top) into quarters. Remove and discard the core; then cut out the pineapple, leaving ½ inch in the shells; set the shells aside. Working over a bowl, cut the pineapple into ½-inch cubes, reserving 3 tablespoons of the juice.

3. In a large bowl, combine the cucumbers with the pineapple cubes and tomatoes.

4. In a small bowl, mix the mint with the remaining ½ teaspoon salt and the sugar. Slowly whisk in the vinegar, lemon juice and reserved pineapple juice; whisk in the oil and stir in the chives.

5. Pour the vinaigrette over the salad, cover and marinate in the refrigerator for at least an hour.

6. To serve, mound the salad into the four reserved pineapple shells and garnish with the mint sprigs. Serve chilled.
—F&W

FRESH AND SUN-DRIED TOMATOES WITH ZUCCHINI AND BALSAMIC VINEGAR
4 Servings

1 pound tomatoes, cut into thin wedges
½ pound zucchini, cut into 2-by-¼-inch julienne strips
6 oil-packed sun-dried tomato halves, cut into thin strips
2 tablespoons extra-virgin olive oil
2 teaspoons balsamic vinegar
½ teaspoon salt
¼ teaspoon freshly ground pepper

Toss all of the ingredients together and let marinate at room temperature for 30 to 60 minutes.
—Anne Disrude

TOMATOES ON BASIL CHIFFONADE
For best results, use the ripest, most flavorful tomatoes you can find. This salad is a nice first course, but also makes a wonderful light lunch. Note that the tomatoes need to sit overnight.
4 Servings

2 cups (packed) fresh basil
1 pound tomatoes, cut into 1-inch wedges
4 garlic cloves, bruised
2 teaspoons capers, chopped
1 cup extra-virgin olive oil
½ teaspoon freshly ground pepper
⅛ teaspoon salt
2 teaspoons white wine vinegar
3 cups curly endive, torn into pieces

1. Chop 1 cup of the basil. In a medium bowl, combine the tomatoes, garlic, capers, olive oil, pepper and chopped basil. Toss to coat the tomatoes with oil. Cover with plastic wrap and refrigerate overnight.

2. Pour off and reserve the marinade from the tomato-basil mixture. Measure out ¼ cup of the marinade. Stir in the salt and vinegar. (The extra marinade may be reserved and used as a salad dressing or to marinate more tomatoes.)

3. Just before serving, divide the endive among 4 salad plates. With a sharp knife, finely shred the remaining 1 cup basil and distribute evenly over the endive. Arrange the tomato wedges on top of the basil. Drizzle 1 tablespoon of the dressing over each plate.
—Marcia Kiesel

MINTED ZUCCHINI AND APPLE SALAD
12 to 16 Servings

½ cup dried currants
¼ cup ruby port or sweet vermouth
2 pounds zucchini, cut into 3-by-¼-inch julienne strips
3 pounds Granny Smith apples, unpeeled, cut into 3-by-¼-inch julienne strips
½ cup fresh lemon juice
1 pound Jarlsberg cheese, cut into 3-by-¼-inch julienne strips
2 medium red onions, minced
½ cup extra-virgin olive oil
¼ cup sherry vinegar
2 teaspoons salt
¼ cup minced fresh mint

1. In a small bowl, soak the currants in the port for at least 2 hours or overnight; drain.

2. In a large bowl, toss the zucchini and apple strips with the lemon juice. Add the cheese and onions and toss well to combine.

3. In a medium bowl, whisk together the olive oil, vinegar and salt. Stir in the currants and mint and pour this dressing over the salad. Toss gently but thoroughly. Taste and add more salt if desired. Serve at room temperature.
—W. Peter Prestcott

ZUCCHINI-RADISH SALAD
Serve this crisp vegetable mixture as a condiment to grilled poultry or fish, or use as a sandwich topping.
Makes About 2 Cups

1 medium zucchini (6 ounces), cut into very long thin julienne strips
16 radishes, cut lengthwise into 1/16-inch julienne strips
1 tablespoon safflower or other flavorless vegetable oil
½ teaspoon sherry vinegar
¼ teaspoon salt
⅛ teaspoon coarsely cracked pepper

Crisp the zucchini and radishes in a bowl of ice water for 15 minutes. Drain and pat dry on paper towels. Just before serving, toss with the oil, vinegar, salt and pepper.
—Anne Disrude

BAGNA CAUDA SALAD
A classic *bagna cauda* is a traditional Italian hot anchovy dip, usually served with crudités. In this takeoff, the vegetables are cooked briefly to enhance color and texture, then arranged attractively on the plate and served with the tangy anchovy sauce as a dressing.
4 Servings

½ pound red potatoes, cut into 1½-by-¼-inch sticks
1 medium zucchini
¼ pound sugar snap peas
¼ pound green beans
½ small head of cauliflower, broken into ½-inch florets
½ small bunch of broccoli, broken into ½-inch florets
3 large mushrooms, quartered
2 tablespoons fresh lemon juice
¾ cup olive oil, preferably extra-virgin
6 large garlic cloves, thinly sliced
8 flat anchovy fillets, minced
1 tablespoon capers
¼ teaspoon coarsely cracked pepper
½ teaspoon grated lemon zest

1. Cook the potatoes in a medium saucepan of boiling salted water until easily pierced with a fork, 5 to 8 minutes. Drain and set aside.

2. Cut the zucchini crosswise into ¼-inch-thick rounds. Stack the rounds and slice again into ¼-inch sticks. Bring a large saucepan of salted water to a boil. Put the zucchini in a strainer and dip in the boiling water for about 2 seconds to bring out the green color. Rinse under cold running water until cool. Drain on paper towels.

3. Add the sugar snap peas to the strainer and again blanch for about 2 seconds. Rinse under cold running water until cool. Drain on paper towels. Repeat separately with the green beans, cauliflower and broccoli, blanching the beans for 1 minute and the cauliflower and broccoli for 30 seconds each.

4. In a small bowl, toss the mushrooms with the lemon juice.

5. To assemble the salad, arrange the vegetables decoratively on 4 plates.

6. Make the sauce by warming the olive oil in a medium skillet over moderate heat. Add the garlic and slowly cook until light brown, about 10 minutes. Remove and discard the garlic. Add the anchovies to the skillet. Cook, mashing with the back of a spoon, until dissolved. Add the capers, pepper and lemon zest. Pour the hot sauce over the vegetables. Serve warm.
—Anne Disrude

ACHAR KUNING

This Indonesian pickled vegetable salad should have four vegetables or more, with different colors and textures to add interest. Other vegetables that may be used to good account are carrots, in julienne strips; string beans, blanched and cut into 1-inch pieces; and sticks of Kirby cucumber.

6 Servings

1 tablespoon corn or peanut oil
1 garlic clove, bruised
¼ teaspoon turmeric
¼ cup cider vinegar
2 teaspoons minced fresh ginger
2 teaspoons sugar
½ teaspoon salt
1 medium red or green bell pepper, cut into ¼-inch-wide strips
1 can (5 ounces) sliced bamboo shoots, rinsed and drained
1 can (15 ounces) baby corn, rinsed and drained
1 medium zucchini, cut into 2-by-⅜-inch julienne strips

1. In a wok or large skillet, heat the oil over moderately high heat. Add the garlic and turmeric and cook, stirring constantly, until the garlic is fragrant but not browned, about 1 minute. Add the vinegar, ginger, sugar, salt and ¼ cup of water. Simmer the dressing for 3 minutes.

2. Add the bell pepper, bamboo shoots, corn and zucchini. Increase the heat to high and cook, stirring, for 3 minutes. Turn out into a bowl. Let cool, then refrigerate, tossing occasionally, for at least 4 hours, and preferably overnight, to chill and develop the flavors. Serve chilled.
—Copeland Marks

INDONESIAN MIXED SALAD WITH PINEAPPLE

This salad, called *asinan*, would also make a nice, light vegetarian entrée.

4 to 6 Servings

2 medium carrots, cut into julienne strips
2 firm Chinese tofu cakes
1 can (8 ounces) unsweetened pineapple chunks, drained and ¼ cup of the juice reserved
3 tablespoons cider vinegar
2 tablespoons crunchy peanut butter
2 teaspoons sugar
½ teaspoon salt
1 to 2 teaspoons crushed red pepper, to taste
3 Kirby cucumbers, cut into ½-inch dice
1 bunch of watercress, tough stems removed (about 4 cups)

1. In a medium saucepan of boiling salted water, blanch the carrots until crisp-tender, about 2 minutes. Drain and rinse under cold running water; drain well.

2. In a large saucepan of water, simmer the tofu for 10 minutes. Remove with a slotted spoon, wrap in a kitchen towel and let drain for 15 minutes. Cut into ½-inch dice. (*The recipe can be made ahead to this point. Refrigerate the carrots and tofu separately.*)

3. In a small bowl, mix the pineapple juice, vinegar, peanut butter, sugar, salt and hot pepper.

4. To assemble, combine the carrots, bean curd, pineapple, cucumbers and watercress in a small salad bowl; toss lightly. Pour the dressing over the salad and toss again to mix. Serve at room temperature or slightly chilled.
—Copeland Marks

VEGETABLE SALAD A LA GRECQUE

The vegetables in this classic preparation are cooked separately to keep their flavors distinct. Begin with the mildest vegetable—mushrooms in this case—and finish with cauliflower, the most strongly flavored.

8 Servings

⅔ cup extra-virgin olive oil
2 medium onions, finely chopped
1 bottle dry white wine
1½ cups fresh lemon juice (from about 6 lemons)
3 imported bay leaves
2 teaspoons thyme
2 teaspoons mustard seeds
1 tablespoon salt
¾ teaspoon freshly ground pepper
¾ pound mushrooms, stems trimmed, quartered if large
¾ pound zucchini, cut into 3-by-½-inch sticks
¾ pound thin green beans, cut in half
¾ pound small Belgian endives, halved lengthwise
¾ pound cauliflower florets
½ cup chopped parsley

1. In a large nonreactive saucepan, heat the olive oil. Add the onions and cook over moderately high heat until translucent, 4 to 5 minutes.

2. Add the wine, lemon juice, bay leaves, thyme, mustard seeds, salt, pepper and 5 cups of water. Bring to a boil. Reduce the heat to moderately low, cover and simmer for 10 minutes.

3. One at a time, add the vegetables to the broth and return the liquid to a boil; then cook (uncovered except for the mushrooms) until just tender: about 15 minutes for the mushrooms, 1½ minutes for the zucchini, 4 minutes for the beans, 6 minutes for the endives and 4 minutes for the cauliflower. Remove the vegetables to individual bowls with a slotted spoon as they are cooked.

4. Cut the endives into bite-size pieces. Group the vegetables on a platter or in a container and refrigerate. Sprinkle with the parsley before serving.
—Bob Chambers

SPICED FRUIT SALAD

Rujak is a popular fruit salad on several of the Indonesian islands. If you can't find papaya or mango, substitute tart apples, peaches, nectarines or your favorite firm seasonal fruit. Just be sure the fruit, jicama and cucumbers add up to eight cups.
12 Servings

½ pound jicama, peeled and cut into thin julienne strips
2 Kirby cucumbers, thinly sliced
1 cup cubed (½-inch) firm-ripe papaya
1 cup cubed (½-inch) pineapple or canned unsweetened chunks
3 kiwis, peeled and cut into ½-inch cubes
1 firm tart apple, cut into ½-inch cubes
1 mango, peeled and sliced
⅓ cup dry-roasted peanuts
1 to 2 fresh hot red chile peppers, seeded and thinly sliced, or ½ to 1 teaspoon crushed red pepper, to taste
⅓ cup (packed) brown sugar
2 tablespoons tamarind paste* dissolved in ⅓ cup water, strained
*Available at Asian markets

1. Put the jicama, cucumbers, papaya, pineapple, kiwis, apple and mango in a serving dish and toss lightly.

2. In a food processor, coarsely chop the peanuts. Add the chile peppers, brown sugar and tamarind liquid. Process to a thick, chunky paste. If it is too thick to act as a dressing, thin with 1 tablespoon hot water.

3. Pour the dressing over the salad and toss well. Refrigerate for at least 1 and up to 3 hours. Serve chilled.
—Copeland Marks

GRAPEFRUIT AND AVOCADO SALAD WITH TOMATO-CUMIN DRESSING

Bitter greens pair magically with avocado and grapefruit, creating a taste that is much more interesting than the sum of these simple ingredients.
8 to 10 Servings

5 large grapefruits
6 ripe avocados, preferably Hass
¼ cup fresh lemon juice
2 medium red onions, very thinly sliced
1 teaspoon coarse (kosher) salt
½ teaspoon freshly ground pepper
6 cups (loosely packed) stemmed arugula or watercress
24 Niçoise or other small black olives
Tomato-Cumin Dressing (p. 260)

1. Using a sharp knife, peel the grapefruits, making sure to remove all the bitter white pith. Working over a bowl, cut in between the membranes to release the sections.

2. Halve the avocados lengthwise and remove the pits. Using a large spoon, scoop out the flesh in one piece. Cut each half lengthwise into ¼-inch slices. Sprinkle with 2 tablespoons of the lemon juice and set aside until ready to use.

3. In a medium bowl, toss the onions with the remaining 2 tablespoons lemon juice and the salt and pepper.

4. Arrange the arugula around the rim of a large platter, then fill in to cover the surface. Fan the avocado slices around the outer edge of the platter, then arrange the grapefruit sections inside the ring of avocado and top with the olives. Mound the onions in the center. Drizzle about ⅓ cup of the Tomato-Cumin Dressing over the grapefruit. Serve the remaining dressing alongside.
—Susan Costner

WARM CABBAGE SLAW WITH BACON

When served with grilled sausages and buttered potatoes, this salad is perfect for a chilly night.
6 to 8 Servings

½ pound lean bacon, thinly sliced and cut into 1-inch pieces
1 tablespoon caraway seeds
3 tablespoons olive oil
1 small head (2 pounds) green cabbage, finely shredded
¾ teaspoon salt
¼ teaspoon freshly ground pepper
1½ cups sour cream
1 tablespoon fresh lemon juice

1. In a large skillet, sauté the bacon over moderately high heat, stirring occasionally, until the bacon is crisp and brown around the edges. Drain the bacon on paper towels; discard the fat in the pan.

2. In a small saucepan, combine the caraway seeds with 1 cup of water and place over moderate heat. Bring the water to a full boil and boil until the liquid has reduced to 1 tablespoon. Remove from the heat and set aside.

3. Add the olive oil to the skillet and place it over moderate heat. Add half the cabbage and cook, stirring occasionally, until it has browned slightly and is tender, about 10 minutes.

4. Reduce the heat to low. Stir in the reserved bacon, the remaining raw cabbage, the caraway seeds with their reduced liquid, the salt, pepper and the sour cream; cook for 1 minute. Remove the pan from the heat and stir in the lemon juice. Serve immediately.
—Jim Fobel

CREAMY GARLIC COLESLAW
Makes About 2 Quarts

1½ pounds green cabbage, shredded
1 tablespoon salt
¾ cup mayonnaise
½ teaspoon finely grated lemon zest
1 tablespoon plus 1 teaspoon fresh lemon juice
1 large garlic clove, minced
¾ teaspoon freshly ground pepper
2 medium carrots—shredded, rinsed and squeezed dry

1. In a large bowl, cover the shredded cabbage with cold water. Stir in the salt and let stand at room temperature for 1 hour. Drain well, cover and refrigerate until cold.

2. In a large bowl, combine the mayonnaise, lemon zest, lemon juice, garlic and pepper. Add the cabbage and carrots and toss well. Season with additional salt to taste. *(The slaw can be refrigerated, covered, for up to 1 day.)*
—Marcia Kiesel

SHREDDED RADISH SLAW
This slaw is a perfect accompaniment to charcoal-grilled, marinated meats such as teriyaki and yakitori.
Makes About 4 Cups

4 cups shredded radishes (about 28 medium)
½ cup rice vinegar
2 teaspoons light soy sauce
1 teaspoon sugar
3 tablespoons Oriental sesame oil

1. In a medium bowl, combine the radishes, vinegar, soy sauce, sugar and sesame oil. Toss the radishes with the dressing until evenly coated.

2. Cover the bowl tightly and refrigerate the salad for about 2 hours before serving. It is best served the same day it's made.
—F&W

FIVE-VEGETABLE SLAW
This light, colorful side dish also makes a delicious sandwich topping. It tastes even better the day after it is made.
Makes About 4½ Cups

4 cups finely shredded green cabbage
½ cup thinly sliced red bell pepper
½ cup thinly sliced green bell pepper
½ of a medium cucumber—peeled, halved lengthwise, seeded and thinly sliced
3 scallions, thinly sliced on the diagonal
⅓ cup mayonnaise
1 tablespoon cider vinegar
¼ teaspoon salt
¼ teaspoon freshly ground black pepper

In a large bowl, toss together the cabbage, red and green bell peppers, cucumber and scallions. Mix in the mayonnaise, vinegar, salt and black pepper. Cover and refrigerate for up to 2 days before serving.
—Marcia Kiesel

COLESLAW WITH BULGUR
This zesty salad is a fine accompaniment to traditional outdoor fare.
6 Servings

½ cup coarse or medium bulgur
½ medium head of green cabbage, finely shredded
2 medium carrots, grated
1 very small onion, grated
¾ cup sour cream
¼ cup mayonnaise
2 tablespoons cider vinegar
1 teaspoon dry mustard
2 teaspoons sugar
¾ teaspoon salt
½ teaspoon freshly ground pepper

1. Place the bulgur in a small bowl and rinse several times to remove any impurities. Add cold water to cover by 1½ inches and let soak for 45 minutes, or until tender. Drain in a fine sieve and press to remove as much moisture as possible.

2. Transfer the bulgur to a large bowl and add the cabbage, carrots and onion.

3. In a small bowl, whisk the sour cream, mayonnaise, vinegar, mustard, sugar, salt and pepper until blended. Toss the dressing with the cabbage until well mixed. Cover and refrigerate for at least 6 hours before serving to blend the flavors.
—Michèle Urvater

RED PEPPER SLAW

I like red peppers so much that I use them two ways—roasted for the flavor and raw for the crunch.

4 Servings

3 small red bell peppers—1 whole, 2 thinly sliced
½ medium head of green cabbage, finely shredded
1 small onion, thinly sliced
½ cup thinly sliced sour gherkins or other pickles
1 garlic clove, minced
⅓ cup mayonnaise
2 teaspoons white wine vinegar
Salt and freshly ground black pepper

1. Roast the whole red pepper directly over a gas flame or under the broiler, turning, until charred all over, about 5 minutes. Place the pepper in a paper bag and set aside for 10 minutes to steam. When cool enough to handle, peel the pepper and remove the core, seeds and ribs. Slice into thin strips.

2. In a large bowl, combine the cooked and uncooked red peppers with the cabbage, onion, gherkins, garlic, mayonnaise and vinegar. Season with salt and black pepper to taste. Cover and refrigerate until ready to serve. *(The recipe can be made up to 1 day ahead.)*

—Marcia Kiesel

SASSY SLAW

Get this pretty, peppery salad ready 24 to 48 hours in advance.

10 to 12 Servings

2 pounds green cabbage, shredded
1 small turnip, shredded
2 carrots, shredded
4 red radishes, chopped
⅔ cup finely chopped red onion
¼ cup minced parsley
3 tablespoons minced fresh dill
1 cup mayonnaise
½ cup cider vinegar
1 teaspoon sugar
¼ teaspoon freshly ground black pepper
⅛ teaspoon freshly ground white pepper
¼ to ½ teaspoon crushed red pepper, to taste
½ teaspoon salt

1. In a large salad bowl, toss together the cabbage, turnip, carrots, radishes, red onion, parsley and dill.

2. In a small bowl, whisk together the mayonnaise, vinegar, sugar, black pepper, white pepper, hot pepper and salt. Add the dressing to the slaw and toss, mixing very thoroughly. Add additional salt to taste and toss again. Cover and refrigerate, stirring occasionally, for 24 to 48 hours. (The longer it stands, the creamier and more peppery the slaw becomes.) Just before serving, stir and adjust the seasonings if necessary.

—Leslie Newman

CUCUMBER AND TOMATO RAITA

An Indian *raita* falls somewhere between a salad and a condiment, since it is intended to complement the flavors of the main dish, but would not be eaten as a separate course.

2 to 4 Servings

1 large cucumber
1 tablespoon salt
1 small onion, minced
2 tomatoes, cut into ¼-inch dice
1 tablespoon minced fresh coriander (cilantro)
⅛ teaspoon cayenne pepper
1 cup plain low-fat yogurt

1. Peel, seed and cut the cucumber into ¼-inch dice. Mix it with the salt and let stand 15 minutes, then squeeze out excess water. Place the cucumber in a medium bowl.

2. Add the onion, tomatoes, fresh coriander, cayenne and yogurt and mix well. Chill for 1 or 2 hours to let the flavors combine.

—F&W

CUCUMBER, WALNUT AND DILL RAITA

The use of walnuts and dill reflects the Persian influence on northern Indian cuisine and gives this refreshing yogurt salad a special texture and taste.

10 Servings

1 cup walnut halves
5 medium cucumbers—peeled, seeded and finely diced
4 cups plain low-fat yogurt
1 cup sour cream
¼ cup minced fresh dill
½ teaspoon salt
¼ teaspoon cayenne pepper
¼ teaspoon freshly ground black pepper

1. Preheat the oven to 250°. Scatter the walnuts on a baking sheet and toast in the oven for 15 minutes to crisp them and bring out their full flavor. Let the walnuts cool and then chop them into pieces about the same size as the diced cucumbers.

2. In a large bowl, combine the yogurt, sour cream, cucumbers, dill, salt, cayenne and black pepper. Mix to blend well. *(The recipe can be prepared to this point several hours in advance. Set the walnuts aside at room temperature; refrigerate the yogurt mixture.)*

3. Shortly before serving, stir the walnuts into the yogurt mixture and adjust the seasoning if necessary.

—Leslie Newman

LEBANESE EGGPLANT AND YOGURT SALAD

Prepared 24 hours in advance, this richly spiced salad-for-all-seasons is irresistible with—or without—lamb, beef or chicken. I have never had any leftovers. Even in Mediterranean countries, where fresh mint is readily available, the dried herb is preferred for its intense flavor in certain stews and salads such as this one.

10 Servings

2 eggplants (about 1 pound each), peeled and sliced ¼ inch thick
2 teaspoons coarse (kosher) salt
4 cups plain yogurt
1½ teaspoons minced garlic
½ teaspoon ground roasted cumin (see Note)
1½ teaspoons crushed dried mint leaves
⅛ teaspoon cayenne pepper
Salt and freshly ground black pepper
½ cup light olive oil
Fresh mint leaves or sprigs of parsley, for garnish

1. Sprinkle the eggplant slices with the coarse salt and let drain in a colander for at least 30 minutes. (This rids the eggplant of excess and sometimes bitter juices and reduces the amount of oil it will absorb.)

2. Meanwhile, in a large colorful serving bowl, mix the yogurt, garlic, cumin, mint, cayenne and salt and black pepper to taste.

3. Preheat the oven to 450°. Pat the eggplant slices dry. Brush with the olive oil and place on baking sheets. Set the sheets in the oven and bake, turning once, until the eggplant slices are tender and lightly browned, 15 to 20 minutes.

4. Drain the eggplant on paper towels. While still slightly warm, fold the slices into the yogurt mixture, immersing them completely (the eggplant will get mashed a bit). Cover and refrigerate for 24 hours, stirring 3 or 4 times.

5. Remove from the refrigerator 1 to 2 hours before serving so that the salad will be cool, not cold. Stir well and garnish with mint leaves.

NOTE: To make ground roasted cumin: Heat about 3 tablespoons cumin seeds in a small heavy skillet over moderate heat, shaking the pan constantly, until the seeds darken and become fragrant. (This takes only a few minutes and happens very suddenly, so be poised to pour the seeds onto a plate as soon as they are toasted.) Let cool slightly. In a spice mill or with a mortar and pestle, grind the seeds to a fine powder.
—Leslie Newman

CAULIFLOWER RAITA

This cool side dish is a great way to use cauliflower in the warmer months.

4 Servings

1 head of cauliflower (about 2 pounds), cut into large florets
1½ teaspoons cumin seeds
½ cup plain low-fat yogurt
¼ cup chopped fresh mint
½ teaspoon salt
¼ teaspoon freshly ground pepper

1. Place the cauliflower florets in a steamer pot, cover and steam over high heat until barely tender, about 4 minutes. Drain, then refrigerate to chill.

2. Meanwhile, in a small dry skillet, toast the cumin seeds over high heat, shaking the pan, until fragrant, about 30 seconds. Pound in a mortar, grind in a spice mill or chop with a knife to form a coarse powder.

3. In a large bowl, mix the cumin with the yogurt, mint, salt and pepper. Fold in the cauliflower and toss lightly to coat evenly. *(The recipe can be made up to 2 hours ahead; cover and refrigerate.)* Serve slightly chilled.
—Marcia Kiesel

CURRIED POTATO AND YOGURT SALAD

4 Servings

2 cups diced (½-inch), peeled potatoes
1½ tablespoons vegetable oil
¼ cup chopped onion
⅛ teaspoon cayenne pepper
1½ teaspoons curry powder
¼ teaspoon ground ginger
½ teaspoon allspice
1 teaspoon finely grated orange zest
1½ teaspoons finely diced jalapeño pepper, or more, to taste
1 tablespoon chopped fresh coriander (cilantro) or minced parsley
½ teaspoon salt
¼ cup dried currants
1 cup plain low-fat yogurt

1. Cook the potatoes in boiling water to cover just until tender. Drain.

2. In a skillet, heat the oil over medium heat. Add the onion and sauté for 1 minute. Off the heat, add cayenne, curry powder, ginger, allspice, and orange zest and stir to blend with the oil.

3. Add the drained potatoes and toss with the spice mixture until they are coated. Add the jalapeño and fresh coriander and toss.

4. In a small bowl, blend the salt, currants and yogurt and pour over the salad. Toss well, transfer to a serving bowl, and chill for an hour or two.
—F&W

WINE-DRENCHED POTATO SALAD WITH WHITE WINE VINAIGRETTE

White wine is splashed over cooked potatoes as they cool to add a seductively elegant aroma to this unusually fresh-tasting potato salad.

8 Servings

3 pounds small red potatoes
½ cup dry white wine
White Wine Vinaigrette (p. 258)
½ cup sliced scallions (including some of the green tops)
⅓ cup plus ¼ cup finely minced parsley
1 cup diced celery
½ cup diced plus ⅓ cup finely minced green bell pepper
¾ teaspoon salt
¾ cup sliced radishes (optional)
1 medium cucumber—peeled, seeded and diced
2 medium tomatoes, sliced

1. Cook the potatoes in a large pot of boiling salted water until just tender; do not overcook. Drain the potatoes.

2. When the potatoes are just cool enough to handle, peel them. Halve the potatoes and then cut into ⅜-inch half-rounds. Place the potatoes in a large bowl and pour the wine and ½ cup of the White Wine Vinaigrette over them and toss gently. Set aside to cool to room temperature.

3. When the potatoes have cooled, add the scallions, ⅓ cup of the parsley, the celery, the ½ cup diced green pepper and the salt. Pour up to ½ cup of the remaining vinaigrette over the potatoes to moisten them (the amount needed will vary according to the moisture content of the potatoes). Add the radishes and cucumber. Gently toss the salad, cover it, and refrigerate for an hour or so or until completely chilled.

4. Place the tomatoes in a shallow bowl and pour ½ cup of the vinaigrette over them. Cover and marinate in the refrigerator for about 30 minutes.

5. Give one final, gentle toss to the salad. Arrange the marinated tomatoes around the outside edge of the bowl. Sprinkle the ⅓ cup minced green pepper over the center of the salad and sprinkle the remaining ¼ cup parsley over the marinated tomatoes. Serve cold.

—F&W

"GREEN" POTATO SALAD

6 Servings

⅓ cup olive oil
2 tablespoons red wine vinegar
1 teaspoon Dijon mustard
1 teaspoon caraway seeds
½ teaspoon salt
½ teaspoon freshly ground black pepper
2 pounds small red potatoes (about 2 inches)
¼ cup minced parsley
1 medium green bell pepper, cut into ⅛-inch dice
¾ cup thinly sliced scallions (about 4 medium)
½ pound bacon, cooked until crisp, drained

1. In a small bowl, mix the oil, vinegar, mustard, caraway seeds, salt and black pepper. Set the dressing aside.

2. Place the potatoes in a medium saucepan of cold water. Bring to a boil and cook over moderate heat until tender but slightly resistant in the center when pierced with a knife, about 20 minutes. Do not overcook. Drain and transfer to a bowl; cover with foil or a towel to keep warm.

3. One at a time, peel the potatoes quickly and replace them in the covered bowl (hold the hot potato in a folded paper towel to protect your hand from the heat). When all the potatoes are peeled, cut them into ¼-inch slices and drop into a shallow serving dish. Sprinkle on the parsley, green pepper and scallions.

4. Whisk the dressing to blend and pour over the vegetables while the potatoes are still warm; toss gently. Crumble the bacon over the salad and serve, or marinate at room temperature for up to 6 hours.

—Diana Sturgis

TOMATO-POTATO SALAD

A refreshing version of potato salad, this one is loaded with cherry tomatoes and flavored with fragrant fresh coriander.

12 Servings

3 pounds small red potatoes
1 teaspoon salt
2 large green bell peppers, cut into ¼-inch dice
4 celery ribs, cut into ¼-inch dice
¼ cup (packed) minced fresh coriander (cilantro)
¼ cup grated onion (optional)
2 pints cherry tomatoes—halved crosswise, squeezed to remove seeds and excess juice, then halved again
1½ cups mayonnaise
2 tablespoons Dijon mustard
½ teaspoon freshly ground black pepper

1. Cook the potatoes in a large saucepan of boiling salted water until just tender, 20 to 30 minutes. Drain, let cool to room temperature and refrigerate until chilled.

2. Peel the potatoes and cut them into ½-inch cubes. Place them in a large bowl and add the green peppers, celery, coriander, onion and tomatoes; toss gently to avoid breaking the potatoes.

3. In a small bowl, mix the mayonnaise, mustard, salt and pepper until blended. Add the dressing to the potato mixture and toss gently until the vegetables are coated. Season with more salt, if desired. Serve chilled.
—*Jim Fobel*

POTATO SALAD A L'ORANGE WITH SWEET PEPPERS
8 Servings

2 pounds red bell peppers
2½ pounds small red potatoes
1 tablespoon red wine vinegar
1 tablespoon fresh orange juice
1 teaspoon grated orange zest
¾ teaspoon freshly ground black pepper
¾ teaspoon salt
⅓ cup safflower oil
1 cup thinly sliced scallions
1 tablespoon chopped parsley

1. Broil the peppers about 3 inches from the heat, turning, until blackened all over, 15 to 20 minutes. Place them in a paper bag for about 10 minutes to loosen the skins. Peel off the blackened skins under running water.

2. Halve the peppers over a bowl to catch the juices. Remove the seeds, cores and ribs. Slice the peppers into strips about 2 by ½ inch. Strain and reserve 1 tablespoon of the juices.

3. Place the potatoes in a large saucepan and cover them with cold water. Bring to a boil and cook until tender, about 20 minutes. Drain, peel if desired and cut into ¼- to ⅜-inch-thick slices. Place in a large bowl.

4. In a medium bowl, whisk together the vinegar, orange juice, orange zest, reserved pepper juices, black pepper and salt. Gradually whisk in the oil.

5. Pour the dressing over the warm potatoes. Add the roasted peppers, scallions and parsley and toss to mix.
—*Diana Sturgis*

LIGHTENED POTATO SALAD

Use one, two or a mixture of the vegetables listed below to lighten this mayonnaise-dressed potato salad, but use a total of 6 cups. For a less traditional flavor, add a good handful of any minced fresh herb, such as basil, tarragon or coriander.
6 to 8 Servings

3 pounds waxy potatoes, peeled and cut into ½-inch dice
½ cup dry white wine
3 tablespoons cider vinegar
1 teaspoon salt
½ teaspoon freshly ground pepper
4 jumbo eggs—hard-cooked, peeled and chopped
¾ cup mayonnaise
2 cups julienned zucchini (1 by ⅛ inch)
1 cup julienned jicama (1 by ⅛ inch)
1 cup julienned daikon radish (1 by ⅛ inch)
1 cup (packed) watercress leaves and small stems
1 cup slivered fresh water chestnuts
⅓ cup minced fresh basil, tarragon or coriander (optional)

1. Bring a large pot of salted water to a boil. Add the potatoes and cook, stirring occasionally, until easily pierced with a fork, about 10 minutes. Drain well and place in a large bowl.

2. In a small nonreactive saucepan, combine the wine, vinegar, ½ teaspoon of the salt and ¼ teaspoon of the pepper. Bring just to a boil. Pour the hot dressing over the potatoes and toss to coat. Let the potatoes cool to room temperature, tossing occasionally.

3. Add the chopped hard-cooked eggs and the mayonnaise to the potatoes and toss to coat. Just before serving, add the vegetables, remaining ½ teaspoon salt and ¼ teaspoon pepper and the herbs, if using. Season with additional salt and pepper to taste.
—*Anne Disrude*

SWEET POTATO SALAD

This is a variation on a traditional Chinese salad preparation in which a vegetable is thinly cut, or julienned, and then marinated overnight in a mixture of vinegar, sugar and salt. This dressing is usually used with turnips, white radishes and young ginger.
6 Servings

1¼ pounds sweet potatoes or yams
3 tablespoons distilled white vinegar
1 teaspoon Oriental sesame oil
1½ tablespoons sugar
1 teaspoon salt
1 teaspoon chopped fresh coriander (cilantro)
6 radicchio leaves, for serving

1. Peel, wash and dry the sweet potatoes. Cut into 2-by-¼-inch julienne strips. Or use the slicing disk of a food processor to slice pieces of the sweet potato; then stack them and cut into ¼-inch strips. Place in a large bowl. Add the vinegar, sesame oil, sugar and salt. Toss the sweet potatoes to mix well, cover and refrigerate overnight.

2. Shortly before serving, remove the salad from the refrigerator, add the chopped coriander and toss. Arrange the radicchio leaves on individual salad

dishes or on a platter and scoop a portion of the sweet potatoes into each. Serve chilled.
—Eileen Yin-Fei Lo

RED POTATO SALAD WITH GREEN BEANS AND RED PEPPER STRIPS

This vegetable salad can be composed to resemble a Christmas wreath. Although this salad feeds a crowd, the recipe is easily halved or quartered.

32 Servings

½ cup minced shallots
⅓ cup Dijon mustard
½ cup tarragon wine vinegar or white wine vinegar
2 garlic cloves, crushed through a press
2 teaspoons tarragon
1 teaspoon salt
1½ teaspoons freshly ground pepper
3 cups olive oil
8 pounds small red potatoes
4 pounds green beans, trimmed and snapped in half
4 large red bell peppers, cut into 2-by-½-inch strips

1. In a medium bowl, whisk the shallots, mustard, vinegar, garlic, tarragon, salt and pepper. Whisk in the olive oil in a thin stream; set the dressing aside.
2. Put the potatoes in a large pot and add enough lightly salted cold water to cover. Bring the water to a boil. Simmer the potatoes until tender, 12 to 15 minutes; drain.
3. When just cool enough to handle, cut the potatoes in half. Toss the warm potatoes with half of the vinaigrette. Cover with plastic wrap and refrigerate overnight.
4. In a large pot of lightly salted boiling water, cook the beans until crisp-tender, 6 to 8 minutes. Drain and rinse under cold running water; drain well.
5. Toss the potatoes with ⅔ cup of the vinaigrette and place in the center of 2 large platters. Toss the beans and pepper strips with the remaining dressing and arrange around the potatoes.
—Bob Chambers

POTATO AND WHITE BEAN SALAD

The combination of potatoes and white beans might seem odd if you've never tried it, but the flavors and textures of the two starches work wonderfully together.

6 Servings

1 large red bell pepper
1¼ pounds small red potatoes
3 cups cooked or canned white beans, drained
2 tablespoons chopped chives
¾ teaspoon salt
¼ teaspoon freshly ground black pepper
2 tablespoons red wine vinegar
1 teaspoon green peppercorn mustard
¼ cup extra-virgin olive oil
2 tablespoons safflower oil

1. Roast the pepper directly over a gas flame or under the broiler as close to the heat as possible, turning, until charred all over. Place the pepper in a bag and let steam for 10 minutes. Peel the pepper; remove the core, seeds and ribs.
2. In a large saucepan of boiling salted water, cook the potatoes until tender, 10 to 15 minutes; drain. Cut the potatoes into quarters.
3. In a large bowl, toss together the potatoes, beans, chives, salt and black pepper.
4. In a food processor, combine the vinegar, mustard, olive oil, safflower oil and roasted pepper. Puree until smooth, about 30 seconds. Pour over the beans and potatoes and toss to coat. Season with additional salt and pepper to taste. Serve at room temperature.
—Lee Bailey

INDIAN FRUIT AND POTATO SALAD

6 Servings

1 medium boiling potato
½ cup fresh lemon or lime juice
1 teaspoon chili powder
½ teaspoon cumin
½ teaspoon freshly ground pepper
½ teaspoon garam masala or curry powder
¼ teaspoon ground ginger
Pinch of salt
1 large Granny Smith apple, peeled and cut into ½-inch cubes
2 firm Bosc or Comice pears, peeled and cut into ½-inch cubes
1 cucumber, peeled and cut into ½-inch cubes
1 orange—peeled with a knife to remove the white pith, quartered and thinly sliced
2 cups cantaloupe cubes (¾-inch) or 1 mango, peeled and cut into ¾-inch cubes
2 bananas, thinly sliced

1. In a medium saucepan of boiling water, cook the potato until tender, about 25 minutes. Drain and let cool to room temperature.
2. In a small bowl, mix together the lemon juice, chili powder, cumin, pepper, garam masala, ginger and salt.
3. Pour half of this dressing into a serving bowl. Add the apple and pears to

the bowl and toss well. Add the cucumber, orange and cantaloupe to the bowl and toss again.

4. Peel the potato and cut into ½-inch cubes. Add the bananas and potato to the bowl, pour on the remaining dressing and toss gently. Cover and refrigerate until very cold, 1 to 2 hours.
—Linda Burum & Linda Merinoff

TANGY CORN AND LIMA BEANS
6 Servings

6 medium ears of corn
1 package (10 ounces) frozen baby lima beans or 1¾ cups shelled fresh lima beans
1½ cups (loosely packed) fresh coriander (cilantro) leaves
¾ teaspoon crushed red pepper
¾ teaspoon salt
½ cup corn oil or other vegetable oil
3 tablespoons fresh lime juice

1. Cook the corn in a large saucepan of water until tender, about 8 minutes (less if the corn is very young and fresh); do not overcook. Drain and refresh at once under cold running water for 2 to 3 minutes to stop the cooking. Drain and set aside.
2. In another saucepan of boiling water, cook the lima beans until just tender, 4 to 5 minutes if frozen, slightly longer if fresh. Drain into a colander and refresh under cold running water. Drain well and place in a bowl.
3. Cut the corn kernels from the cobs. Add the kernels to the lima beans.
4. In a blender or food processor, process the fresh coriander, hot pepper, salt, oil and lime juice until the coriander is minced. Pour the dressing over the vegetables and toss to blend well. Let the vegetables marinate at room temperature for at least 1 and up to 6 hours.
—Diana Sturgis

BLACK BEAN AND CELERY SALAD
This flavorful salad, with its crunch of fresh vegetables, is tossed with a tangy lime dressing. You can make this a no-cook dish by substituting three 1-pound cans of black beans for the dried; if you do so, rinse and drain the beans and begin at Step 2.
16 Servings

1 pound dried black beans, rinsed and picked over
2¼ teaspoons salt
5 celery ribs, finely diced
1 large green bell pepper, finely chopped
3 red onions (about ¾ pound)—2 finely chopped, 1 thinly sliced
⅔ cup olive oil or other vegetable oil
⅓ cup fresh lime juice
¾ teaspoon freshly ground black pepper
Grated zest of 1 lime
Red onion rings, for garnish

1. Place the beans in a large saucepan or flameproof casserole and add enough cold water to cover by 2 inches. Let soak overnight. Alternatively, bring the water to a boil. Remove from the heat, cover and let soak for 1 hour.
2. Drain the beans, return them to the pot and add fresh cold water to cover by 2 inches. Bring to a boil, reduce the heat to low and simmer until the beans are almost tender, about 45 minutes. Add 1 teaspoon of the salt and cook until tender, 30 to 45 minutes longer. Drain the beans and set aside to cool. *(The beans can be cooked up to 3 days ahead. Refrigerate, covered.)*
3. In a large bowl, combine the black beans, celery, green pepper, chopped red onions and sliced red onion. Toss to mix.
4. In a small jar, combine the oil, lime juice, black pepper, lime zest and remaining 1¼ teaspoons salt. Cover with a tight lid and shake to blend.
5. Pour the dressing over the bean salad and toss well. *(The salad can be made up to 3 hours ahead and tossed again just before serving. Let stand at room temperature.)* Garnish with onion rings just before serving.
—Diana Sturgis

BLACK BEAN AND BELL PEPPER SALAD
This simple salad can be tossed together in no time at all and is perfect for last-minute preparations.
4 Servings

1 can (16 ounces) black beans, rinsed and drained
1 large green bell pepper, diced
½ cup peeled, seeded and diced fresh tomato or drained and chopped canned tomato
2 scallions, thinly sliced
1 jalapeño pepper—seeded, deribbed and minced
1 tablespoon chopped fresh coriander (cilantro)
1 tablespoon corn oil
1 tablespoon fresh lime juice
¼ teaspoon salt
⅛ teaspoon freshly ground black pepper

1. Place the beans in a medium bowl. Add the bell pepper, tomato, scallions, jalapeño pepper and coriander. Toss to mix.
2. In a small bowl, combine the corn oil, lime juice, salt and black pepper. Pour the dressing over the bean mixture and toss well to coat.
—Diana Sturgis

Potato Crisps (p. 173).

Tempura Onion Rings (p. 165).

Corn Chili (p. 158).

BLACK BEAN, CORN AND POBLANO PEPPER SALAD WITH CUMIN VINAIGRETTE

This side salad could complement any southwestern meal, but seems especially appropriate alongside a grilled entrée. It can be assembled one day ahead, and the quantities can be increased if a crowd is on the way. Fresh coriander (cilantro) fans will recognize this salad as a perfect vehicle for the herb. Chop the leaves and stir them into the salad just before serving.

4 to 6 Servings

1½ cups dried black beans, rinsed and picked over
1 tablespoon salt
4 poblano peppers
1 tablespoon cumin seeds
¼ cup sherry vinegar
1 tablespoon Dijon mustard
2 teaspoons freshly ground black pepper
½ cup olive oil
2 cups corn kernels—fresh, defrosted or drained canned
18 cherry tomatoes, cut in half
3 scallions, thinly sliced

1. In a large bowl, combine the beans with enough cold water to cover by at least 3 inches. Let stand for 24 hours at room temperature. Alternatively, cover the beans with water in a saucepan and bring to a boil. Remove from the heat and let stand, covered, for 1 hour.

2. Drain the beans and transfer to a large saucepan. Add enough fresh cold water to cover by at least 3 inches and bring to a boil over high heat. Reduce the heat and simmer, partially covered, for 30 minutes. Stir in 2 teaspoons of the salt and continue cooking until the beans are just tender, 35 to 45 minutes longer; drain.

3. Meanwhile, roast the poblanos directly over a gas flame or under a broiler as close to the heat as possible until the skins are lightly charred. Seal in a paper bag and let steam for 10 minutes. Rub away the charred skin. Stem and seed the peppers and cut into ¾-inch squares.

4. In a small dry skillet, toast the cumin seeds over moderately high heat, tossing, until fragrant and golden brown, about 1 minute. Grind in a spice mill or finely chop on a cutting board.

5. In a small bowl, whisk together the vinegar, mustard, cumin, black pepper and remaining 1 teaspoon salt. Slowly whisk in the oil.

6. In a large bowl, combine the black beans, poblano peppers and corn. Pour the dressing over the salad and toss well. *(The recipe can be prepared to this point up to 24 hours ahead. Cover and refrigerate. Let return to room temperature before proceeding.)*

7. Add the tomatoes and scallions to the salad, toss well and season with additional salt and pepper to taste. Serve at room temperature.

—Michael McLaughlin

THREE-BEAN MACARONI SALAD

Although this salad joins two classics, it doesn't use either of the traditional dressings. Instead, it is seasoned with orange zest and Chinese flavorings, such as star anise, ginger and sesame oil. Let the salad marinate at room temperature for at least one hour before serving to develop the flavors. Add the vinegar just before serving, so as not to discolor the green beans.

4 to 6 Servings

1½ teaspoons Szechuan peppercorns
4 star anise pods*
1 small dried hot red pepper
⅓ cup safflower or other light vegetable oil
1 tablespoon Oriental sesame oil
6 scallions—white part minced, green part cut into 1½-by-⅛-inch julienne strips
2 tablespoons minced fresh ginger
1½ tablespoons minced garlic
Minced zest of 1 navel orange
3 tablespoons soy sauce
½ pound shaped pasta, such as orecchiette or small shells
½ pound green beans, cut on the diagonal into ½-inch pieces
½ pound yellow wax beans, cut on the diagonal into ½-inch pieces
2 cups cooked small red or black beans, such as small kidney beans or turtle beans (from 1 cup dried), or canned
⅓ cup chopped fresh mint
3 tablespoons white wine vinegar
Salt

*Available at Asian markets

1. In a small skillet, toast the Szechuan peppercorns over high heat, tossing, until fragrant, about 1 minute. Remove from the pan and let cool slightly.

2. In an electric spice grinder, grind the star anise, hot pepper and roasted Szechuan peppercorns to a fine powder. If you do not have a spice grinder, use a large knife to coarsely chop the spices, then use the flat of the blade or a mortar and pestle to grind them to a powder.

3. In a small skillet, heat the safflower and sesame oils over moderately low heat. Add the minced scallion white, ginger, garlic, orange zest and spice mixture. Cook over moderately low heat, stirring occasionally, until the garlic is softened but not browned, about 5 minutes. Stir in the soy sauce, remove from the heat and set aside. *(The flavored oil can be made several hours or even days ahead. Cover and refrigerate.)*

4. In a large pot of boiling salted water, cook the pasta until tender but still firm, 12 to 15 minutes. Drain well.

5. In a large pot of boiling salted water, cook the green and wax beans until crisp-tender, 3 to 5 minutes. Rinse under cold running water; pat dry.

6. At least 1 hour before serving, toss the red or black beans and the green and wax beans with the pasta, flavored oil and julienned scallion green. Let stand at room temperature for 1 to 2 hours to allow the flavors to develop. Just before serving, add the mint, vinegar and salt to taste.
—*Anne Disrude*

BEAN, CORN AND
RICE SALAD WITH
CHILI VINAIGRETTE

This spicy salad is a perfect accompaniment for grilled lamb or barbecued chicken. The flavors improve if the salad sits at room temperature for about an hour before serving.
6 to 8 Servings

3½ cups cooked converted rice, cooled (from 1 cup raw rice)
1 can (16 ounces) pink beans, rinsed and drained
1½ cups cooked fresh corn or 1 can (12 ounces) no-salt-added corn niblets
⅓ cup chopped scallions
2 pickled jalapeño peppers— seeded, deribbed and minced
⅓ cup safflower or corn oil
2 tablespoons fresh lime juice
1 tablespoon cider vinegar
1 tablespoon (packed) brown sugar
1 teaspoon chili powder
¾ teaspoon salt
½ teaspoon cumin

1. In a large bowl, combine the rice, beans, corn, scallions and jalapeño peppers. Toss to mix.

2. In a small bowl, combine the oil, lime juice, vinegar, brown sugar, chili powder, salt and cumin. Whisk until the sugar dissolves and the mixture is well blended.

3. Pour the dressing over the salad and toss to coat. Let stand at room temperature, tossing occasionally, for up to 4 hours before serving, or cover and refrigerate for up to 3 days.
—*Susan Wyler*

BLACK-EYED PEA
SALAD WITH
ORANGE-JALAPEÑO
DRESSING
8 Servings

1 pound dried black-eyed peas, rinsed and picked over
1 bay leaf
1 large red onion, finely diced
2 teaspoons grated orange zest
¼ cup fresh orange juice
5 garlic cloves, minced
2 tablespoons unsulphured molasses
⅓ cup red wine vinegar
¼ teaspoon hot pepper sauce
½ teaspoon thyme
1 teaspoon salt
½ teaspoon freshly ground black pepper
¾ cup light olive oil
1 large cucumber—peeled, seeded and cut into ½-inch dice
1 small jalapeño pepper with its seeds, minced

1. In a large saucepan, cover the peas with 4 inches of cold water. Add the bay leaf and bring to a boil over moderate heat. Reduce the heat to low and simmer, stirring occasionally, until the peas are tender, about 50 minutes; add water if the peas look dry.

2. Drain the peas in a colander and rinse well under cold water; discard the bay leaf. Transfer the peas to a large bowl and toss with the red onion.

3. In a medium bowl, combine the orange zest, orange juice, garlic, molasses, vinegar, hot sauce, thyme, salt and black pepper. Whisk in the olive oil. Add the dressing to the peas and toss well. Set aside at room temperature until serving time. *(The recipe can be prepared up to 1 day ahead. Cover and refrigerate.)*

4. Just before serving, fold in the cucumber and jalapeño.
—*Marcia Kiesel*

BLACK-EYED PEA
AND CORN SALAD

Use fresh black-eyed peas if you can find them. Otherwise, frozen peas (cooked according to the package directions), cooked dried peas or canned peas are a fine substitute.
4 Servings

1 cup cooked black-eyed peas
1 cup cooked corn, fresh or frozen
3 medium celery ribs, thinly sliced
1 medium red bell pepper, cut into ¼-inch dice
3 scallions, thinly sliced
3 tablespoons balsamic vinegar
2 teaspoons honey
1 teaspoon Dijon mustard
½ teaspoon salt
⅛ teaspoon freshly ground black pepper
¼ cup plus 3 tablespoons safflower oil
1 bunch of young dandelion leaves, washed and dried (optional)

1. In a large bowl, combine the peas, corn, celery, bell pepper and scallions.

2. In a bowl, whisk together the vinegar, honey, mustard, salt and black pepper. Whisk in the oil until blended. *(The recipe can be prepared to this point up to 4 hours ahead. Refrigerate the vegetables and dressing separately.)*

3. Pour the dressing over the vegetables and toss well. Divide the dandelion leaves, if using, among 4 chilled plates. Spoon the salad onto the leaves and serve.
—Melanie Barnard & Brooke Dojny

CHICKPEA SALAD

I first sampled this dish at a casual family restaurant on the outskirts of Aix-en-Provence. In the summer months, I keep this salad on hand to serve as an appetizer or lunch dish or as a side dish with grilled fish or meats.

8 to 10 Servings

½ pound dried chickpeas, rinsed and picked over
2 tablespoons red wine vinegar
5 garlic cloves, minced
½ teaspoon salt
⅛ teaspoon freshly ground pepper
2 tablespoons finely chopped mixed fresh herbs, such as rosemary, thyme, tarragon and parsley
½ cup extra-virgin olive oil
½ cup (2 ounces) oil-cured black olives, preferably from Nyons, pitted and halved
1 medium red onion, minced

1. Place the dried chickpeas in a medium saucepan and add enough cold water to cover by 2 inches. Set aside to soak overnight.

2. Drain the chickpeas, return them to the saucepan, cover again with cold water and bring to a boil over high heat. Reduce the heat, cover and simmer over low heat until tender, about 2 hours and 15 minutes.

3. In a medium bowl, whisk together the vinegar, garlic, salt, pepper and fresh herbs. Gradually whisk in the oil.

4. Drain the chickpeas. Add to the dressing along with the olives and onion. Toss to mix. Serve warm, at room temperature or chilled. *(The recipe can be made up to 3 days in advance and stored, covered, in the refrigerator.)*
—Patricia Wells

CHICKPEA AND BAKED BELL PEPPER SALAD WITH CUMIN

Start this recipe a day ahead since the chickpeas need to soak overnight. Baking bell peppers with oil and garlic creates a flavorful liquid to use as a dressing.

6 Servings

2 cups dried chickpeas
2 large red bell peppers
6 garlic cloves, unpeeled
¼ cup plus 2 tablespoons extra-virgin olive oil
1 sprig of rosemary (optional)
2½ teaspoons cumin seeds
2 tablespoons fresh lemon juice
1 tablespoon red wine vinegar
¾ teaspoon salt
½ teaspoon freshly ground black pepper
1 small red onion, finely chopped
2 small tomatoes, cut into ½-inch dice

1. In a large saucepan, cover the chickpeas with cold water and let soak overnight at room temperature. The next day, strain off the water. Cover the chickpeas with 6 inches of fresh water and bring to a boil over high heat. Using a ladle, skim the water to remove any impurities that rise to the surface. Reduce the heat to moderately low. Simmer, adding 2 cups of hot tap water to the pan every 30 minutes or so as necessary, until the chickpeas are tender, about 3 hours. Remove from the heat and let cool in the liquid. Drain and transfer to a large bowl. *(The recipe can be made to this point up to 1 day ahead; cover and refrigerate.)*

2. Preheat the oven to 425°. In a small roasting pan, combine the bell peppers, garlic, 2 tablespoons of the olive oil, the rosemary, if desired, and 2 tablespoons of water. Cover with aluminum foil and bake for 20 minutes, or until the peppers and garlic are soft when pierced with a fork. Remove from the oven and set aside, covered, for 5 minutes. Then transfer the peppers to a large plate and set aside to cool slightly.

3. Meanwhile, peel the garlic cloves and place them in a small bowl. Mash them with a fork. Stir in the juices from the roasting pan; discard the rosemary.

4. In a small skillet, toast the cumin seeds over high heat until fragrant, about 30 seconds. Pound in a mortar, grind in a spice mill or coarsely chop with a knife to form a coarse powder. Stir the cumin into the garlic dressing along with the lemon juice, vinegar, salt, black pepper and the remaining ¼ cup olive oil.

5. Pour the dressing over the chickpeas, add the red onion and mix well. Set aside at room temperature for 30 minutes to blend the flavors.

6. Working over a bowl, peel the peppers; remove and discard the cores, seeds and ribs. Cut into ½-inch dice. Reserve all the juices from the peppers. Shortly before serving, fold the peppers and their liquid into the chickpeas along with the tomatoes. Season with additional salt to taste. Serve the salad at room temperature or slightly chilled.
—Marcia Kiesel

FRENCH LENTIL SALAD WITH GLAZED PEARL ONIONS

The secret to this salad is dressing the lentils while they're still warm so that they will absorb all the flavors. Imported French green lentils, which are tiny and hold their shape, are my choice for this salad, but ordinary lentils would work just as well.

6 to 8 Servings

3 medium smoked ham hocks
3 small onions, stuck with 8 whole cloves
6 garlic cloves
1 imported bay leaf
½ cup plus 2 tablespoons extra-virgin olive oil
1 pound pearl onions, peeled
¼ cup tomato paste
3 tablespoons white wine vinegar
1 cup dry white wine
½ cup dried currants
¾ teaspoon salt
½ teaspoon freshly ground pepper
1 pound lentils, preferably French green lentils*
2 tablespoons Dijon mustard
½ cup chopped flat-leaf parsley
*Available at specialty food shops

1. In a large saucepan, combine the ham hocks, clove-studded onions, garlic cloves and bay leaf. Add 2 quarts of cold water and bring to a boil. Reduce the heat to low, cover and simmer for 1 hour.

2. Meanwhile, heat 2 tablespoons of the olive oil in a large nonreactive skillet, preferably nonstick. Add the pearl onions and sauté over moderately high heat, stirring frequently, until lightly browned all over, 5 to 7 minutes.

3. Add the tomato paste, 1 tablespoon of the vinegar, the wine, currants, ½ teaspoon of the salt, ¼ teaspoon of the pepper and enough water just to cover. Bring to a boil, reduce the heat to low and simmer until the onions are tender and glazed and the liquid has reduced to a thick sauce, about 30 minutes. Remove from the heat.

4. Strain the ham hock stock, reserving the garlic cloves and hocks. Return the stock to the pot. Add the lentils and bring to a boil. Reduce the heat to moderately low and simmer, stirring occasionally, until the lentils are tender but still firm, 30 to 35 minutes. Drain off any excess liquid.

5. Meanwhile, remove the meat from the ham hocks and coarsely chop.

6. Crush the reserved garlic cloves into a small bowl. Whisk in the mustard and the remaining 2 tablespoons vinegar, ¼ teaspoon salt, ¼ teaspoon pepper and ½ cup olive oil. Stir in the chopped parsley.

7. In a large bowl, combine the lentils with the onions and their sauce and chopped ham. Add the dressing and toss well. Let cool and refrigerate. Serve at room temperature.
—*Bob Chambers*

TABBOULEH

Probably the most popular bulgur dish in this country, tabbouleh is a wonderfully fresh tasting salad, perfumed with parsley and fresh mint.

6 to 8 Servings

1 cup bulgur, preferably fine
2 cups minced parsley
½ cup thinly sliced scallions
⅓ to ½ cup minced fresh mint, to taste
½ cup fresh lemon juice
½ teaspoon grated lemon zest
¾ teaspoon salt
¼ teaspoon freshly ground pepper
⅓ cup olive oil
2 tomatoes—peeled, seeded and cut into ½-inch dice (optional)
Romaine lettuce leaves

1. Place the bulgur in a small bowl and rinse several times to remove any impurities. Add cold water to cover by 1½ inches and soak for 45 minutes, or until tender. Drain in a fine sieve; then squeeze out excess moisture with your hands; the bulgur should be as dry as possible.

2. In a medium bowl, combine the bulgur, parsley, scallions and mint.

3. In a small bowl, whisk the lemon juice, lemon zest, salt, pepper and oil. Stir this dressing into the bulgur mixture. Chill, covered, for 2 hours.

4. Just before serving, mix in the tomatoes, if desired. Serve on lettuce leaves.
—*Michèle Urvater*

CRACKED WHEAT-SPINACH SALAD

Prepare the mushrooms and pear while the cracked wheat is cooking to prevent them from discoloring.

4 to 6 Servings

⅓ cup plus 2 teaspoons olive oil
2 large shallots, finely chopped
1 garlic clove, minced
½ cup cracked wheat
¼ teaspoon salt
¼ pound mushrooms
1 large pear
1 tablespoon Dijon mustard
2 tablespoons white wine vinegar
6 scallions, thinly sliced
4 plum tomatoes or 2 medium tomatoes, cut into ⅜-inch dice
3 tablespoons raisins
⅓ cup cashews or walnuts, broken or coarsely chopped
Freshly ground pepper
½ pound spinach, stemmed and torn into large pieces

1. In a heavy medium saucepan, heat 2 teaspoons of the olive oil. Add the shallots and garlic and cook over moderate heat, stirring frequently, for 1

minute. Add the cracked wheat and continue to cook, stirring, until the kernels are covered with oil and the bottom of the pan is dry, about 1 minute. Add the salt and 1 cup of water, stir once and bring to a boil. Cover, reduce the heat to low and cook until the liquid is absorbed, 10 to 15 minutes. Remove from the heat, fluff with a fork and transfer to a bowl to cool to room temperature.

2. Trim and thinly slice the mushrooms. Cut the pear into small dice.

3. In a large bowl, whisk together the mustard and vinegar. Slowly beat in the remaining ⅓ cup olive oil, whisking until the dressing is blended and creamy. Add the mushrooms, pear, scallions, tomatoes, raisins and nuts; toss to coat with dressing.

4. Add the cooled cracked wheat and toss again. Season with additional salt and pepper to taste. Arrange the spinach around the edge of the bowl and fill the center with the tossed grain salad.
—Dorie Greenspan

QUINOA SALAD

Quinoa, originally from the Andes, is now being grown in the mountains of Colorado and New Mexico. This delicate grain is light and easy to digest and is higher in protein than most other grains. Quinoa should be rinsed well before cooking, or it will have a noticeably bitter edge. This salad is best served warm or at room temperature when first made, but it can also be refrigerated for up to one day. If you're making it ahead, add the pine nuts just before serving.
4 Servings

1 cup quinoa
¾ teaspoon salt
¼ cup dried currants
3 tablespoons pine nuts
1 tablespoon fresh lemon juice
1 teaspoon finely grated lemon zest
¼ teaspoon paprika
¼ teaspoon cumin
¼ teaspoon ground coriander
1 tablespoon minced fresh coriander (cilantro)
¼ cup fruity olive oil
6 dried apricot halves, finely diced
2 tablespoons snipped chives or 3 small scallions, green part only, thinly sliced
3 tablespoons finely diced yellow or green bell pepper
Boston or Bibb lettuce leaves

1. Rinse the quinoa very well in a bowl of cool water. Pour into a strainer and rinse briefly.

2. In a medium saucepan, bring 2 cups of water to a boil. Add ½ teaspoon of the salt and stir in the quinoa. Cover and cook over low heat until slightly chewy, about 10 minutes. If all the water has not been absorbed, drain well. Transfer the quinoa to a large bowl and stir in the currants.

3. Meanwhile, toast the pine nuts in a small dry skillet over moderate heat, shaking the pan from time to time, until the nuts are golden brown, about 5 minutes. Set aside.

4. In a small bowl, combine the lemon juice, lemon zest, paprika, cumin, ground coriander, 2 teaspoons of the fresh coriander and the remaining ¼ teaspoon salt. Whisk in the olive oil.

5. Add the apricots, chives, bell pepper, toasted pine nuts and the dressing to the warm quinoa and toss well. To serve, arrange the lettuce leaves on a small platter and mound the salad on top. Sprinkle with the remaining 1 teaspoon fresh coriander before serving.
—Deborah Madison

CURRIED RICE SALAD

In this salad, the spice of curry powder is tempered by the sweetness of fresh papaya and preserved mango chutney.
8 Servings

2 cups rice
2 teaspoons salt
1 large yellow bell pepper, finely diced
1 large red bell pepper, finely diced
1 large green bell pepper, finely diced
1 ripe papaya—peeled, seeded and finely diced
2 tablespoons finely diced mango pieces from Major Grey chutney
1 tablespoon Dijon mustard
4 teaspoons curry powder
Pinch of cayenne pepper
2 tablespoons white wine vinegar
¼ teaspoon freshly ground black pepper
⅓ cup extra-virgin olive oil
½ pint cherry tomatoes, halved

1. In a large saucepan, bring 2 quarts of water to a boil over high heat. Add the rice and 1½ teaspoons of the salt. Stir frequently until the water returns to a boil. Cook until the rice is tender, 12 to 15 minutes. Drain the rice in a colander and rinse under cold water; drain well. Let stand for 15 minutes.

2. Meanwhile, in a large bowl, combine the diced bell peppers, the papaya and the mango.

3. In a small bowl, whisk together the mustard, curry powder, cayenne, vinegar, remaining ½ teaspoon salt and the black pepper. Gradually whisk in the olive oil in a thin stream; the dressing should be thick.

4. Add the dressing and the rice to the bell peppers and toss well. Cover and refrigerate overnight if you like. Toss with the cherry tomato halves before serving. Taste and add more salt if desired. Serve at room temperature.
—*Bob Chambers*

RICE SALAD WITH MINT AND TOMATILLOS

Two large green peppers, cut into small dice, can be substituted for the fresh tomatillos; the recipe will be different, but good nonetheless. Do not use canned tomatillos as they are too soft to be used in this recipe. This salad is best served at room temperature.

6 to 8 Servings

1¼ cups converted rice
1 teaspoon mild vegetable oil
1¼ teaspoons salt
5 fresh tomatillos
4 medium garlic cloves, unpeeled
1 tablespoon plus 1 teaspoon minced fresh ginger
2 tablespoons sugar
¼ cup fresh lime juice
2 teaspoons Oriental sesame oil
2 teaspoons soy sauce
⅛ teaspoon hot pepper sauce
1 large papaya—peeled, seeded and cut into ¼-inch dice
3 tablespoons finely chopped fresh mint

1. In a medium saucepan, bring 2¾ cups of water to a boil. Add the rice, vegetable oil and 1 teaspoon of the salt and stir well. Cover and simmer over low heat until the water is absorbed and the rice is tender, about 20 minutes. Remove from the heat and let stand, covered, for 10 minutes. Spoon the rice into a large serving bowl and fluff well. Refrigerate, uncovered, until cooled.

2. Remove the husks from the tomatillos. Rinse to remove the sticky coating and dry well. Pierce the tomatillos in 1 or 2 places with a sharp knife.

3. Line a large heavy skillet or a griddle with foil. Add the tomatillos and roast over moderately high heat, turning, until blackened in spots and barely soft, about 10 minutes. Halve and discard the white cores, then dice the tomatillos.

4. Roast the garlic the same way but over moderate heat, turning frequently, until barely soft but not blackened, about 12 minutes. Peel the garlic.

5. In a food processor, puree the garlic. Add the ginger, sugar and the remaining ¼ teaspoon salt. Puree until smooth. Add the lime juice, sesame oil, soy sauce and hot pepper sauce. Process until the dressing is well blended.

6. Add the tomatillos, papaya and mint to the cooked rice and toss to mix. *(The rice and dressing can be prepared to this point up to 1 day ahead and refrigerated separately.)*

7. Two to 4 hours before serving, toss the rice with the dressing. Mix again just before serving.
—*Linda Burum & Linda Merinoff*

8 DESSERTS

CITRUS SALAD WITH HONEY-LIME DRESSING

This refreshing fruit salad has the right balance of tartness and sweetness. The fruit here can be prepared a day ahead without damaging taste or texture.
8 Servings

4 large pink grapefruits
4 navel oranges
2 tablespoons honey
1 tablespoon fresh lime juice
½ teaspoon finely grated lime zest
Pinch of salt

1. Using a sharp knife, peel the grapefruits, removing all the bitter white pith. Working over a bowl, cut in between the membranes to release the sections. Discard the membranes.

2. Using a small sharp knife, peel the oranges, removing all the bitter white pith. Slice the oranges crosswise ¼ inch thick and add to the bowl of grapefruit sections. *(The recipe can be prepared to this point up to 1 day ahead. Cover and refrigerate the fruit and its juice. Let return to room temperature for about 1 hour before proceeding.)*

3. In a small bowl, whisk together the honey, lime juice, lime zest and salt until smoothly combined.

4. Arrange the fruit on a large platter: Lay the orange slices in the center in an overlapping circle and surround with the grapefruit sections. Stir 2 tablespoons of the juice that has collected at the bottom of the fruit bowl into the honey-lime dressing and spoon it over the fruit just before serving.
—Tracey Seaman

BERRY FRUIT SALAD

Macerating the berries overnight in the refrigerator with sugar and lemon zest yields a lovely syrup and lends a subtle lemon flavor to the berries. Serve them with sweetened whipped cream flavored with fresh lemon juice or with lemon sherbet.
6 to 8 Servings

3 cups strawberries (about 1 pint), quartered
3 cups blueberries (about 1 pint)
1½ cups raspberries (about ½ pint)
½ cup sugar, preferably superfine
½ teaspoon grated lemon zest

In a large glass bowl, toss together the strawberries, blueberries and raspberries with the sugar and lemon zest. Cover the bowl with plastic wrap and refrigerate overnight before serving.
—Diana Sturgis

STRAWBERRY AND FRESH FIG SALAD

Strawberries sprinkled with balsamic vinegar have been around for a while. This version, using tender, locally grown berries, sweet juicy figs and a sprinkling of shredded basil, is good as a light refreshing dessert, first-course salad or intermezzo. This recipe can be doubled or tripled.
2 Servings

1½ cups halved or quartered strawberries (about 1 pint)
3 fresh figs, quartered
2 teaspoons balsamic vinegar
2 tablespoons shredded fresh basil leaves

In a small bowl, combine the strawberries, figs, vinegar and basil. Gently toss together. Divide the salad evenly between 2 small plates.
—Diana Sturgis

STRAWBERRY, KIWI AND ORANGE SALAD

The tart, tangy citrus dressing on this salad would work equally well with almost any other combination of fruits.
8 Servings

1½ cups fresh orange juice (from about 5 oranges)
⅓ cup fresh lemon juice
¼ cup sugar
2 teaspoons arrowroot
2 tablespoons orange liqueur
4 navel oranges
4 kiwis, peeled and cut lengthwise into eighths
5 pints of ripe strawberries, halved lengthwise if large
Fresh mint sprigs, for garnish

1. In a small nonreactive saucepan, combine the orange juice, lemon juice and sugar and bring to a boil. Reduce the heat to low and simmer until reduced to 1⅓ cups, about 10 minutes.

2. Dilute the arrowroot in 1 tablespoon of cold water. Stir into the syrup and return to a boil, stirring. Remove from the heat and let cool to room temperature. Stir in the orange liqueur. Cover and refrigerate for at least 2 hours.

3. Using a sharp knife, peel the oranges, removing all the bitter white pith. Cut in between the membranes to release the sections.

4. Toss the fruit with the dressing up to 2 hours before serving. Garnish with mint sprigs.
—Bob Chambers

SWEET BEAN CRISPS

You'll be pleasantly surprised when you taste these crunchy bean cookies. Serve them for dessert alongside vanilla ice cream or fill the cookie jar and watch them disappear.

Makes About 5 Dozen

1¼ cups all-purpose flour
2 teaspoons baking powder
½ teaspoon cinnamon
½ teaspoon ground cloves
½ teaspoon salt
½ cup golden raisins
1 teaspoon freshly grated orange zest
½ cup sweet Marsala
1½ tablespoons Pernod
1 can (19 ounces) chickpeas—drained, rinsed and patted dry
1 cup milk
1 teaspoon vanilla extract
1 stick (4 ounces) unsalted butter, softened
¾ cup (packed) dark brown sugar
1 egg, at room temperature
Confectioners' sugar, for dusting

1. Preheat the oven to 375°. Sift together the flour, baking powder, cinnamon, cloves and salt.
2. In a small saucepan, combine the raisins, orange zest, Marsala and Pernod. Bring to a boil over high heat. Remove from the heat and let cool to room temperature.
3. In a food processor or blender, combine the chickpeas, milk and vanilla. Puree until smooth, 4 to 5 minutes.
4. In a large bowl, cream together the butter and brown sugar until fluffy. Beat in the egg. Blend in the cooled raisin mixture and the chickpea puree. Add the dry ingredients and mix until thoroughly combined.
5. Spoon heaping teaspoonfuls of batter about 2 inches apart onto lightly greased cookie sheets. Bake for 25 to 30 minutes, or until the cookies are lightly browned around the edges. Transfer to wire racks and let cool completely. Dust lightly with confectioners' sugar.
—*Anne Disrude*

SWEET CHESTNUT CROQUETTES

8 Servings

1 cup milk
1 vanilla bean, split, or 1 teaspoon vanilla extract
3 egg yolks
¼ cup sugar
2½ tablespoons all-purpose flour
1 teaspoon unsalted butter, at room temperature
1 cup ground roasted chestnuts* (about 5 ounces)
½ cup coarsely chopped glazed chestnut pieces*
1 cup all-purpose flour
2 whole eggs, beaten
2 cups fresh bread crumbs (made from 5 or 6 slices of firm-textured white bread, crusts removed)
1½ to 2 quarts peanut oil, for deep-frying
Confectioners' sugar, for garnish
*Available at specialty food shops

1. In a heavy medium saucepan, bring the milk with the vanilla bean to a boil over moderately high heat; immediately reduce the heat to low and simmer for 5 minutes.
2. Meanwhile, in a medium bowl, whisk the egg yolks and sugar together until thickened and lemon colored. Beat in the flour until blended.
3. Remove the vanilla bean from the milk (see Note) and, whisking constantly, gradually pour the hot milk into the egg yolk mixture. Return to the saucepan and bring to a boil over moderately high heat, stirring constantly. Cook, stirring constantly, until the pastry cream is thick and smooth and begins to pull away from the sides of the pan, about 3 minutes. Scrape into a small bowl; dot the surface with the butter to prevent a skin from forming and let cool to room temperature.
4. Fold the ground and chopped chestnuts into the pastry cream. Cover tightly and refrigerate until well chilled, about 3 hours or overnight.
5. Place the flour in a shallow dish, the whole beaten eggs in a second dish and the bread crumbs in a third shallow dish. To make each croquette, scoop out 1 tablespoon of the pastry cream and quickly roll it into a small ball between your palms. Roll each ball in flour to coat all over; shake lightly on your fingers to remove any excess. Dip into the beaten eggs, letting the excess drip back into the dish. Roll in the bread crumbs until completely coated. Roll lightly between your palms to remove excess crumbs.
6. Heat 2½ inches of oil in a deep-fat fryer or deep heavy saucepan to 375°. Fry the croquettes in batches without crowding for about 3 minutes, or until golden brown. Remove and drain on paper towels. Serve hot, sprinkled with confectioners' sugar, if desired.

NOTE: The vanilla bean can be rinsed off, dried and used again.
—*John Robert Massie*

SWEET RICE WITH APRICOT CROQUETTES

The croquette mixture should be chilled for at least 3 hours, so plan accordingly.
8 Servings

½ cup rice
1½ cups milk
1 vanilla bean, split, or 1 teaspoon vanilla extract
¼ cup sugar
1 tablespoon unsalted butter
Pinch of salt
3 egg yolks, lightly beaten
½ cup apricot or other fruit preserves, strained through a sieve
1 cup all-purpose flour
2 whole eggs, beaten
2 cups fresh bread crumbs (made from 5 or 6 slices of firm-textured white bread, crusts removed)
1½ to 2 quarts peanut oil, for deep-frying
Apricot Sauce (recipe follows) or confectioners' sugar, for serving

1. Preheat the oven to 300°. Place the rice in a medium heatproof saucepan or casserole and add cold water to cover. Bring to a boil over moderately high heat. Drain, rinse with warm tap water, drain again and return to the pan.

2. In another medium saucepan, combine the milk, vanilla bean and sugar. Bring to a boil over moderate heat; immediately remove from the heat. Strain the hot milk over the rice; reserve the vanilla bean (see Note). Stir in the butter and salt and return to a boil. Cover and place in the oven. Bake for about 30 minutes, or until all the liquid is absorbed. The rice will be sticky.

3. Mix the egg yolks into the hot rice. Transfer to a bowl and let cool slightly. Cover and refrigerate until well chilled, about 3 hours or overnight.

4. To make each croquette, scoop out 1 tablespoon of the chilled rice and quickly roll it into a ball between your palms. Use the handle of a wooden spoon or a chopstick to make a hole through to the center of each ball. Using a small pastry bag with a small plain tip or a cone made from a triangle of parchment, squeeze a dab of apricot preserves into the center of each croquette. Pinch to close and roll the ball back into shape.

5. Place the flour in a shallow dish, the whole beaten eggs in a second dish and the bread crumbs in a third shallow dish. Roll each ball of rice in the flour to coat all over; shake lightly on your fingers to remove any excess. Dip into the beaten eggs, letting the excess drip back into the dish. Roll in the bread crumbs until completely coated. Roll lightly between your palms to remove excess crumbs.

6. Fry the croquettes: Heat 2½ inches of oil in a deep-fat fryer or deep heavy saucepan to 375°. Fry the croquettes in batches without crowding for about 3 minutes, or until golden brown. Remove and drain on paper towels.

7. Serve hot with Apricot Sauce or dust with confectioners' sugar.

NOTE: The vanilla bean can be rinsed off, dried and used again.
—*John Robert Massie*

APRICOT SAUCE
Makes About 1 Cup

1 jar (12 ounces) apricot preserves
2 tablespoons brandy

Strain the preserves through a fine-mesh sieve into a small saucepan. Bring to a boil over moderate heat. Remove from the heat and stir in the brandy. Serve hot or warm.
—*John Robert Massie*

CHOCOLATE-BEET BUNDT CAKE WITH PECANS

Adding ground beets to the batter makes this cake very moist. The beets must be chopped very fine—two or three 10-second runs in the food processor should do it. Slice the cake thin (it's as rich as pound cake) and serve with whipped cream.
12 to 16 Servings

½ cup plus 2 tablespoons unsweetened cocoa powder
2¾ cups sifted all-purpose flour
1 cup finely chopped pecans
1 teaspoon baking powder
3 sticks (¾ pound) unsalted butter, at room temperature
2 cups granulated sugar
5 jumbo eggs
2 teaspoons vanilla extract
2 medium beets—peeled, cut into pieces and very finely chopped
¾ cup milk, at room temperature
1 tablespoon confectioners' sugar

1. Preheat the oven to 300°. Butter a 12-cup bundt pan well, paying special attention to the crevices and central tube. Dust the pan with 2 tablespoons of the cocoa, tapping out any excess.

2. In a small bowl, combine ¼ cup of the flour with the pecans; toss well to coat and set aside. Into a medium bowl, sift the remaining 2½ cups flour and ½ cup cocoa powder with the baking powder 3 times; set aside.

3. In an electric mixer, cream the butter until silvery and light, 3 to 4 minutes. Beat in the granulated sugar, 1 tablespoon at a time, then beat on high speed until fluffy, 2 to 3 minutes.

4. With the mixer on low, beat in the eggs, 1 at a time. Beat in the vanilla. Add the beets and mix until incorporated. By hand, stir in the sifted dry ingredients alternately with the milk, using about 4 additions of the dry and 3

of milk and beating just enough to mix after each addition. Fold in the pecans and any dredging flour.

5. Spoon the batter into the pan and bake for 1 hour and 30 minutes, or until the cake has pulled away from the sides and a skewer inserted in the center comes out clean; the top will crack slightly.

6. Let the cake cool in the pan on a rack. To unmold, loosen around the edge and central tube with a thin spatula, then invert onto the rack and let cool for 1 hour. Just before serving, dust the cake lightly with confectioners' sugar.
—Jean Anderson

RED BEET DEVIL'S
FOOD CAKE
Serve this cake with whipped cream or à la mode with coffee ice cream.
Makes One 13-by-9-Inch Cake

6 medium beets
3 ounces unsweetened chocolate
1½ cups granulated sugar
3 eggs
1 cup vegetable oil
1¾ cups all-purpose flour
1½ teaspoons baking soda
Confectioners' sugar, for dusting

1. In a medium saucepan, combine the beets and enough water to cover. Bring to a boil over high heat, reduce the heat to moderately low and simmer until tender, about 1 hour. Drain, rinse with cold water and peel the beets. In a food processor, puree until completely smooth.

2. Preheat the oven to 350°. Lightly grease a 13-by-9-by-2-inch baking pan. In a small double boiler, melt the chocolate. Set aside to cool.

3. In a medium bowl, whisk the granulated sugar with the eggs and oil until blended. Stir in the pureed beets and the cooled chocolate.

4. Sift the flour with the baking soda. Stir the dry ingredients into the chocolate-beet mixture. Spoon the batter into the prepared pan and bake for 25 to 30 minutes, or until the cake springs back slightly when touched. Transfer to a rack to cool completely. Dust the top of the cake with confectioners' sugar.
—Marcy Goldman

MACE-SCENTED
CARROT TORTE
Using a food processor with a steel blade to grind the nuts and mince the carrots makes this torte quick and easy.
8 Servings

1 cup pecans (4 ounces), ground
2 tablespoons all-purpose flour
2 teaspoons baking powder
½ teaspoon ground mace
4 eggs, separated
⅔ cup granulated sugar
2 tablespoons fresh lemon juice
⅔ cup minced carrots (about 2 medium)
Confectioners' sugar, for dusting

1. Preheat the oven to 350°. Butter an 8-inch springform pan. Line the bottom with parchment or wax paper and butter the paper. Dust the pan with flour; tap out any excess. In a small bowl, combine the pecans, flour, baking powder and mace.

2. In a large bowl, whisk together the egg yolks and granulated sugar until pale. Add the lemon juice, carrots and pecan-flour mixture. Stir until blended.

3. Beat the egg whites until they form stiff peaks. Fold into the carrot mixture; do not overmix. Turn the batter into the prepared pan and smooth the top.

4. Bake in the center of the oven for 60 to 65 minutes, until golden, this type of torte will naturally sink slightly in the center. Let cool on a rack for 10 minutes. Run a blunt knife around the sides of the pan and unmold the cake. Turn right-side up and let cool completely on a rack. Slide onto a platter and garnish with a sprinkling of confectioners' sugar.
—Dorie Greenspan

PLYMOUTH BAY
BERRY CAKE
One of my most pleasant summer memories is of an August spent on Cape Cod, where we hiked through the pines collecting buckets of wild blueberries. Once home, we ground hard yellow corn into a fine grain and made hearty blueberry corn muffins and pancakes. The rich, moist Plymouth Bay Berry Cake evokes those blue and yellow summer days by the sea.
12 Servings

Cornmeal Cake (recipe follows)
2½ cups heavy cream, chilled
¾ cup plus 3 tablespoons superfine sugar
1 pint blueberries (about 2 cups)
1 cup diced fresh fruit, such as nectarines, peaches, sliced bananas and strawberries or raspberries
6 egg whites, at room temperature
Pinch of salt

1. Preheat the oven to 500°. Using a long serrated knife, trim the top of the Cornmeal Cake to make an even surface. Slice the cake horizontally into 4 even layers. Place the bottom layer, crusty-side down, on a springform pan base.

2. In a large chilled bowl, beat the cream until it begins to thicken, about 1 minute. Gradually add 3 tablespoons of the sugar and beat until stiff, about 1 minute.

3. Using a spatula, spread ⅔ cup of the sweetened whipped cream over the bottom cake layer to within ¼ inch of the edge. Top with half of the blueberries. Cover with 1 cup of the whipped cream, spreading it evenly and keeping a ¼-inch border. Place the second cake layer on top and cover with a layer of whipped cream and ¾ cup of the diced fruit. Top with the third cake layer and the remaining blueberries and whipped cream. Top with the last cake layer, cover with plastic wrap and refrigerate.

4. Meanwhile, in a large mixer bowl, beat the egg whites on medium speed until frothy. Add the salt and beat until soft peaks form, about 2 minutes. Gradually beat in the remaining ¾ cup sugar, 1 tablespoon at a time, increasing the speed to high before adding the last 3 tablespoons. Beat until the meringue forms stiff peaks, about 1 minute.

5. Cover the top and sides of the cake with the meringue, swirling it on top into a decorative pattern. Place the cake (on its springform base) on a cookie sheet and bake for 4 to 5 minutes, or until the top and sides are golden brown. Transfer to a serving platter and decoratively garnish the top of the cake with the remaining ¼ cup diced fruit.
—Peggy Cullen

CORNMEAL CAKE
Makes One 9-Inch Round Cake

¾ cup yellow cornmeal
½ cup plus 3 tablespoons all-purpose flour
2 teaspoons baking powder
¼ teaspoon salt
2 sticks (8 ounces) unsalted butter, softened to room temperature
1¼ cups superfine sugar
4 whole eggs, at room temperature
6 egg yolks, at room temperature
1 teaspoon vanilla extract

1. Preheat the oven to 350°. Butter a 9-inch springform pan. Sprinkle 2 tablespoons of the cornmeal into the pan and tilt and shake to coat the bottom and sides; tap out any excess.

2. In a medium bowl, sift together the flour, baking powder, salt and remaining ½ cup plus 2 tablespoons cornmeal.

3. In a large mixer bowl, beat the butter on medium-high speed until light and fluffy, about 1 minute. Gradually add the sugar and beat until the sugar is completely dissolved, about 3 minutes. Add the whole eggs and egg yolks, 1 at a time, beating well after each addition. Beat in the vanilla.

4. Gradually fold in the flour mixture until well blended. Scrape the batter into the prepared pan and bake for 40 to 45 minutes, or until the cake is golden brown and a toothpick inserted into the center comes out clean. Let cool on a rack for 10 minutes. Invert to unmold onto the rack and let the cake cool completely. (*The recipe can be prepared up to 1 day in advance, wrapped in plastic wrap and stored at room temperature or frozen for up to 2 months.*)
—Peggy Cullen

LEMON-GLAZED PARSNIP TEA BREAD

Parsnips add moistness and sweetness to this quick bread, and freshly ground pepper heightens its piquancy. The top of the loaf may crack slightly, but this is characteristic of rich vegetable breads.
12 Servings

2 tablespoons fine dry bread crumbs
1 pound parsnips, peeled and cut into 2-inch chunks
2¼ cups sifted all-purpose flour
½ cup dried currants
½ cup coarsely chopped walnuts
1 tablespoon baking powder
1 teaspoon cinnamon
1 teaspoon ground ginger
½ teaspoon ground cloves
½ teaspoon freshly grated nutmeg
½ teaspoon ground cardamom
¼ teaspoon freshly ground pepper
¼ teaspoon salt
¾ cup (packed) light brown sugar
1 tablespoon grated orange zest
2 jumbo eggs, lightly beaten
½ cup flavorless vegetable oil
½ cup milk
¼ cup fresh lemon juice
1 cup confectioners' sugar

1. Preheat the oven to 350°. Butter a 9-by-3-inch loaf pan. Coat the bottom and sides of the pan with the bread crumbs; tap out the excess.

2. Finely chop the parsnips in a food processor.

3. In a small bowl, combine ¼ cup of the flour with the currants and walnuts; toss well.

4. Into a large bowl, sift the remaining 2 cups flour 3 times with the baking powder, cinnamon, ginger, cloves, nutmeg, cardamom, pepper and salt. Add the brown sugar and orange zest and stir well to mix. Blend in the nuts, currants and all the dredging flour.

5. In another bowl, whisk together the eggs, oil, milk and 2 tablespoons of the lemon juice; add the parsnips.

6. Make a well in the dry ingredients and add the parsnip mixture; stir just to combine. Spoon the batter (it will be very stiff) into the pan, spreading it evenly and smoothing the top. Bake for 1 hour and 15 minutes, or until a skewer inserted in the center comes out clean.

7. Let the loaf cool on a rack for 15 minutes. Loosen around the edges with a thin spatula, turn out onto the rack and let cool for 1 hour.

8. In a small bowl, combine the confectioners' sugar with the remaining 2 tablespoons lemon juice. Drizzle the glaze over the loaf, letting the excess drip down the sides. Smooth the top, then let set for 1 hour before slicing.
—Jean Anderson

PUMPKIN-GINGER CAKE

In this exquisitely tender cake, ginger is blended with another spice—allspice—to produce a deep and complex flavor that is especially pleasing when combined with pumpkin and molasses.
8 Servings

1 stick (4 ounces) unsalted butter, softened
½ cup maple syrup
½ cup (packed) light brown sugar
2 eggs, lightly beaten, at room temperature
2 tablespoons unsulphured molasses
½ teaspoon vanilla extract
1 cup canned unsweetened pumpkin puree
1¾ cups all-purpose flour
2¼ teaspoons ground ginger
1 teaspoon baking soda
½ teaspoon salt
½ teaspoon ground allspice

1. Preheat the oven to 350°. Butter and flour a shallow 4-cup ring mold or a deep, 4-cup kugelhopf pan. In a large bowl, beat together the butter, maple syrup and brown sugar. Stir in the eggs, molasses and vanilla until combined. Stir in the pumpkin puree.

2. In a sifter or sieve set over the bowl, combine the flour, ginger, baking soda, salt and allspice. Gradually sift the dry ingredients into the pumpkin mixture, folding them in with a rubber spatula until just blended.

3. Pour the batter into the prepared pan and smooth the surface with the spatula. Bake the cake for 30 minutes for the shallow pan and up to 45 minutes for the kugelhopf pan, or until a cake tester inserted in the center comes out dry. Let cool for about 10 minutes in the pan before unmolding onto a rack. Serve warm or at room temperature.
—Marcia Kiesel

PUMPKIN BOURBON POUND CAKE
10 to 12 Servings

2 sticks (8 ounces) unsalted butter, softened
1 cup granulated sugar
1 cup light brown sugar
4 eggs, at room temperature, beaten
¼ cup milk
2 tablespoons bourbon
1 teaspoon vanilla extract
2 cups fresh or canned unsweetened pumpkin puree
3 cups all-purpose flour
2 teaspoons baking soda
2 teaspoons baking powder
1¾ teaspoons cinnamon
½ teaspoon ground ginger
¼ teaspoon ground allspice
¼ teaspoon freshly grated nutmeg
Small pinch of salt
1 cup plus 2 tablespoons pecan halves, coarsely chopped
Bourbon Caramel Glaze (recipe follows)

1. Preheat the oven to 350°. Generously grease and flour a 12-cup angel food cake pan or bundt pan.

2. In a large bowl, using an electric mixer, cream the butter, granulated sugar and brown sugar until fluffy. Gradually beat in the eggs on low speed. Beat in the milk, bourbon, vanilla and pumpkin puree.

3. In a bowl, sift the flour with the baking soda, baking powder, cinnamon, ginger, allspice, nutmeg and salt. Add the dry ingredients to the pumpkin mixture and mix just until blended. Stir in 1 cup of the pecans.

4. Spoon the batter into the prepared pan and bake for about 1 hour, or until a cake tester inserted in the center comes out clean. Let the cake cool in the pan for 10 minutes before inverting onto a rack to cool completely. Transfer to a plate.

5. Drizzle the warm Bourbon Caramel Glaze over the top of the cake. Garnish with the remaining 2 tablespoons pecans.
—Marcy Goldman

BOURBON CARAMEL GLAZE
Makes About 1¼ Cups

1 stick (4 ounces) unsalted butter
½ cup (packed) light brown sugar
About 3 tablespoons heavy cream
1½ cups confectioners' sugar, sifted
3 tablespoons bourbon
¼ teaspoon vanilla extract

In a small heavy saucepan, melt the butter over moderately low heat. Stir in the brown sugar until incorporated. Stir in the 3 tablespoons cream and simmer, stirring, for 3 minutes. Using an electric mixer, beat in the confectioners' sugar, bourbon and vanilla. If necessary, add more cream to achieve a heavy glaze consistency. Use warm.
—Marcy Goldman

BUTTERNUT SQUASH, APPLESAUCE AND DATE CAKE
Makes Two 9-Inch Loaves

2 cups (loosely packed) pitted dates
1 small butternut squash
2 cups sugar
1½ cups vegetable oil
4 eggs
1 cup unsweetened applesauce
1 teaspoon vanilla extract
1 teaspoon finely minced lemon zest
3⅔ cups all-purpose flour
2 teaspoons baking powder
2 teaspoons baking soda
2 teaspoons cinnamon
¼ teaspoon ground ginger
½ cup finely chopped walnuts

1. Preheat the oven to 350°. Generously grease two 9-inch loaf pans. Soak the dates in a bowl of very hot water for 5 minutes. Drain, halve lengthwise and set aside.

2. Halve the squash lengthwise and scoop out and discard the seeds. Place on a baking sheet cut-side down and bake for about 30 minutes, until soft. Let cool, then scoop the pulp into a food processor; puree until smooth. Measure 1 cup of the puree for the cake and save the remainder for another use.

3. In a large bowl, beat the sugar, oil, eggs, squash puree, applesauce, vanilla and lemon zest with an electric mixer until well combined.

4. In a medium bowl, sift the flour with the baking powder, baking soda, cinnamon and ginger. On low speed, gradually beat the dry ingredients into the wet ingredients just until blended. Stir in the dates and walnuts.

5. Spoon the batter into the prepared pans. Bake the cakes for 1 hour, or until golden brown and a cake tester inserted in their centers comes out clean. Transfer to a rack to cool slightly, then invert the cakes onto the rack to cool completely.
—Marcy Goldman

ORANGE-RHUBARB BREAKFAST CAKE
Makes One 13-by-9-Inch Cake

⅓ cup granulated sugar
1 tablespoon unsalted butter, softened
½ cup finely chopped walnuts
1½ teaspoons finely grated orange zest
1 teaspoon cinnamon
1⅓ cups (packed) light brown sugar
⅔ cup vegetable oil
1 cup buttermilk
1 egg
1 teaspoon vanilla extract
2¼ cups all-purpose flour
¼ cup oat bran
1 teaspoon baking soda
Pinch of salt
2 cups fresh or partially thawed frozen rhubarb, cut into ½-inch pieces
1 tablespoon confectioners' sugar

1. Preheat the oven to 350°. Lightly butter a 13-by-9-by-2-inch baking pan.

2. In a small bowl, using your fingers, rub the granulated sugar and butter together until blended. Mix in the walnuts and ½ teaspoon each of the orange zest and cinnamon. Set the streusel topping aside.

3. In a large bowl, blend the brown sugar and oil. Add the buttermilk, egg and vanilla and stir to combine. In a medium bowl, toss the flour, oat bran, baking soda, salt and remaining 1 teaspoon orange zest and ½ teaspoon cinnamon. Stir the dry ingredients into the buttermilk mixture until combined. Fold in the rhubarb.

4. Spoon the batter into the prepared pan. Sprinkle the reserved streusel topping over the top. Bake for about 50 minutes, or until a cake tester inserted in the center comes out clean. Let cool on a rack for at least 10 minutes. Sift the confectioners' sugar on top and serve.
—Marcy Goldman

CARROT CHEESECAKE WITH NUT-CRUMB CRUST

The important thing to remember in this recipe is to cook the carrots until they are soft enough to puree to silk.
8 Servings

¾ pound carrots, cut into 2-inch chunks
⅔ cup fine graham cracker crumbs
⅔ cup fine gingersnap crumbs
⅔ cup finely ground pecans or walnuts plus ¼ cup finely chopped pecans or walnuts
⅓ cup granulated sugar
4 tablespoons unsalted butter, at room temperature
½ cup (packed) light brown sugar
4 teaspoons minced preserved stem ginger
1 tablespoon fresh lemon juice
1 teaspoon grated orange zest
¼ teaspoon cinnamon
¼ teaspoon mace
¼ teaspoon allspice
1 pound cream cheese, cut into small cubes and softened
4 eggs

1. In a medium saucepan, bring 4 cups of water to a boil. Add the carrots, cover and cook over moderate heat until very tender, about 45 minutes.

2. Meanwhile, preheat the oven to 400°. In a medium bowl, combine the graham cracker and gingersnap crumbs with the ⅔ cup ground nuts and the granulated sugar; toss well. Work in the butter until the mixture is uniformly crumbly.

3. Pat the crust mixture over the bottom and up the sides of a buttered 9-inch springform pan. Bake the crust for 5 to 7 minutes, until firm. Remove from the oven and set aside. Reduce the oven temperature to 350°.

4. When the carrots are done, drain well. Return them to the pan and cook over moderate heat, shaking the pan, to cook off any excess moisture. Transfer the carrots to a food processor and puree for 30 seconds; scrape down the sides of the bowl and puree for 30 seconds longer, until absolutely smooth.

5. Add the brown sugar, ginger, lemon juice, orange zest, cinnamon, mace and allspice to the processor and puree for 30 seconds. Scrape down the sides and puree for 30 seconds longer. Let the mixture stand until cool.

6. Add the cream cheese and puree for 1 minute, stopping to scrape down the sides of the bowl every 20 seconds.

7. Beat in the eggs, 1 at a time, by turning the machine on and off 5 times after each addition. Pour the mixture into the crust and scatter the ¼ cup chopped nuts over the top.

8. Bake for about 50 minutes, until softly set and a cake tester inserted midway between the center and the edge comes out clean. Remove from the oven and let cool. Cover loosely and refrigerate for 1 to 2 hours before serving.
—Jean Anderson

APPLE AND GREEN TOMATO TART WITH CALVADOS CREAM

Neatness counts in this recipe. Cut the apple and tomato slices to a uniform size. Eat or discard the small ends.

6 to 8 Servings

1 cup all-purpose flour
⅓ cup confectioners' sugar
1 stick (4 ounces) unsalted butter, softened
1 large tart apple—peeled, cored and cut lengthwise into thin rounds
1 large green tomato, cored and cut lengthwise into thin rounds
1 tablespoon fresh lemon juice
3 tablespoons granulated sugar
1 tablespoon cornstarch
½ cup blackberry preserves
1 cup heavy cream
2 tablespoons Calvados or applejack

1. Sift the flour and confectioners' sugar into a medium bowl. Add the butter and blend well to form a stiff dough. Press and pat the dough into the bottom of a rectangular tart pan, preferably black, 9 by 3 inches. Don't line the sides of the pan, but leave a little lip around the edges of the dough. Cover with plastic wrap and refrigerate for at least 20 minutes.

2. Preheat the oven to 350°. Prebake the pastry for 30 minutes, or until dry.

3. Arrange alternating, overlapping slices of the apple and green tomato down the length of the prebaked pastry. Sprinkle with the lemon juice. Mix the granulated sugar and cornstarch and sift evenly over the apple and tomato slices.

4. Return to the 350° oven and bake for 25 minutes, or until the fruit is tender and the crust lightly browned. Slide the tart onto a serving plate.

5. In a small nonreactive saucepan, heat the preserves over moderately low heat, stirring occasionally, until melted, about 3 minutes. Strain to remove the seeds. Spread over the fruit.

6. Whip the cream until soft peaks form. Beat in the Calvados. Serve the tart warm or at room temperature, with the whipped cream on the side.
—Lee Bailey

RHUBARB PIE WITH LATTICED WALNUT CRUST

6 Servings

PASTRY
1¾ cups all-purpose flour
2 tablespoons sugar
1 teaspoon cinnamon
¼ teaspoon freshly grated nutmeg
¼ teaspoon salt
6 tablespoons cold unsalted butter, cut into small pieces
4 tablespoons chilled vegetable shortening, cut into small pieces
½ cup ground walnuts
1 egg
2 tablespoons heavy cream

FILLING
3½ cups rhubarb, cut into ½-inch pieces
3 tablespoons cornstarch
½ cup (packed) dark-brown sugar
½ to ⅔ cup granulated sugar, to taste

1. Make the pastry: Sift the flour, sugar, cinnamon, nutmeg and salt into a bowl.

2. Add the butter and shortening and blend them in with a pastry blender or your fingertips until you have small, even-sized particles. Add the ground nuts.

DESSERTS

3. Beat together the egg and 2 tablespoons heavy cream in a small bowl. Add the mixture, a bit at a time, to the flour-fat mixture, tossing with a fork until all the liquid is incorporated. Form the pastry into a ball, wrap it in plastic, and refrigerate it for at least 30 minutes.

4. Preheat the oven to 400°. Cut off two-thirds of the pastry and refrigerate the remainder. Roll the dough out on a floured surface into a round about 11 inches in diameter. Fit it into a buttered 9½-inch fluted tart pan with a removable bottom, making the sides slightly thicker than the bottom and fitting the pastry carefully into each curve of the pan. Trim the top edge neatly.

5. Prick the bottom of the pastry all over with a fork. Line the pan with buttered aluminum foil, buttered-side down, and fill it with beans or pie weights to reach the rim.

6. Set the pan on a baking sheet and bake it on the bottom shelf of the oven for 5 minutes. Remove the foil and weights and bake the shell for 2 or 3 minutes longer, or until it barely begins to color. Place the pan on a rack and cool the shell.

7. Make the filling: In a bowl, combine the rhubarb, cornstarch and brown and granulated sugars; toss well. Let stand for 30 minutes to 1 hour, stirring a few times. Meanwhile, preheat the oven to 425°.

8. Roll out the remaining dough into a round about 10 inches in diameter. Cut it into 12 strips with a plain or fluted pastry wheel or a knife.

9. Pour the rhubarb and its syrup into the shell. Moisten the ends of the longest pastry strip and place it over the center. Press down the ends. Criss-cross the pastry strips until you have a latticed top. Trim overhanging ends of strips.

10. Brush the lattice sparingly with some cream and sprinkle the pie with sugar.

11. Bake the pie in the lower level of the oven for 15 minutes. Place it on a shelf in the middle level, lower the heat to 375°, and bake for about 30 minutes more, or until the filling is bubbling and syrupy. If the edge of the pie begins to brown too much, cover it with foil.

12. Cool the pie for 10 minutes. Release the sides of the pan. Set the pie, still on the pan bottom, on a rack to cool.
—F&W

FRESH RHUBARB TART

Rhubarb is readily available throughout the year in the Paris markets. In the early spring months it comes from hothouses in Holland and is generally more tender and sweeter than the field-grown kind, which appears by late spring. When buying rhubarb, look for young stalks. Always discard the leaves (they are full of poisonous oxalic acid) and, if necessary, peel the stalks to remove the coarse strings before cooking. Although rhubarb is naturally tart, I like to add lemon zest and candied ginger to emphasize its unusual acidic flavor.

8 Servings

1½ cups unbleached flour
Pinch of salt
1 cup plus 2 tablespoons sugar
1 stick (4 ounces) plus 2 tablespoons cold unsalted butter, cut into small dice
1 teaspoon almond extract
2 pounds rhubarb stalks, trimmed and cut into ½-inch dice
2 tablespoons finely chopped candied ginger, plus fine slivers of candied ginger for garnish
1 teaspoon finely grated lemon zest
1 cup crème fraîche, for serving
Fresh mint sprigs, for garnish

1. In a food processor, combine the flour, salt and 2 tablespoons of the sugar. Pulse briefly to combine. Add the butter and process until crumbly, about 15 seconds. With the machine on, pour in the almond extract and 1 tablespoon of cold water and process until the dough just barely forms a ball.

2. Press and pat the dough evenly into a 9½-inch tart pan with a removable bottom; make sure to press the dough into the grooved sides of the pan. With a fork, prick the bottom of the tart shell at 1-inch intervals. Wrap well and freeze for at least 30 minutes or overnight.

3. Preheat the oven to 400°. Remove the tart shell from the freezer and bake in the middle of the oven for 15 minutes. Reduce the temperature to 375° and bake for about 15 minutes longer, until light golden brown all over. Remove the tart shell from the oven and reduce the temperature to 325°.

4. Meanwhile, in a heavy medium nonreactive saucepan, combine the rhubarb, chopped candied ginger and the remaining 1 cup sugar. Cover tightly and cook over moderate heat until the rhubarb is very soft, about 15 minutes. Uncover, increase the heat to high and cook, stirring constantly, until a thick puree forms, about 5 minutes longer. Let the puree cool for 10 minutes, then stir in the lemon zest.

5. Spread the rhubarb puree in the tart shell and bake for 20 minutes. Let cool to room temperature on a rack. Garnish the tart with the slivered candied ginger.

6. Remove the sides of the pan and serve the tart in wedges, garnished with a large dollop of crème fraîche and fresh mint sprigs.
—Ann Chantal Altman

Classic Squid Salad (p. 42).

*At left, Scalloped Potatoes with Sweet Marjoram and Parmesan Cheese (p. 177).
Above, Rosemary-Scented Sweet Red Pepper Bisque (p. 81).*

Pumpkin-Ginger Cake (p. 245).

BUTTERMILK RHUBARB COBBLER

Buttermilk makes an exceptionally light cobbler batter, and rhubarb is the right fruit match.

6 to 8 Servings

¾ cup plus
2 tablespoons sugar
1 cup plus 1 tablespoon all-purpose flour
½ teaspoon finely grated orange zest
¼ teaspoon cinnamon
1½ pounds fresh rhubarb, cut into ¾-inch pieces or 1½ pounds frozen rhubarb, thawed (a scant 6 cups)
1 teaspoon baking powder
1 teaspoon baking soda
¼ teaspoon salt
4 tablespoons cold unsalted butter, cut into small pieces
⅔ cup buttermilk
Softly whipped cream, pouring cream or vanilla ice cream, for serving (optional)

1. Preheat the oven to 400°. Generously butter a shallow 1½- to 2-quart oval or rectangular nonreactive baking dish.

2. In a large bowl, stir together ¾ cup of the sugar, 1 tablespoon of the flour, the orange zest and the cinnamon. Add the rhubarb and toss to combine. Spread the mixture in the prepared baking dish and bake for 10 minutes.

3. Meanwhile, in a large bowl, stir the remaining 2 tablespoons sugar and 1 cup flour with the baking powder, baking soda and salt. Add the butter and, using your fingers, rub it into the flour mixture until the mixture is the size of very small peas. Make a well in the center and pour in the buttermilk. Stir with a fork until all the ingredients are moistened and a soft dough forms.

4. Remove the fruit from the oven. Using a spoon, drop the dough in 6 or 8 evenly spaced mounds on top of the hot fruit. Return to the oven and bake for 25 to 30 minutes, until the fruit is bubbling and the topping is a rich golden brown. Serve hot with cream or ice cream.

—Melanie Barnard & Brooke Dojny

CRACKED-WHEAT PUDDING

This has all the flavor of an Indian pudding but more texture because of the wheat.

8 Servings

1 quart milk
½ to ¾ cup finely cracked wheat (see Note)
⅓ cup molasses
¼ teaspoon ground cloves
¼ teaspoon freshly grated nutmeg
1 small cinnamon stick
½ cup raisins
2 tablespoons dark rum
2 eggs, lightly beaten
4 teaspoons chopped blanched almonds

1. Preheat the oven to 325°. In a large saucepan, slowly heat the milk with the cracked wheat, molasses, cloves, nutmeg and cinnamon stick. Reduce the heat and simmer, stirring constantly, for about 5 minutes, or until thickened.

2. Transfer to a medium bowl and let cool for 10 minutes. Lightly butter eight 1-cup ramekins or custard cups. When the mixture has cooled slightly, remove the cinnamon stick and whisk in the raisins, rum and eggs.

3. Spoon the mixture into the ramekins. Place them in a large baking pan and pour enough boiling water into the pan to reach three-fourths of the way up the sides of the ramekins. Bake for 25 minutes, or until set. Serve with a sprinkling of almonds in the center of each pudding.

NOTE: If you like a custard-like pudding, use ½ cup of wheat; if you prefer a denser pudding, use ¾ cup.

—Michèle Urvater

OLD-FASHIONED BAKED RICE PUDDING

For a special treat, serve this homey rice pudding with a little half-and-half.

6 to 8 Servings

¾ cup raisins
1½ quarts milk
1 cup sugar
¼ teaspoon salt
1 vanilla bean, split lengthwise or 1½ teaspoons vanilla extract
⅔ cup rice
1 whole egg
2 egg yolks
Cinnamon

1. Put the raisins in a bowl. Add hot water to cover and set aside until the raisins are plumped, about 30 minutes. Drain the raisins.

2. In a heavy medium saucepan, combine 1 quart of the milk with the sugar, salt and vanilla bean (if using vanilla extract, do not add it until Step 3). Bring to a boil over moderately high heat. Stir in the rice. Reduce the heat and simmer, covered, stirring occasionally, until the rice is tender, about 40 minutes. Remove the vanilla bean and scrape the seeds into the rice.

3. Preheat the oven to 400°. In a medium bowl, whisk together the whole egg, egg yolks and remaining 2 cups milk. Stir the egg mixture and raisins

into the rice. (If using vanilla extract, add it at this point.)

4. Spoon the pudding into a large, shallow, lightly buttered baking dish. Place in a larger roasting pan and pour in enough warm water to reach halfway up the sides of the baking dish. Bake in the center of the oven for 25 minutes.

5. Stir the pudding and bake for another 25 minutes. Lightly sprinkle the surface of the pudding with cinnamon.

6. Bake for 25 minutes longer, or until the top is crusty and lightly browned but the pudding is not completely set. Remove the pudding from the water bath and let cool on a rack. Serve warm, at room temperature or slightly chilled.
—*Richard Sax*

COCONUT-BASMATI RICE PUDDING WITH SATIN CHOCOLATE SAUCE

You can use another type of aromatic rice, such as the domestic Texmati, in place of the basmati here.
4 Servings

1 quart milk
1 cup shredded sweetened coconut
½ cup basmati rice*
¼ cup sugar
Pinch of salt
1 vanilla bean, halved crosswise (optional)
1 teaspoon vanilla extract
Satin Chocolate Sauce (recipe follows)
*Available at Indian and specialty food shops and some supermarkets

1. In a heavy medium saucepan, combine the milk, coconut, rice, sugar, salt and vanilla bean. Bring to a simmer over moderate heat. Reduce the heat to very low, setting the pan slightly off the burner if necessary to maintain a gentle simmer. Cook, stirring frequently to prevent a skin from forming, until the rice is very tender, the milk has reduced and the mixture is thick and creamy, about 1½ hours. Set aside to cool slightly.

2. Remove the vanilla bean and squeeze the seeds out of the open ends into the pudding; stir well. (Save the vanilla bean for another use.) Stir in the vanilla extract. Serve the pudding hot, warm or cold, with the Satin Chocolate Sauce.
—*Marcia Kiesel*

SATIN CHOCOLATE SAUCE

The better the quality of the chocolate used, the better this sauce will be. Valrhona is the richest flavored chocolate I've ever tasted. Vanilla adds a subtle perfume to this luxurious sauce. Pour it over ice cream or any sweet treat that could use a touch of inspiration.
Makes About 1 Cup

¼ cup sugar
½ of a vanilla bean, split lengthwise
1½ teaspoons instant espresso powder
½ cup heavy cream
3 ounces bittersweet chocolate, such as Valrhona,* chopped
1 tablespoon unsalted butter
*Available at specialty food shops

1. In a small heavy saucepan, combine the sugar with ⅓ cup of water. Using the tip of a small knife, scrape the seeds from the vanilla bean into the sugar water; add the bean. Bring to a boil over high heat and boil for 3 minutes. Whisk in the espresso powder, add the heavy cream and return to a boil, stirring frequently. Remove from the heat.

2. Add the chocolate and let stand until melted, about 2 minutes. Whisk the sauce until smooth. Then reheat over high heat, whisking, just until hot, about 30 seconds. Remove from the heat and whisk in the butter. Serve the sauce hot, warm or at room temperature. (*The sauce can be refrigerated, covered, for up to 1 week. Reheat in a saucepan or in a microwave oven just until melted and warm.*)
—*Marcia Kiesel*

PEACHES-AND-CREAM RICE PUDDING
8 Servings

3 cups milk
⅔ cup rice, preferably short-grain
½ teaspoon salt
1 cup heavy cream
3 whole eggs
2 egg yolks
⅓ cup plus 2 tablespoons sugar
½ teaspoon vanilla extract
⅛ teaspoon freshly grated nutmeg
⅛ teaspoon ground cloves
⅛ teaspoon cinnamon
⅛ teaspoon ground ginger
4 medium peaches, peeled and diced
1½ teaspoons fresh lemon juice
1½ teaspoons raspberry vinegar or distilled white vinegar
2 teaspoons kirsch

1. In a medium saucepan, bring 2 cups of the milk to a simmer over moderate heat. Stir in the rice and salt, reduce the heat to low and cook, covered, for 20 minutes; set aside.

2. Preheat the oven to 325°. Butter eight 1-cup custard cups. In a small saucepan, scald the remaining 1 cup milk and the heavy cream. Remove from the heat and set aside.

3. Meanwhile, in a large bowl, whisk the whole eggs, egg yolks and ⅓ cup of the sugar together until thick and fluffy, 3 to 5 minutes.

4. Whisking constantly, add the scalded milk-cream mixture in a slow, steady stream to the egg mixture. Add

the vanilla, nutmeg, cloves, cinnamon and ginger, and mix for another minute. Fold in the rice and one-fourth of the peaches.

5. Divide the custard evenly among the eight cups, filling each to within ½ inch of the top. Set the cups in a baking pan and add 1 inch of hot water. Bake in the center of the oven for about 1 hour, or until a knife inserted in the center comes out clean. Remove the cups from the water and set aside for an hour to cool. To unmold, run a knife tip around the inside of each cup, invert it over a dessert plate and shake it sharply.

6. Meanwhile, in a food processor, puree the remaining peaches with the lemon juice, vinegar, kirsch and remaining 2 tablespoons sugar, adding more kirsch and sugar if desired. Spoon around each custard and serve.
—Jim Fobel

RICE PUDDING
WITH BLUEBERRY
AND BAY COMPOTE
Basmati rice adds a delicious, nutty flavor to this pudding, and the whipped cream that is folded in at the last moment makes this traditional dessert uncommonly light.
6 Servings

1 cup basmati rice*
4 bay leaves
Pinch of salt
3 cups milk
¾ cup sugar
1 teaspoon grated orange zest
1 teaspoon vanilla extract
⅓ cup heavy cream
2 tablespoons confectioners' sugar
Blueberry and Bay Compote
 (recipe follows)
*Available at Indian and specialty
 food markets and some
 supermarkets

1. In a large saucepan, combine the rice with 2½ cups of water. Add the bay leaves and salt and bring to a boil over high heat. Reduce the heat to low, cover and cook, without removing the lid, for 20 minutes. Remove from the heat and keep covered for 10 minutes longer.

2. Stir the milk and sugar into the rice and simmer over low heat, stirring occasionally, until the mixture is thick and creamy and the rice is very soft, about 45 minutes.

3. Spoon the pudding into a bowl and stir in the orange zest and vanilla. Keep the bay leaves in the pudding and cool to room temperature, stirring occasionally. Place plastic wrap directly over the pudding and refrigerate overnight.

4. Place a stainless steel bowl and beaters in the freezer for about 10 minutes. Place the cream and confectioners' sugar in the bowl and beat until stiff peaks form. Remove the bay leaves from the pudding and fold in the whipped cream. Serve slightly chilled, with the Blueberry and Bay Compote.
—Marcia Kiesel

BLUEBERRY AND
BAY COMPOTE
Serve this compote with ice cream, pound cake or rice pudding.
Makes About 3 Cups

2 cups fresh orange juice
½ cup sugar
2 bay leaves, broken in half
1 pint fresh blueberries

In a medium nonreactive saucepan, combine the orange juice, sugar and bay leaves. Bring to a boil over high heat until reduced to about ⅔ cup, about 15 minutes. Stir in the blueberries and remove from the heat. Serve warm or cold.
—Marcia Kiesel

RICE CARLOTTA
This dessert must be refrigerated overnight, so plan accordingly.
8 to 10 Servings

1 cup golden raisins
½ cup peach brandy
1½ envelopes (¼ ounce each)
 unflavored gelatin
½ cup rice
5 cups milk
3 tablespoons unsalted butter
1½ teaspoons almond extract
1¼ cups sugar
6 egg yolks, at room temperature
3 tablespoons peach preserves
1 cup heavy cream, chilled
Peach Sauce (recipe follows)

1. In a small bowl, combine the raisins with the brandy; stir in the gelatin and set aside to soften.

2. Bring 3 quarts of water to a boil and stir in the rice. Cover and boil for 5 minutes; remove from the heat and let stand for 10 minutes. Drain the rice.

3. In a large saucepan, bring 3½ cups of the milk, the butter, almond extract and ½ cup of the sugar to a boil. Add the rice, cover and reduce the heat to low. Simmer gently, stirring occasionally, for 1¼ hours, or until the liquid has been absorbed and the rice is very tender. Set aside and let cool completely.

4. Meanwhile, in a large bowl, beat the egg yolks until thick and pale, about 5 minutes. Gradually beat in the remaining ¾ cup sugar and continue beating until the mixture is lemon colored and falls from the beater in a slowly dissolving ribbon.

5. In a small saucepan, bring the remaining 1½ cups milk to a boil over high heat. Very slowly whisk the boiling milk into the egg yolk mixture in a thin stream. Transfer the mixture to a heavy medium saucepan and cook over low heat, stirring constantly, until the custard thickens enough to coat a wooden spoon, about 10 minutes.

6. Add the raisin and brandy mixture to the custard and cook, stirring constantly, until the gelatin is completely dissolved, about 1 minute longer. Stir in the peach preserves; remove from the heat and let the custard cool completely.

7. Fold the rice into the cooled custard and refrigerate until the mixture is the consistency of stiffly whipped cream, about 1 hour; do not allow to set completely.

8. Meanwhile, whip the heavy cream until it forms stiff peaks. Gently fold the whipped cream into the rice custard.

9. Generously oil an 8-cup mold. Turn the rice custard into the mold, cover with wax paper and refrigerate overnight.

10. Before serving, run a wet knife around the rim of the mold, then dip the mold into hot water for a few seconds. Unmold onto a round platter and surround with some of the Peach Sauce; pass the remaining sauce separately.
—*W. Peter Prestcott*

PEACH SAUCE
Makes About 3 Cups

2 packages (16 ounces each) frozen peach slices, defrosted
3 tablespoons peach brandy
2 tablespoons heavy cream
1 teaspoon almond extract

Place the peaches, brandy, cream and almond extract in a blender or food processor and puree until smooth.
—*W. Peter Prestcott*

RICE PUDDING
ICE CREAM
Italians actually make rice ice cream; it's a bit crunchy and as white as snow. However, this recipe is more a twist on the rice puddings of old. It's every bit as creamy and rich as they were, although no one ever soaked raisins in rum for those simple desserts.
Makes About 2 Quarts

2 quarts milk
1½ cups sugar
½ teaspoon cinnamon
3 tablespoons long-grain white rice
½ cup raisins
5 tablespoons rum
One 4-inch piece of vanilla bean, split
1 cinnamon stick
5 egg yolks
⅔ cup heavy or whipping cream

1. Preheat the oven to 300°. In a buttered 2-quart baking dish or soufflé dish, combine 1 quart of the milk, ¼ cup of the sugar, the ground cinnamon and the rice; mix thoroughly. Bake uncovered for about 2 hours, or until the pudding is creamy, the rice is very soft and the top is lightly browned. Set aside to cool to room temperature.

2. Meanwhile, in a small bowl, cover the raisins with the rum and let macerate until plump, at least 30 minutes.

3. In a heavy medium saucepan, combine the remaining 1 quart milk and 1¼ cups sugar with the vanilla bean and cinnamon stick. Cook over moderate heat, stirring, until the mixture is hot and the sugar dissolves, 4 to 5 minutes.

4. In a medium bowl, lightly whisk the egg yolks. Whisk about 1 cup of the hot milk into the yolks, then whisk the mixture back into the remaining milk in the saucepan. Cook over moderate heat, stirring constantly, until the custard reaches 175°, thickens slightly and coats the back of a spoon, 8 to 10 minutes. (Do not boil or the mixture will curdle.) Strain the custard through a fine sieve into a medium bowl. Stir in the cream and let cool to room temperature. Refrigerate until completely chilled, then stir in the rice pudding and the raisins.

5. Freeze the mixture in two batches. Pour half of the mixture into an ice cream maker and freeze according to the manufacturer's instructions. To store, transfer the ice cream to a clean, chilled container; cover tightly and freeze for up to 1 week. Repeat with the remaining ice cream mixture.
—*Carol Field*

9

DRESSINGS, CONDIMENTS & PICKLES

EXOTIC OIL VINAIGRETTE
Makes About 1 Cup

3 tablespoons white wine vinegar
Salt and freshly ground pepper
¼ cup peanut oil
¼ cup walnut oil
1 tablespoon plus 1 teaspoon Oriental sesame oil

Place the vinegar and salt and pepper to taste in a small bowl. Add the oils and, using a fork or a small whisk, beat until well combined.
—F&W

RED WINE VINAIGRETTE
Makes About 1⅓ Cups

¾ teaspoon coarse (kosher) salt
½ teaspoon freshly ground pepper
1 garlic clove or shallot, finely minced
3 teaspoons Dijon mustard
¼ cup red wine vinegar
3 tablespoons lemon juice
½ cup olive oil
½ cup vegetable oil

1. In a small bowl, vigorously whisk together the salt, pepper, garlic, mustard, vinegar, and lemon juice. Whisk until frothy, 2 to 3 minutes.
2. Gradually add the oils in a slow stream, whisking vigorously all the while. (Wrapping a towel around the base of the bowl will prevent it from moving while you add the oil.) When you have added all the oil, whisk for another moment—the dressing should be quite fluffy. If the vinaigrette is not to be used immediately, refrigerate it in a covered jar. Let the dressing warm to room temperature and whisk it again just before using it.
—F&W

WHITE WINE VINAIGRETTE
Makes About 1½ Cups

¾ teaspoon salt
½ teaspoon freshly ground pepper
1 garlic clove, minced
1 tablespoon Dijon mustard
¼ cup white wine vinegar
3 tablespoons fresh lemon juice
1 cup olive oil (or ½ cup olive and ½ cup vegetable oil)

1. In a medium bowl, combine the salt, pepper and garlic. Use the handle of a wooden spoon to pound the garlic into the salt and pepper until it forms a paste.
2. Add the mustard, vinegar, lemon juice, and oil. Stir with a wire whisk until well blended.
—F&W

PORT DRESSING
Makes About 1 Cup

1 tablespoon Dijon mustard
½ cup fragrant olive oil
1 tablespoon walnut oil
1 tablespoon safflower oil
1 tablespoon tarragon wine vinegar
1 tablespoon fresh lemon juice
3 tablespoons tawny port
¼ teaspoon salt
⅛ teaspoon finely ground pepper

Place the mustard in a food processor or blender. With the machine running, slowly add the olive, walnut and safflower oils; then add the vinegar, lemon juice, port, salt and pepper.
—W. Peter Prestcott

DIJON VINAIGRETTE
Serve this intense mustard dressing with a contemporary Niçoise salad. Use grilled fresh tuna, romaine lettuce, tomatoes, hard-cooked eggs, boiled potatoes and slender crisply cooked green beans; garnish with Niçoise olives.
Makes About 1 Cup

¼ cup Dijon mustard
¼ cup red wine vinegar
¼ teaspoon freshly ground pepper
½ cup mild olive oil

In a small bowl, stir together the mustard, vinegar and pepper. Whisk in the oil in a thin stream until incorporated. (The dressing can be made up to 1 week ahead and refrigerated in a sealed jar.)
—Tracey Seaman

OIL-AND-EGG VINAIGRETTE
Try this dressing on a salad made with fresh spinach, homemade bacon bits, shiitake mushrooms sautéed in a dry skillet, cherry tomatoes and thinly sliced red onion. It is natural for this dressing to separate slightly.
Makes About 1¼ Cups

2 tablespoons white wine vinegar
2 teaspoons Dijon mustard
¾ teaspoon salt
½ teaspoon fresh lemon juice
¾ cup mild olive oil
4 hard-cooked eggs, finely chopped
¼ cup minced fresh chives
½ teaspoon freshly ground pepper

In a bowl, whisk together the vinegar, mustard, salt and lemon juice. Slowly whisk in the oil until incorporated. Stir in the eggs, chives and pepper.
—Tracey Seaman

NO-YOLK CAESAR DRESSING

For a simple salad, toss this dressing with torn romaine lettuce leaves. Sprinkle homemade croutons on top.

Makes About 1 Cup

2 anchovy fillets, rinsed, or 2 teaspoons anchovy paste
1 large garlic clove, minced
1 tablespoon fresh lemon juice
2 teaspoons Dijon mustard
3 tablespoons red wine vinegar
⅓ cup olive oil
Freshly grated Parmesan cheese
Freshly ground pepper

1. In a medium bowl, mash the anchovies to a paste with a fork. Whisk in the garlic, lemon juice and mustard until combined. Whisk in the vinegar.

2. Gradually add the oil in a thin stream, whisking vigorously until the dressing is thick and glossy. If the oil begins to separate out, stop pouring and keep whisking until it is thoroughly incorporated; then whisk in the remaining oil in a steady stream. *(The dressing can be made to this point up to 2 weeks ahead and refrigerated in a sealed jar. Shake before using.)*

3. When serving, add salad to the dressing in a bowl and toss. Season to taste with Parmesan cheese and pepper and toss again.

—*Tracey Seaman*

BEET VINAIGRETTE

This sweet and earthy, carmine-colored sauce breaks into graceful bubbles and splashes on the plate. (If you prefer a slightly more homogenized effect, emulsify the mixture in the blender before serving.) It suits squab, quail, beef, lamb, chicken, pork or potatoes.

Makes About 2 Cups

1 large beet (about 6 ounces), peeled and cut into 8 pieces
¼ cup sherry vinegar
Ginger Oil (recipe follows)
Salt and freshly ground pepper
Pinch of sugar (optional)

1. In a small nonreactive saucepan, combine the beet with 1½ cups water. Bring to a simmer over moderate heat and cook until soft, about 20 minutes. Drain, reserving ¾ cup of the cooking liquid.

2. In a food processor or blender, combine the beet with the reserved cooking liquid and puree until smooth.

3. Transfer the puree to a medium bowl and whisk in the vinegar, Ginger Oil, and salt and pepper to taste. Add a pinch or two of sugar, if necessary, to balance the acid. Use immediately or refrigerate the vinaigrette in a covered jar or bottle for up to 1 week. Let return to room temperature and stir or shake lightly before serving.

—*Jean-Georges Vongerichten*

GINGER OIL

Makes About ¾ Cup

4 tablespoons ground ginger
1 cup grape-seed or safflower oil

1. In a medium bowl, whisk the ginger with 3 tablespoons water to form a paste. Gradually whisk in the oil. Pour into a jar and let stand at room temperature, covered, for at least 2 days.

2. Carefully pour the clear oil into a jar and discard the ginger sediment left behind. The oil can be kept covered in the refrigerator indefinitely.

—*Jean-Georges Vongerichten*

CARAMELIZED SHALLOT VINAIGRETTE

A fairly dense and hefty vinaigrette, this sauce suits sturdy foods such as potatoes, mushrooms and polenta and meats such as calf's liver, beef and squab.

Makes About 1 Cup

½ cup plus 2 tablespoons hazelnut oil
5 medium shallots (about 4 ounces), very thinly sliced
¼ teaspoon salt
¼ cup sherry vinegar
½ teaspoon minced fresh marjoram or thyme
Freshly ground pepper

1. In a medium skillet, heat 2 tablespoons of the oil over moderate heat. Add the shallots and sprinkle with the salt. Cook, tossing frequently, until deeply browned, 7 to 8 minutes. Add 2 tablespoons of water and stir for a few seconds until the water evaporates. Remove from the heat and let cool slightly.

2. In a medium bowl, whisk the remaining ½ cup oil with the vinegar and marjoram. Stir in the shallot mixture and pepper to taste. Refrigerate in a covered jar or bottle for up to 3 weeks. The vinaigrette is best if allowed to sit for 2 to 3 days. Let return to room temperature and stir or shake lightly before serving.

—*Jean-Georges Vongerichten*

DRESSINGS, CONDIMENTS & PICKLES

VIRGIN MARY VINAIGRETTE

Use this zesty dressing with a salad of grilled vegetables and meats. For example, poach and grill Italian sausage links; then cut them into chunks. Toss yellow and red bell pepper pieces, zucchini disks and halved mushrooms in the dressing, arrange on skewers and grill until charred. Combine with the sausages, add more dressing and toss well.

Makes About 1¼ Cups

1 garlic clove
¾ teaspoon salt
½ cup tomato juice or vegetable juice cocktail
3 tablespoons red wine vinegar
1 teaspoon Worcestershire sauce
1 teaspoon tomato paste
½ teaspoon freshly ground pepper
½ cup olive oil
¼ cup minced celery leaves
2 dashes of hot pepper sauce

1. On a work surface, coarsely chop the garlic. Using a fork, mash the garlic and salt together to form a paste. Transfer to a medium bowl.

2. Whisk in the tomato juice, vinegar, Worcestershire sauce, tomato paste and pepper until combined.

3. Whisk in the oil in a thin stream until incorporated. Stir in the celery leaves and hot pepper sauce. *(The dressing can be made up to 3 days ahead and refrigerated in a sealed jar.)*
—Tracey Seaman

TOMATO-CUMIN DRESSING

Makes About 1½ Cups

¼ cup fresh lemon juice (2 lemons)
¼ cup tomato juice
2 teaspoons cumin
1 garlic clove, minced
½ cup coarsely chopped fresh coriander (cilantro)
½ teaspoon salt
½ teaspoon freshly ground pepper
¾ cup olive oil

In a medium bowl, whisk together the lemon juice, tomato juice, cumin, garlic and coriander. Then whisk in the salt and pepper and, if necessary, adjust the seasoning. Gradually whisk in the olive oil until incorporated. *(The dressing can be made 1 to 2 days ahead and refrigerated, covered.)*
—Susan Costner

GREEN PEPPERCORN TERIYAKI DRESSING

This Asian-inspired dressing marries well with a multitude of salad combinations. Mix with thinly sliced cooked flank steak—from last night's dinner perhaps—cut crosswise into strips, tossed with romaine lettuce, radicchio and endive and garnished with enoki mushrooms. You can substitute cooked shrimp or chicken for the steak.

Makes About 1¾ Cups

¼ cup soy sauce
¼ cup mirin (see Note)
¼ cup sake (see Note)
¼ cup plus 1 tablespoon rice vinegar
2 tablespoons drained green peppercorns, minced or lightly crushed
1 tablespoon minced fresh ginger
1 medium garlic clove, minced
½ cup plus 2 tablespoons vegetable oil

In a medium bowl, stir together the soy sauce, mirin, sake and rice vinegar. Whisk in the peppercorns, ginger and garlic, then whisk in the oil in a thin stream until incorporated. *(The dressing can be made up to 5 days ahead and refrigerated in a sealed jar.)*

NOTE: Mirin, or sweet sake, is available at Asian markets. Sake can be purchased at most liquor stores.
—Tracey Seaman

CREAMY ROASTED GARLIC DRESSING

The sublime roasted-garlic flavor of this dressing would be a crowning touch on salads and sandwiches alike. Slather it on leftover meats or poultry, or mix it in with cold cooked sliced potatoes for another version of potato salad.

Makes About 1 Cup

1 medium head of garlic (3 ounces)
2 teaspoons mild olive oil
½ cup sour cream
¼ cup mayonnaise
1 medium scallion, thinly sliced
1 tablespoon white wine vinegar
¾ teaspoon freshly ground pepper
½ teaspoon salt

1. Preheat the oven to 375°. Cut approximately ½ inch off the top of the garlic head. Place the garlic on a 6-inch square of aluminum foil and drizzle the oil on top. Wrap the garlic in the foil to enclose completely. Roast for 45 minutes, or until soft. Set aside to cool.

2. In a medium bowl, stir together the sour cream and mayonnaise. Pinch the root end of the roasted garlic and squeeze the flesh into the bowl. Using a rubber spatula, mash the garlic against the side of the bowl and then stir it into the sour cream and mayonnaise. Stir in

the scallion, vinegar, pepper and salt. *(The dressing can be made up to 1 week ahead and refrigerated in a sealed jar.)*
—*Tracey Seaman*

CREAMY TARRAGON VINAIGRETTE
Makes About 1⅓ Cups

½ cup dry white wine
½ cup white wine vinegar
10 peppercorns
2 large shallots, finely chopped (about 3 tablespoons)
2 tablespoons dried tarragon
1 hard-cooked egg yolk, at room temperature
1 teaspoon Dijon mustard
¼ teaspoon salt
⅛ teaspoon freshly ground white pepper
1 cup safflower oil

1. In a small, nonreactive saucepan, combine the wine, vinegar, peppercorns, shallots and tarragon. Boil until the liquid is reduced to ⅓ cup, about 10 minutes. Strain through several layers of dampened cheesecloth and set aside to cool to lukewarm.

2. In a large bowl, mash the egg yolk with a fork. Mash in the mustard, salt and white pepper and 2 tablespoons of the wine reduction.

3. Start whisking in the oil in a very thin stream; after about one-third of the oil has been added and the mixture starts to thicken, add 2 more tablespoons of the reduction. Whisk in another third of the oil, then add the remaining reduction. Whisk in the remaining oil.
—*F&W*

VINAIGRETTE DRESSING WITH GARLIC AND FRESH HERBS
Every good cook has a great salad dressing; I like to spike mine with fresh herbs and garlic. Because this creamy vinaigrette is based on an egg yolk, the total yield is more than you will need to serve 6 or 8. Simply refrigerate the remainder in a tightly covered jar and shake to blend well before using.
Makes About ¾ Cup

1 egg yolk
1 garlic clove, crushed through a press
1½ teaspoons Dijon mustard
2 tablespoons red wine vinegar
½ cup mild olive oil
1 tablespoon minced fresh basil
1 tablespoon minced parsley
¾ teaspoon coarse (kosher) salt
½ teaspoon freshly ground pepper

In a medium bowl, whisk together the egg yolk, garlic, mustard and vinegar until well blended. Gradually whisk in the oil. Blend in the minced basil, parsley, salt and pepper.
—*Lydie Marshall*

BUTTERMILK-HERB DRESSING
Drizzle this dressing on salads of vegetables and greens arranged in pleasing color combinations. For instance, combine mixed crisp lettuce leaves, tomato wedges, roasted yellow bell pepper strips and thinly sliced carrots, cucumbers and red onion.
Makes About 2 Cups

¾ cup buttermilk
¾ cup mayonnaise
1 large shallot or 1 small onion, minced
2 medium garlic cloves, minced
½ teaspoon salt
½ teaspoon freshly ground pepper
⅛ teaspoon curry powder
1 tablespoon minced fresh basil
1 teaspoon minced fresh thyme

In a medium bowl, whisk together the buttermilk, mayonnaise, shallot and garlic. Whisk in the salt, pepper, curry powder, basil and thyme. *(The dressing can be made up to 3 days ahead and refrigerated in a sealed jar.)*
—*Tracey Seaman*

CHUNKY-CREAMY BLUE CHEESE DRESSING
This dressing goes well with crudités or with a salad of red leaf lettuce, bite-sized chunks of Red Delicious apples, chopped toasted walnuts and pieces of cooked chicken.
Makes About 2 Cups

1 cup mayonnaise
½ cup plain low-fat yogurt
2 teaspoons white wine vinegar
¼ teaspoon freshly ground pepper
4 ounces crumbled Danish blue cheese (about 1 cup)
2 medium scallions, minced

In a medium bowl, whisk together the mayonnaise and yogurt until smooth. Whisk in the vinegar and the pepper. Fold in the cheese and scallions. Season to taste with more pepper. *(The dressing can be made up to 2 weeks ahead and refrigerated in a sealed jar. It will thicken slightly as it sits.)*
—*Tracey Seaman*

DRESSINGS, CONDIMENTS & PICKLES

PEANUT DRESSING

You can make a quick slaw for this dressing in a food processor fitted with a shredding disk. Shred carrots, jicama and peeled broccoli stalks. Alternatively, grate the vegetables by hand. In a large bowl, combine the shredded vegetables with finely diced red bell pepper and bean sprouts. Toss the vegetables with the dressing just before serving. Peanut dressing also makes a tasty dip for grilled chicken, or mix it with leftover cooked noodles.
Makes About 1 Cup

¼ cup creamy peanut butter
2 tablespoons Oriental sesame oil
2 tablespoons fresh lemon juice
2 teaspoons minced fresh ginger
1 garlic clove, minced
½ teaspoon salt
¼ teaspoon cayenne pepper
1 tablespoon distilled white vinegar
1 tablespoon soy sauce
¼ cup olive oil
¼ cup peanut or corn oil

   1. In a medium bowl, whisk together the peanut butter, sesame oil, lemon juice, ginger, garlic, salt and cayenne until well blended. Whisk in the vinegar and soy sauce.
   2. Whisk in the olive and peanut oils in a thin stream until incorporated. *(The dressing can be made up to 5 days ahead and refrigerated in a sealed jar.)*
—Tracey Seaman

SALSA VINAIGRETTE

Serve this fresh tomato salsa with Black-Eyed Pea Cakes (p. 107).
Makes About 2¾ Cups

3 medium tomatoes—peeled, seeded and coarsely chopped
¼ cup finely diced red onion
2 tablespoons thinly sliced scallions
1½ teaspoons minced garlic
1 jalapeño pepper, seeded and finely minced
2 tablespoons chopped fresh coriander (cilantro)
¾ teaspoon cumin
½ teaspoon freshly ground black pepper
½ teaspoon salt
2 tablespoons olive oil
1 tablespoon cider vinegar

Combine all the ingredients in a large bowl and toss well. *(The recipe can be prepared up to 1 day in advance, covered and refrigerated. Drain off any excess liquid before serving.)*
—Carolina's, Charleston, South Carolina

TOMATO-CORIANDER SALSA
Makes About 2 Cups

2 medium tomatoes, finely diced
½ teaspoon salt
1 small onion, minced
¼ cup minced fresh coriander (cilantro)
2 fresh hot chiles, preferably serrano—seeded, deribbed and minced

Place the tomatoes in a small nonreactive bowl and sprinkle with the salt; let stand for 10 minutes. Mix in the onion, coriander, chiles and 2 tablespoons of cold water. Serve at room temperature.
—Jim Fobel

RED TOMATO SALSA

Red tomato salsa is a staple in southwestern cooking; with the addition of fresh coriander (cilantro), it is known as *pico de gallo*, a popular, all-purpose salsa.
Makes About 2 Cups

2 garlic cloves, unpeeled
1 pound small tomatoes, seeded and cut into ¼-inch dice
3 tablespoons finely diced red onion
2 tablespoons finely diced red bell pepper
1 teaspoon fresh lime juice
1 fresh red jalapeño pepper, seeded and finely diced
½ teaspoon salt

   1. Preheat the oven to 400°. Loosely wrap the garlic cloves in foil and roast until very soft, about 25 minutes. Squeeze the garlic pulp from the skins into a small bowl and mash to a paste.
   2. In a medium bowl, combine all of the remaining ingredients. Add the garlic pulp and stir to blend.
—Stephan Pyles

TOMATILLO SALSA
Makes About 2 Cups

4 large garlic cloves, unpeeled
4 scallions, finely chopped
1 pound fresh tomatillos—husked, rinsed, cored and cut into ¼-inch dice
2 tablespoons chopped fresh coriander (cilantro)
3 to 4 serrano chiles, seeded and finely diced
2 teaspoons fresh lime juice
½ teaspoon salt

   1. Preheat the oven to 400°. Loosely wrap the garlic cloves in foil and roast for about 25 minutes, until very soft. Squeeze the garlic pulp from the skins into a small bowl and mash to a paste.

2. In a medium bowl, combine all of the remaining ingredients. Add the garlic pulp and stir to blend.
—Stephan Pyles

YELLOW TOMATO SALSA

Yellow tomatoes are softer and sweeter than red tomatoes, and this salsa makes a wonderful accompaniment to grilled fish.

Makes About 2 Cups

1 pound yellow tomatoes or 1 pint yellow plum tomatoes, seeded and diced
2 serrano chiles, seeded and finely diced
2 tablespoons finely diced yellow bell pepper
3 tablespoons finely diced mango
2 teaspoons fresh orange juice
½ teaspoon salt

In a medium bowl, combine the tomatoes, chiles, bell pepper, mango, orange juice and salt.
—Stephan Pyles

YOGURT AND GRILLED EGGPLANT RELISH

Serve slightly chilled as a side dish for grilled meats.
6 Servings

1 tablespoon finely sliced scallions (white and tender green)
1 medium (¾- to 1-pound) eggplant
2 cups plain low-fat yogurt
1 medium tomato, blanched and cut into ¼-inch pieces
¼ teaspoon grated fresh ginger
2 tablespoons finely minced fresh mint
¾ teaspoon salt
⅛ teaspoon cayenne pepper
Freshly ground black pepper

1. Soak the scallions in cold water to cover for 30 minutes. Drain well and pat dry.
2. Meanwhile, grill the eggplant over hot coals until it begins to char. When one side gets quite charred, roll it over slightly. Keep rolling it until it is charred on all sides.
3. Peel off and discard the darkened skin, quickly rinsing the eggplant if necessary. Chop the skinned eggplant to a pulp.
4. Put the yogurt into a serving bowl and beat lightly with a fork or whisk until creamy. Stir in the scallions, eggplant, tomato, ginger, mint, salt, cayenne and black pepper to taste. Chill until serving time.
—Madhur Jaffrey

SPICED FRESH TOMATO CHUTNEY

There is no cooking involved in this refreshing chutney. Flavored with orange, mint and fresh coriander, it makes a tasty relish for charcoal-grilled hamburgers or steaks and is also good on sandwiches.
Makes About 1½ Cups

¼ cup minced fresh mint
¼ cup minced fresh coriander (cilantro)
2 tablespoons minced onion
½ teaspoon salt
¼ teaspoon freshly ground pepper
2 tablespoons tomato paste
¼ cup fresh orange juice
2 medium tomatoes (about ¾ pound), cut into ½-inch dice

In a medium bowl, combine the mint, coriander, onion, salt and pepper. With a wooden spoon or a pestle, lightly pound the mixture to bruise the herbs and extract some moisture. Blend in the tomato paste and orange juice. Stir in the tomatoes. Cover and refrigerate for at least 1 hour before serving.
—Jim Fobel

PAPAYA-TOMATILLO CHUTNEY

Chutneys and salsas are used extensively in southwestern cuisine and are common accompaniments to grilled meats and fish.
Makes About 5 Cups

¾ pound tomatillos—husked, rinsed, cored and quartered
1 large papaya, peeled and cut into ½-inch dice
2 small red bell peppers, cut into ½-inch dice
1 small green bell pepper, cut into ½-inch dice
8 scallion whites, thinly sliced
2 small jalapeño peppers, seeded and minced
¾ cup red wine vinegar
½ cup diced fresh pineapple
½ cup (packed) light brown sugar
1 tablespoon chopped fresh coriander (cilantro)
1 teaspoon salt
1 garlic clove, minced
¼ teaspoon cumin

In a large, heavy, nonreactive saucepan, combine all of the ingredients. Bring to a boil over high heat, stirring frequently. Reduce the heat to moderately low and boil gently, stirring occasionally, until thick, 30 to 35 minutes. Transfer to a bowl and let cool completely. Cover and refrigerate for up to 5 days.
—Stephan Pyles

FRESH CORIANDER CHUTNEY
Makes About ½ Cup

½ teaspoon cumin seeds
3 cups fresh coriander (cilantro) sprigs, chopped
½ cup fresh mint, chopped
1 small green Thai chile, minced
1 garlic clove, minced
2 tablespoons fresh lime juice
¼ teaspoon salt
½ teaspoon freshly ground black pepper

1. In a small dry skillet, toast the cumin seeds over high heat until fragrant, about 30 seconds. In a mortar, pound the seeds to a coarse powder. Or chop the seeds finely with a knife.

2. In a food processor, combine the cumin with all the remaining ingredients. Process to a paste, stopping to scrape down the sides of the bowl once or twice. Scrape the chutney into a bowl and press plastic wrap directly on the surface. Refrigerate for up to 1 day.
—*Madhur Jaffrey*

TOMATO CHUTNEY WITH LEEKS AND CARROTS

Depending on the availability of good, vine-ripened tomatoes, you may use either all fresh, all canned, or a combination (our preference).
Makes About 5 Half-Pint Jars

2 large carrots, peeled and cut into ¼-inch dice
1 pound fresh tomatoes—peeled, seeded and coarsely chopped
1 pound canned tomatoes in thick puree (see Note), coarsely chopped
2 leeks (white portions only), cut into ¾-inch pieces, finely diced
1 celery rib, finely diced
1 small onion or ½ of a medium onion
2 teaspoons grated orange zest
1 teaspoon salt
½ teaspoon cayenne pepper
1¾ cups red wine vinegar
1 cup granulated sugar
½ cup (packed) brown sugar
½ teaspoon allspice
3 garlic cloves, finely chopped
1 tablespoon minced fresh ginger
1 cup golden raisins

1. In a medium saucepan of boiling water, cook the carrots until they just lose their raw firmness, 5 to 8 minutes. Drain well.

2. In a large nonreactive saucepan, combine the carrots, fresh tomatoes, canned tomatoes, leeks, celery, onion, orange zest, salt and cayenne. Cover and heat until the vegetables begin to sizzle very slightly. Stir, then re-cover and cook for 4 or 5 minutes, until the vegetables begin to release their juices.

3. Add the vinegar, the granulated and brown sugars, allspice, garlic and ginger. Stir over moderate heat until the sugar has dissolved. Adjust the heat to a steady simmer and cook until the vegetables are tender and the liquid is thick enough to coat a spoon evenly, about 1½ hours. Stir often to prevent sticking or burning.

4. Add the raisins to the chutney. Check the seasoning and add more salt and cayenne if needed. Fill the 5 half-pint screw-band canning jars, leaving ¼ inch of headspace, with the chutney. Put on the two-piece lids, and process 15 minutes in a boiling-water bath. Cool and store.

NOTE: If the canned tomatoes are in a thin liquid, not tomato puree, do not use this liquid. Instead, use 2 tablespoons tomato paste mixed with ¼ cup of the canning liquid or water.
—*F&W*

ROASTED PEPPERS WITH OLIVES AND THYME

These peppers will keep for 2 to 3 months in the refrigerator.
Makes About 1 Quart

6 pounds red or yellow bell peppers
6 brine-cured black olives
2 sprigs of fresh thyme
2 garlic cloves, peeled
1 tablespoon red or white wine vinegar
About ¾ cup extra-virgin olive oil

1. Roast the peppers directly over a gas flame or under the broiler, as close to the heat as possible, turning frequently, until blackened all over. Place the peppers in a paper or plastic bag and let steam for 10 minutes. When cool enough to handle, peel off the skins under running water. Place the peeled peppers in a large bowl.

2. Halve the peppers and remove the cores, seeds and ribs over the bowl to catch the juices. Place the peppers in a wide-mouth quart jar. Strain the juices over the peppers.

3. Add the olives, thyme and garlic to the peppers. Add the vinegar and enough oil to cover the peppers. Stir to release any air bubbles, then cover and refrigerate.
—*Anne Disrude*

ROASTED PEPPERS WITH ANCHOVIES AND CAPERS

These peppers will keep for 2 to 3 months in the refrigerator.
Makes About 1 Quart

6 pounds red or yellow bell peppers
1 can flat anchovy fillets—drained, rinsed and patted dry
1½ tablespoons drained capers
2 imported bay leaves
4 garlic cloves
1 tablespoon red or white wine vinegar
About ¾ cup extra-virgin olive oil

1. Roast the peppers directly over a gas flame or under the broiler, as close to the heat as possible, turning frequently, until blackened all over. Place the peppers in a paper or plastic bag and let steam for 10 minutes. When cool enough to handle, peel off the skins under running water. Place the peeled peppers in a large bowl.

2. Halve the peppers and remove the cores, seeds and ribs over the bowl to catch the juices. Place the peppers in a wide-mouth quart jar. Strain the juices over the peppers.

3. Add the anchovies, capers, bay leaves and garlic to the peppers. Add the vinegar and enough oil to cover the peppers. Stir to release any air bubbles, then cover and refrigerate.
—*Anne Disrude*

ROASTED GARLIC PUREE

I find this puree to be an indispensable ingredient. The sweet, mellow flavor makes it a delicious addition to salad dressings, baked vegetables, soups, stews and sauces. This recipe can be doubled or tripled and will keep for up to two weeks in the refrigerator.
Makes About ¼ Cup

2 large heads of garlic (about ½ pound)
1 teaspoon salt
1 teaspoon freshly ground pepper
2 tablespoons extra-virgin olive oil

1. Preheat the oven to 350°. Cut the heads of garlic in half horizontally, being careful that the cloves remain intact. Season the cut surfaces with the salt and pepper and drizzle on the olive oil. Reassemble the heads, set them in a small roasting pan and bake until the cloves are soft and golden, about 1 hour. Remove and let cool to room temperature.

2. Strain the roasted garlic through a fine-mesh sieve over a bowl; discard any skin that remains. Scrape any paste from the back of the sieve and add to the paste in the bowl. *(To store, scrape the garlic paste into a glass jar, cover the surface with olive oil and refrigerate.)*
—*Bob Chambers*

NEW DILL SLICES

I love these crunchy dill slices on hamburgers with mustard, mayonnaise and ketchup. They are also good alone or with almost any sandwich, such as chicken salad.
Makes About 2 Cups

1 pound (4 to 6 medium) pickling cucumbers, such as Kirbys, or regular cucumbers (2 medium)
1 tablespoon plus 1 teaspoon coarse (kosher) salt
3 tablespoons cider vinegar
½ teaspoon dill seed
1 bay leaf
1 garlic clove, thinly sliced
Several small sprigs of fresh dill (optional)

1. If using Kirbys, trim the ends away and cut into ⅛-inch slices. If using regular cucumbers, peel first and then cut into ⅛-inch slices. In a bowl, toss the cucumbers with 1 tablespoon of the salt. Let stand for 1 hour, tossing occasionally. Rinse and drain 3 times with cold water.

2. Meanwhile, in a small nonreactive saucepan, combine the cider vinegar, dill seed, bay leaf, garlic, remaining 1 teaspoon salt and 1 cup of water. Bring to a boil over moderately high heat. Boil for 1 minute, remove from the heat and let cool to room temperature.

3. With your hands, squeeze out the excess moisture from the cucumber slices. Place in a small bowl, add the fresh dill sprigs, if using, and the cooled brine. Cover and refrigerate for at least 1 hour, or overnight.
—*Jim Fobel*

ABBREVIATED BREAD-AND-BUTTER PICKLES

These pickles will satisfy an old-fashioned pickle craving at a moment's notice. Their taste and crunch are traditional; the only thing that's changed is the amount of time that it takes to make them. They are not too sweet and not too sour and are wonderful in tuna or egg salads or alongside baked beans or potato salad.
Makes About 4 Cups

8 medium pickling cucumbers, such as Kirbys (1½ to 2 pounds)
2 medium onions (about ¼ pound each), thinly sliced
6 ice cubes
¼ cup coarse (kosher) salt
1 cup cider vinegar
½ cup sugar
1 tablespoon mustard seed
½ teaspoon celery seed
½ teaspoon turmeric

1. Trim the ends of the cucumbers and cut them into ¼-inch slices (for a truly authentic look, cut the cucumbers with a crinkle-edged cutter).
2. In a large bowl, combine the cucumbers, onions, ice cubes and salt. Let stand for 1 hour, tossing occasionally. Fill the bowl with cold water and then drain in a colander. Rinse and drain 3 times to rinse off all the salt.
3. In a large nonreactive saucepan, combine the vinegar, sugar, mustard seed, celery seed and turmeric. Bring to a boil over high heat. Add the vegetables. When the liquid barely begins to simmer again, remove from the heat.
4. Transfer the vegetables and their liquid to a bowl and let cool to room temperature. Cover with plastic wrap placed directly on the surface and refrigerate overnight.
—*Jim Fobel*

OLD-FASHIONED BREAD-AND-BUTTER PICKLES
Makes About 1 Quart

5 pickling (Kirby) cucumbers, each about 4 inches long, thinly sliced
5 small white onions (about ½ pound), thinly sliced and separated into rings
3 tablespoons coarse (kosher) salt
1 cup distilled white vinegar
¾ cup sugar
2 teaspoons mustard seed
¼ teaspoon turmeric
¼ teaspoon celery seed

1. In a large bowl, toss the cucumbers and onion rings with the salt. Cover with water and let rest for about 3 hours. Drain and, without rinsing, place in a quart jar or nonreactive bowl.
2. In a small nonreactive saucepan, bring the vinegar, sugar, mustard seed, turmeric, celery seed and 3 tablespoons of water to a boil. Remove from the heat, pour over the cucumbers, cover and refrigerate for 3 days.
—*F&W*

DILLED CARROT STICKS

Try these pickles in place of dilled cucumber pickles. Serve as an hors d'oeuvre or alongside burgers and other grilled meats.
Makes 2 Pints

1½ cups cider vinegar
6 garlic cloves
3 tablespoons dill seeds, lightly crushed
2 tablespoons honey
1 teaspoon salt
1 pound medium carrots, cut into 4-by-⅜-inch sticks

1. Sterilize two 1-pint preserving jars and their lids.
2. In a small nonreactive saucepan, combine the vinegar, garlic, dill seeds, honey, salt and ¾ cup of water. Bring to a boil over moderate heat, then set aside to steep for at least 10 minutes.
3. Meanwhile, pack the carrot sticks upright in the hot jars.
4. Return the brine to a boil. Pour the hot brine over the carrots to within ½ inch of the rim of the jars, dividing the dill seeds and garlic evenly between the jars. Run a small knife or spatula around the inside of the jars to release trapped air bubbles. Add more brine if necessary to top off the jars. Wipe the rims clean and seal with the lids. Let cool slightly, then refrigerate.
—*Jonathan Locke & Sally Gutiérrez*

VERMOUTH-PICKLED ONIONS
Makes About 2½ Cups

1 pound pearl onions, peeled and trimmed (about 4 cups)
4 tablespoons coarse (kosher) salt
1 cup dry vermouth
½ cup distilled white vinegar

1. Bring a large pot of water to a boil, add the onions and 1 tablespoon of the salt. When the boiling resumes, cook for 3 minutes. Drain and transfer to a large bowl; add 1 cup cold water and the remaining salt, stirring until the salt is dissolved. Let rest, covered, overnight.
2. Drain and rinse under cold water. Add the vermouth and vinegar and let rest, covered, overnight.
—*F&W*

CAULIFLOWER-CARROT PICKLES IN HONEY-MUSTARD

Blanched in sherry and dressed with sherry vinegar, these elegant pickles are also flavored with mustard, honey and olive oil. They're a good addition to any tapas selection and are great with ham or corned beef.

Makes About 4 Cups

1 medium head of cauliflower (about 1½ pounds), separated into 1½-inch florets
4 medium carrots (about ½ pound)
1½ cups dry sherry
⅓ cup plus 1 tablespoon sherry wine vinegar or cider vinegar
3 tablespoons coarse (kosher) salt
1 tablespoon sugar
1 teaspoon tarragon
¼ cup Dijon mustard
2 tablespoons honey
⅓ cup olive oil

1. Cut the cauliflorets into halves or quarters (their stems should be about ¼ inch thick). Cut the carrots into 3-by-¼-inch sticks.

2. In a large nonreactive saucepan, combine the sherry, ⅓ cup of the vinegar, the salt, sugar, tarragon and 2 cups of water. Bring to a boil over high heat. Add the cauliflower and carrots and stir for 1 minute (the water will not return to a simmer).

3. Using a slotted spoon (to leave behind as much of the tarragon as possible), remove the vegetables to a colander to drain; do not rinse. Discard the cooking liquid. Place the vegetables on several layers of paper towels and top with several more; blot well to dry.

4. In a medium bowl, combine the mustard and honey with the remaining 1 tablespoon vinegar. Whisking constantly, add the oil drop by drop and then slowly in a steady stream until all has been added and the dressing is creamy and thick.

5. Add the vegetables, toss well, cover with plastic wrap placed directly on the surface and refrigerate overnight. Toss again before serving.

—*Jim Fobel*

KIM CHEE

Although this classic Korean pickle can take up to three weeks to pickle, this version is ready overnight (and happens to taste good after only an hour). It is delicious with grilled beef (such as its traditional partner, Korean *bul goki*) or chicken, and perfect with charcoal-broiled hamburgers. Be sure to use the curly-leafed nappa cabbage and not the longer, thinner Chinese celery cabbage (which is more fibrous and takes longer to cure).

Makes About 5 Cups

1 medium head of nappa cabbage (about 1¾ pounds)
2 medium pickling cucumbers, such as Kirbys (about ¼ pound each)
¼ cup plus 2 teaspoons coarse (kosher) salt
1 tablespoon grated fresh ginger
1 large garlic clove, crushed through a press
1 tablespoon sugar
1 tablespoon sweet paprika
¼ teaspoon cayenne pepper
1 to 2 tablespoons soy sauce
¼ cup rice vinegar
½ cup sliced scallions (4 to 5 medium)

1. Trim away any wilted or bruised outer leaves from the head of cabbage. Cut the head crosswise into 1½-inch slices. Discard the core end.

2. Trim the ends of the cucumbers, halve them lengthwise and scoop out the seeds. Cut them crosswise into ¼-inch slices.

3. Place the cabbage and cucumbers in a large bowl. Sprinkle on ¼ cup of the salt and toss well. Let rest for 1 hour, tossing occasionally.

4. Fill the bowl with cold water and drain the vegetables into a colander. Repeat this process 3 times to rinse off the salt. Drain well.

5. In a medium bowl, combine the ginger, garlic, sugar, paprika, cayenne, 1 tablespoon of the soy sauce and the remaining 2 teaspoons salt. Stir in the vinegar and 1½ cups of cold water.

6. Add the cabbage, cucumbers and scallions. Toss together, place plastic wrap on the surface and refrigerate overnight. Taste and add 1 more tablespoon soy sauce if desired.

—*Jim Fobel*

PICKLED BEETS

This Scandinavian-inspired recipe pairs well with pâté and a variety of meat sandwiches.

Makes 2 Pints

1⅓ pounds beets
1⅓ cups apple cider vinegar
1⅓ cups honey
¾ teaspoon whole cloves
1 cinnamon stick, broken in half

1. In a large pot, cover the beets with water and bring to a boil over high heat. Reduce the heat to moderately low and simmer until tender, about 45 minutes.

2. Meanwhile, sterilize two 1-pint preserving jars and their lids.

3. In a small nonreactive saucepan, combine the vinegar, honey, cloves and cinnamon. Bring to a boil over moderately high heat, then set aside to steep.

4. Drain the beets, rinse under cold water and slip off the skins. Halve any large beets, then slice the beets ¼ inch thick and pack them in the hot jars.

5. Return the brine to a boil. Pour the hot brine over the beets to within ½ inch of the rim of the jars, dividing the cloves and cinnamon stick evenly between the jars. Run a small knife or spatula around the inside of the jars to release trapped air bubbles. Wipe the rims clean and seal with the lids. Let cool slightly, then refrigerate.
—*Jonathan Locke & Sally Gutiérrez*

FRAGRANT PICKLED BEETS WITH PEARL ONIONS

The sophisticated flavor combination of raspberry vinegar and Zinfandel wine adds fragrance and flavor to these deep red pickles. Try them as a tangy accompaniment to poached salmon, broiled swordfish or grilled pork chops with rosemary.
Makes About 3 Cups

1 cup red Zinfandel wine
3 tablespoons sugar
1 teaspoon whole allspice, cracked
1 teaspoon coarse (kosher) salt
6 ounces tiny, white pearl onions (about 1½ cups), peeled
8 medium beets (about 1 pound), peeled and cut into ¼-inch slices
¼ cup raspberry or red wine vinegar

1. In a medium nonreactive saucepan, combine the wine, sugar, allspice, salt and ¼ cup of water. Bring to a boil over moderately high heat. Add the onions and beets. Partially cover and simmer over low heat until the beets are just tender, about 15 minutes.
2. Transfer the onions, beets and their liquid to a bowl and add the vinegar. Let cool to room temperature. Cover and refrigerate for at least 1 hour, or overnight.
—*Jim Fobel*

PICKLED EGGPLANT AND MUSHROOMS

Good as an antipasto before dinner or lunch, these luscious pickles are also good for picnics and brunch or as a counterpoint to more substantial dishes such as breaded veal cutlets or chicken cacciatore.
Makes About 4 Cups

1½ cups plus 1 tablespoon dry white wine
¼ cup plus 1 tablespoon white wine vinegar
2 large garlic cloves, thinly sliced
1 cinnamon stick
1 bay leaf
1 tablespoon plus ¼ teaspoon oregano
1 tablespoon basil
3 whole cloves
2 tablespoons coarse (kosher) salt
2 tablespoons sugar
1 medium eggplant (about 1 pound), unpeeled and cut into 3-by-½-inch sticks
½ pound small mushrooms or large mushrooms, halved or quartered
⅓ cup olive oil
Chopped parsley, for garnish

1. In a medium nonreactive saucepan, combine 1½ cups of the wine, ¼ cup of the vinegar, the garlic, cinnamon, bay leaf, 1 tablespoon of the oregano, the basil, cloves, salt, sugar and 1 cup of water. Bring to a boil over high heat. Reduce the heat to low and simmer for 5 minutes.
2. Add the eggplant and mushrooms to the pickling solution, return to a simmer and cook, stirring occasionally, for 10 minutes.
3. Using a slotted spoon (to leave behind as much of the dried herbs and spices as possible), transfer the eggplant, mushrooms and garlic to a colander; do not rinse. Discard the cooking liquid in the pan and any whole spices in the colander. Place the vegetables in a bowl and let cool to room temperature.
4. Add the olive oil, remaining 1 tablespoon wine, 1 tablespoon vinegar and ¼ teaspoon oregano. Stir gently. Cover with plastic wrap placed directly on the surface and refrigerate overnight. To serve, garnish with chopped parsley.
—*Jim Fobel*

OLIVE OIL PICKLED BEANS

The olive oil in this brine congeals once the beans are refrigerated. Allow a little time before serving for the oil to soften. These pickles are delicious, brine and all, in marinated salads or as a sandwich garnish.
Makes 2 Pints

1½ cups white wine vinegar
¼ cup olive oil
¼ cup brown sugar
1 teaspoon celery seeds
1 teaspoon mustard seeds
⅛ teaspoon crushed red pepper
¾ pound unblemished green beans
1 small yellow onion, sliced lengthwise

1. Sterilize two 1-pint preserving jars and their lids.
2. In a medium nonreactive saucepan, combine the vinegar, olive oil, brown sugar, celery seeds, mustard seeds, red pepper and 1 cup of water. Bring to a boil over high heat, then cover and set aside to steep for at least 10 minutes.
3. Trim the stems off the beans and cut any long beans to stand within ¾ inch of the rim of the preserving jars.

Stand the beans upright in the hot jars and pack in the onion.

4. Return the brine to a boil. Pour the hot brine over the beans and onion to within ½ inch of the jar rim. Run a small knife around the inside of the jars to release trapped air bubbles. Wipe the rims clean and seal with the lids. Let cool slightly, then refrigerate.
—*Jonathan Locke & Sally Gutiérrez*

PICKLED VEGETABLES WITH PINK PEPPERCORNS
Makes About 1 Quart

½ pound carrots, cut into ¼-by-3-inch julienne strips
½ pound green beans
¼ pound snow peas
8 sprigs of fresh coriander (cilantro), lightly crushed
⅔ cup distilled white vinegar
2 tablespoons sherry vinegar or cider vinegar
2 teaspoons salt
2 teaspoons sugar
2 teaspoons pink peppercorns, lightly crushed

1. In rapidly boiling water, blanch the carrots and green beans separately for 2 minutes each and the snow peas for 30 seconds. Remove with a slotted spoon.

2. Place the vegetables, standing upright, in a quart jar. Distribute the coriander evenly among the vegetables.

3. In a small bowl, combine the vinegars, salt and sugar with 1 cup water, stirring until the salt and sugar have dissolved. Pour half the mixture over the vegetables, add the peppercorns and the remaining liquid and refrigerate, covered, for 3 days.
—*F&W*

PICKLED JALAPENOS

Serve this incendiary delight with scrambled eggs, hamburgers and fajitas. These pickles are more flavorful and firmer than any you can find in a jar.
Makes 2 Pints

1½ cups cider vinegar
2 tablespoons honey
1 tablespoon pickling spice*
1 teaspoon salt
4 garlic cloves
1 pound green jalapeño peppers, stemmed and sliced ⅛ inch thick
*Available at most supermarkets

1. Sterilize two 1-pint preserving jars and their lids.

2. In a nonreactive saucepan, combine the vinegar, honey, pickling spice, salt, garlic and ¾ cup of water. Bring to a boil over high heat, then set aside to steep for 10 minutes.

3. Pack the jalapeños to within 1 inch of the rim of the hot jars.

4. Return the brine to a boil. Pour the hot brine over the peppers to within ½ inch of the rim of the jars, dividing the spices and garlic cloves evenly between the jars. Run a small knife or spatula around the inside of the jars to release trapped air bubbles. Wipe the rims clean and seal with the lids. Let cool slightly, then refrigerate.
—*Jonathan Locke & Sally Gutiérrez*

ORIENTAL PICKLED VEGETABLES
Makes About 1 Quart

½ small head of cabbage (about ½ pound)
2 large celery ribs, sliced ¼ inch thick on the bias (about 1 cup)
1 large carrot, cut into paper-thin rounds (about 1 cup)
1 tablespoon coarse (kosher) salt
5 tablespoons soy sauce
3 tablespoons sugar
4 garlic cloves, thinly sliced
3 to 4 small dried hot chiles, split
1 tablespoon distilled white vinegar

1. Cut the cabbage in half through the core and then into 1-inch cubes, core included. In a large bowl, toss the cabbage, celery and carrot with the salt, 1 tablespoon of the soy sauce and ½ cup of water. (Cover and let stand overnight, tossing occasionally.)

2. Drain the vegetables (do not rinse) and return to the bowl. Combine the remaining ¼ cup soy sauce with the sugar, stirring to partially dissolve the sugar. Add to the vegetables along with the garlic, chiles, vinegar and 1 cup water. Toss and transfer to a quart jar or similar container, submerge by weighting with a heavy glass jar or saucer and let stand for 48 hours.
—*F&W*

INDEX

A

Acorn squash and turnip soup, 67
Ancho chile cream, 77
APPETIZERS. *See also* Chips;
    Dips; First courses; Nachos
    artichoke fritters with remoulade sauce, 20
    avocado tempura with coriander and lime, 20
    brussels sprouts hors d'oeuvres, 17
    carrot and cauliflower loaf, 155
    double radish sandwiches, 16
    eggplant cookies with goat cheese and tomato-basil sauce, 17
    fried polenta with gorgonzola, 18
    gingered eggplant mousse, 28
    grilled eggplant bruschetta with prosciutto, 22
    herbed pita crisps, 23
    marinated green beans, 21
    marinated red and green peppers, 21
    peppered pecans, 200
    red potatoes with gorgonzola cream, bacon and walnuts, 18
    roasted peppers and celery on toasted croutons, 22
    spicy pumpkin seeds, 16
    spinach-stuffed mushrooms with cheddar and chervil, 17
    10-ingredient vegetable fritters, 20
    tomato bread crisp, 22
    triangles with two-mushroom filling, 16
    vegetarian spring roll, 19
APPLES
    apple and green tomato tart with calvados cream, 247
    minted zucchini and apple salad, 219
APRICOTS
    apricot sauce, 242
    apricot-shrimp salad, 41
ARTICHOKES
    artichoke fritters with remoulade sauce, 20
    artichoke hearts with mushrooms, 148
    artichoke, corn and oyster gratin, 47
    braised artichoke hearts, 47
    braised artichokes with spinach, 148
    warm artichoke salad with bacon and mustard vinaigrette, 40
ARUGULA
    arugula and scallop salad, 43
    arugula and watercress salad with swiss cheese beignets, 37
ASPARAGUS
    asparagus and beet salad with tarragon, 207
    asparagus bundles, 47
    asparagus pancake with mozzarella, 99
    asparagus soup, 64
    asparagus tart à la bicyclette, 89
    asparagus with butter vinaigrette, 153
    asparagus with hazelnut-orange sauce, 153
    cool asparagus soup, 61
    cream of asparagus and morel soup, 70
    scallop, mussel and asparagus salad with orange-saffron dressing, 43
AVOCADO
    avocado and roasted corn guacamole with toasted corn tortillas, 29
    avocado and veal salad with walnuts, 138
    avocado frittata, 98
    avocado salad with grapefruit and bacon, 32
    avocado soup with papaya-pepper relish, 62
    avocado tempura with coriander and lime, 20
    beet and avocado salad with tarragon vinaigrette, 208
    corn and avocado torta, 159
    crab and avocado salad with coriander-tomatillo vinaigrette, 122
    grapefruit and avocado salad with tomato-cumin dressing, 221
    grilled swordfish and avocado salad baja-style, 120
    guacamole, 29

B

Bagna cauda salad, 219
BARLEY
    barley-ham salad with honey mustard dressing, 141
    mixed grain trio, 193
    mushroom-barley soup, 70
BASIL
    basil chiffonade, tomatoes on, 218
    basil mashed potatoes, 178
    basilico potatoes, 167
BEAN SALADS
    bean, corn and rice salad with chili vinaigrette, 234
    black bean and bell pepper salad, 228
    black bean and celery salad, 228
    black bean, corn and poblano pepper salad with cumin vinaigrette, 233
    black-eyed pea and corn salad, 234
    black-eyed pea salad with orange-jalapeño dressing, 234
    chickpea and baked bell pepper salad with cumin, 235
    chickpea salad, 235
    french lentil salad with glazed pearl onions, 236
    potato and white bean salad, 227
    tangy corn and lima beans, 228
    three-bean macaroni salad, 233
BEANS, DRIED OR CANNED. *See also* Bean salads and names of specific beans
    baked beans with onions and orange marmalade, 185
    baked beans with salsa, 185

frijolemole, 27
Lafayette's spicy baked red beans, 186
sweet bean crisps, 241
three-bean casserole, 101

BEANS, FRESH
 celery root, green bean and potato puree, 158
 garden green bean and tomato salad with warm bacon and pine nut dressing, 206
 green bean and benne salad, 206
 green beans and roasted peppers, 153
 green beans mimosa, 205
 green beans with coconut and popped rice, 153
 green beans in aquavit, 206
 marinated green beans, 21
 olive oil pickled beans, 268
 red potato salad with green beans and red pepper strips, 227
 warm string bean salad, 213

BEEF
 cooked beef salad with mustard-anchovy dressing, 144
 madrigal salad, 144
 oriental beef salad, 144
 south sea beef salad, 146
 thai beef salad, 146
 thai-style beef and lettuce salad with chile-lime dressing, 145

BEETS
 asparagus and beet salad with tarragon, 207
 baked beet and endive salad, 208
 beet and avocado salad with tarragon vinaigrette, 208
 beet and chicory salad, 207
 beet frites, 25
 beet vinaigrette, 259
 chocolate-beet bundt cake with pecans, 242
 fragrant pickled beets with pearl onions, 268
 hearts of palm, beet and endive salad, 207
 pickled beets, 267
 pineapple and beet salad, 208
 red beet devil's food cake, 243
 shredded raw beet and celery salad, 213
 sweet-and-sour beets, 154

BELGIAN ENDIVE
 baked beet and endive salad, 208
 braised pear, celery and endive with parmesan and basil, 49
 Chardenoux's salad of blue cheese, nuts and belgian endive, 203
 hearts of palm, beet and endive salad, 207
 shrimp and endive salad, 127

BLACK BEANS
 black bean and bell pepper salad, 228
 black bean and celery salad, 228
 black bean, corn and poblano pepper salad with cumin vinaigrette, 233
 frijolemole, 27
 southwestern black bean soup, 71
 three-bean casserole, 101

BLACK-EYED PEAS
 baked beans with salsa, 185
 black-eyed pea and corn salad, 234
 black-eyed pea and mint marigold salad, 39
 black-eyed pea salad with orange-jalapeño dressing, 234
 black-eyed-pea cakes with salsa, 107
 collard greens and black-eyed-pea soup with cornmeal croustades, 65
 crab hoppin' john, 123
 spicy black-eyed peas, 187

BLUE CHEESE. See also Roquefort; Stilton
 Chardenoux's salad of blue cheese, nuts and belgian endive, 203
 chunky-creamy blue cheese dressing, 261
 mixed green salad with soured cream and blue cheese dressing, 201

Bread, cornmeal yeast, 66
Bread-and-butter pickles, abbreviated, 266
Bread-and-butter pickles, old fashioned, 266

BROCCOLI
 broccoli, onion and cheese soup, 69
 jade flower broccoli, 154
 warm broccoli salad with toasted pine nuts, 213
Bruschetta, grilled eggplant, with prosciutto, 22

BRUSSELS SPROUTS
 brussels sprouts hors d'oeuvres, 17
 spicy brussels sprouts, 154

BULGUR
 beefsteak tomatoes with minted cracked wheat salad, 116
 bulgur wheat, peppers and bean curd, 104
 coleslaw with bulgur, 222
 cracked wheat-spinach salad, 236
 dilled tabbouleh, 121
 tabbouleh, 236

BUTTERNUT SQUASH
 butternut squash consommé with leek ravioli, 63
 butternut squash gratin, 179
 butternut squash, applesauce and date cake, 246

C

CABBAGE. See also Nappa cabbage
 cabbage salad with garlic, 213
 coleslaw with bulgur, 222
 creamy garlic coleslaw, 222
 red cabbage soup, 64
 red cabbage with granny smith apples, 155
 sassy slaw, 223
 warm cabbage slaw with bacon, 221
Caesar dressing, no-yolk, 259
Caesar salad, warm, 114

CAKES
 butternut squash, applesauce and date cake, 246
 carrot cheesecake with nut-crumb crust, 246
 chocolate-beet bundt cake with pecans, 242
 cornmeal cake, 244
 lemon-glazed parsnip tea bread, 244
 mace-scented carrot torte, 243
 orange-rhubarb breakfast cake, 246

plymouth bay berry cake, 243
pumpkin bourbon pound cake, 245
pumpkin-ginger cake, 245
red beet devil's food cake, 243
CARROTS
baby carrot and turnip salad, 214
carrot and cauliflower loaf, 155
carrot and orange salad with black mustard seeds, 214
carrot cheesecake with nut-crumb crust, 246
carrot chips, 24
carrots glazed with lime butter, 155
cauliflower-carrot pickles in honey-mustard, 267
charred carrot puree, 156
charred carrot soup, 80
cream of carrot soup with brandy and chervil, 80
creamy carrot soup, 80
dilled carrot sticks, 266
mace-scented carrot torte, 243
sage-glazed carrots, 156
CAULIFLOWER
carrot and cauliflower loaf, 155
cauliflower cress soup with corn wafers, 68
cauliflower-carrot pickles in honey-mustard, 267
curried cauliflower with peas, 156
mexican cauliflower in cheese sauce, 157
shrimp salad with cauliflower, 127
Caviar, eggplant, 27
Caviar, gratin of eggplant, 160
CELERY
black bean and celery salad, 228
braised pear, celery and endive with parmesan and basil, 49
cream of celery soup, 77
shredded raw beet and celery salad, 213
CELERY ROOT
celery root chips, 24
celery root, green bean and potato puree, 158
deviled celery root, 157
root vegetable cakes, 57
scalloped celery root and potatoes, 157

Cèpes bordelaise, 164
Chardenoux's salad of blue cheese, nuts and belgian endive, 203
Cheesecake, carrot, with nut-crumb crust, 246
Chestnuts, sweet, croquettes, 241
CHICKEN
broiled thai chicken salad, 137
chinese-style chicken salad, 136
crunchy chicken salad, 136
triple-mustard chicken salad, 137
Chicken livers, apples and mustard seed, wilted salad with, 45
CHICKPEAS
chickpea and baked bell pepper salad with cumin, 235
chickpea salad, 235
chickpea stew with greens and spices, 102
garlic-braised eggplant and chickpea casserole, 101
sweet bean crisps, 241
CHIPS
carrot chips, 24
celery root chips, 24
classic nachos, 23
herbed pita crisps, 23
parsnip chips, 24
rutabaga chips, 25
salad nachos vinaigrette, 23
CHOCOLATE
chocolate-beet bundt cake with pecans, 242
red beet devil's food cake, 243
satin chocolate sauce, 254
CHUTNEY. *See Relish/chutney*
CHOWDER
great lakes corn chowder, 81
new jersey tomato and sweet potato chowder, 82
roasted garlic and leek chowder, 82
Citrus's crab coleslaw, 44
Cobbler, buttermilk rhubarb, 253
COLESLAW. *See Slaws*
Collard greens and black-eyed-pea soup with cornmeal croustades, 65
COOKIES
sweet bean crisps, 241
CORIANDER, FRESH

cilantro cream, 77
fresh coriander chutney, 264
tomato-coriander salsa, 262
CORN. *See also Hominy*
bean, corn and rice salad with chili vinaigrette, 234
black bean, corn and poblano pepper salad with cumin vinaigrette, 233
black-eyed pea and corn salad, 234
corn and avocado torta, 159
corn chili, 158
corn pudding with cheese and chiles, 158
corn soup with spicy pumpkin seeds, 61
corn-and-leek soup, 72
great lakes corn chowder, 81
grilled corn soup with cilantro and ancho chile creams, 77
simple stew of lima beans, corn and tomatoes, 102
tangy corn and lima beans, 228
CORNMEAL
corn wafers, 68
cornmeal cake, 244
cornmeal pizza with greens and fontina, 86
cornmeal yeast bread, 66
fried polenta with gorgonzola, 18
plymouth bay berry cake, 243
polenta with peppers, mushrooms and onions, 106
COUSCOUS
couscous dumplings, curried, rutabaga soup with, 67
couscous salad, 135
couscous-stuffed tomatoes, 181
mixed grain trio, 193
orange pistachio couscous, 198
CRAB
Citrus's crab coleslaw, 44
crab and avocado salad with coriander-tomatillo vinaigrette, 122
crab hoppin' john, 123
jade crab salad, 122
soft-shell crab and hazelnut salad, 123
CRACKED WHEAT

beefsteak tomatoes with minted cracked wheat salad, 116
cracked wheat-spinach salad, 236
cracked-wheat pudding, 253
CROQUETTES
   potato and onion croquettes with deep-fried parsley, 168
   sweet chestnut croquettes, 241
   sweet rice with apricot croquettes, 242
CUCUMBERS
   abbreviated bread-and-butter pickles, 266
   cold cucumber fettuccine salad, 214
   cucumber and chive salad, 214
   cucumber and tomato raita, 223
   cucumber, walnut and dill raita, 223
   ginger-spiced cucumber, 160
   marinated cucumber salad, 215
   new dill slices, 265
   old-fashioned bread-and-butter pickles, 266
   sautéed cucumbers, 160
   spicy cucumber and fruit salad, 40
   spicy potato-cucumber soup with dill, 60
   tomato, cucumber and pineapple salad, 218

D
Dandelion and bacon salad with hard cider vinaigrette, 117
DESSERTS. *See also Cakes; Fruit salads; Rice pudding*
   buttermilk rhubarb cobbler, 253
   cracked-wheat pudding, 253
   rice carlotta, 255
   sweet bean crisps, 241
   sweet chestnut croquettes, 241
   sweet rice with apricot croquettes, 242
Deviled celery root, 157
DIPS
   avocado and roasted corn guacamole with toasted corn tortillas, 29
   crispy shallot dip, 28
   eggplant caviar, 27
   frijolemole, 27
   golden madras dip, 28
   guacamole, 29
   middle-eastern eggplant dip, 25
   provençale spinach dip with herb toasts, 27
   warm mediterranean eggplant dip with herbed pita crisps, 26

E
EGGPLANT
   baked stuffed tomatoes with eggplant and zucchini, 181
   braised eggplant and tomato with sage, 161
   eggplant and mushrooms with balsamic glaze, 48
   eggplant and red pepper pizza with rosemary, 85
   eggplant and spinach timbales, 161
   eggplant caviar, 27
   eggplant cookies with goat cheese and tomato-basil sauce, 17
   eggplant crêpes, 105
   fennel and eggplant salad, 215
   garlic-braised eggplant and chickpea casserole, 101
   gingered eggplant mousse, 28
   gratin of eggplant caviar, 160
   grilled eggplant bruschetta with prosciutto, 22
   lebanese eggplant and yogurt salad, 224
   middle-eastern eggplant dip, 25
   millet-and-eggplant casserole, 101
   pickled eggplant and mushrooms, 268
   red pepper and eggplant terrine, 48
   warm mediterranean eggplant dip with herbed pita crisps, 26
   yogurt and grilled eggplant relish, 263
EGGS
   asparagus pancake with mozzarella, 99
   avocado frittata, 98
   huevos rancheros new york style, 99
   pepper and spaghetti frittata, 98
   potato frittata sandwiches with roasted red peppers, 98
   warm egg and rice salad, 117

Escarole, wilted, salad with pancetta and garlic, 46

F
FENNEL
   baked fennel and boston lettuce, 162
   fennel and eggplant salad, 215
   fennel salad, 215
   green salad with lemon and fennel, 205
   salad of flageolet beans with fennel and walnuts, 38
FIRST COURSES. *See also Salads, first-course; Risotto; Soups*
   artichoke, corn and oyster gratin, 47
   asparagus bundles, 47
   braised artichoke hearts, 47
   braised pear, celery and endive with parmesan and basil, 49
   braised wild mushrooms, 51
   caramelized onion with salsa, 49
   eggplant and mushrooms with balsamic glaze, 48
   noodle galettes with basil water and tomatoes, 57
   papillote of leeks with parmesan cheese, 49
   peas and mushrooms in croustades with chervil, 52
   potato cases with shiitake and morel filling, 52
   red pepper and eggplant terrine, 48
   root vegetable cakes, 57
   stuffed white mushrooms in phyllo, 50
   swiss cheese beignets, 37
   wild mushrooms en papillote, 51
   zucchini ribbons with arugula and creamy goat cheese sauce, 50
FISH/SHELLFISH
   coriander swordfish salad with jicama and thin green beans, 119
   crab and avocado salad with coriander-tomatillo vinaigrette, 122
   crab hoppin' john, 123
   curried salmon salad with lemon grass vinaigrette, 120

fresh tuna salad niçoise, 119
grilled fresh tuna and lima bean salad, 118
grilled swordfish and avocado salad baja-style, 120
jade crab salad, 122
lemon shrimp salad, 126
molded seafood salad with shiitake mushrooms and avocado sauce, 134
mussel and potato salad, 124
mussel and red bean salad, 124
seafood salad with watercress, red pepper and fresh herbs, 134
shellfish boil, 133
shrimp and endive salad, 127
shrimp salad with cauliflower, 127
smoked seafood salad with roasted peppers and capers, 121
smoked trout niçoise, 122
soft-shell crab and hazelnut salad, 123
spicy southeast asian shrimp salad, 127
squid salad with lemon and fresh mint, 133
stir-fried scallop and red pepper salad, 125
swedish west coast salad, 45
thai papaya scallop salad, 126
tuna and white bean salad, 118
warm mussel salad with lemon grass and fresh coriander, 125
wild rice and mussel salad, 124

FLAGEOLETS
flageolet and leek soup, 71
salad of flageolet beans with fennel and walnuts, 38

Frijolemole, 27

FRITTATA. *See Eggs*

FRITTERS
artichoke fritters with remoulade sauce, 20
avocado tempura with coriander and lime, 20
swiss cheese beignets, 37
10-ingredient vegetable fritters, 20

FRUIT SALADS
berry fruit salad, 240

citrus salad with honey-lime dressing, 240
spiced fruit salad, 221
spicy cucumber and fruit salad, 40
strawberry and fresh fig salad, 240
strawberry, kiwi and orange salad, 240

G
Gado-gado (layered salad with spicy peanut dressing), 114

GARLIC
creamy roasted garlic dressing, 260
garlic-olive mayonnaise, 99
roasted garlic and leek chowder, 82
roasted garlic puree, 265

Ginger oil, 259

GOAT CHEESE
eggplant cookies with goat cheese and tomato-basil sauce, 17
green gulch lettuces with sonoma goat cheese, 202
salad of pears and mixed greens and chèvre, 37
zucchini ribbons with arugula and creamy goat cheese sauce, 50

GORGONZOLA
fried polenta with gorgonzola, 18
mixed greens with polenta croutons and gorgonzola, 32
red potatoes with gorgonzola cream, bacon and walnuts, 18

GRAINS. *See also Bulgur*
cracked wheat-spinach salad, 236
kasha with wild mushrooms and toasted walnuts, 187
millet-and-eggplant casserole, 101
mixed grain trio, 193
orange pistachio couscous, 198
quinoa salad, 237
saffroned millet pilaf with roasted red peppers, 188
tabbouleh, 236

Grandma's potato, red pepper and zucchini gratin, 176

GRAPEFRUIT
avocado salad with grapefruit and bacon, 32
grapefruit and avocado salad with tomato-cumin dressing, 221

GRATINS
artichoke, corn and oyster gratin, 47
butternut squash gratin, 179
grandma's potato, red pepper and zucchini gratin, 176
gratin of eggplant caviar, 160
plantain gratin, 166
potato and jerusalem artichoke gratin, 176
potato and wild mushroom gratin, 176

Green tomato tart, apple and, with calvados cream, 247

GREEN BEANS. *See Beans, fresh*

GREENS
(relatively) quick greens in potlikker, 164
braised mustard greens, 163
chickpea stew with greens and spices, 102
collard greens and black-eyed-pea soup with cornmeal croustades, 65
cornmeal pizza with greens and fontina, 86
mixed greens with cracklins and hot pepper vinegar, 163
stewed kale, 163
turnips and turnip greens with ginger and garlic, 183

GUACAMOLE
avocado and roasted corn guacamole with toasted corn tortillas, 29
guacamole, 29

H
HAM
barley-ham salad with honey mustard dressing, 141
ham and cheese salad, 142
ham salad with raspberries and pearl onions, 142

Hearts of palm, beet and endive salad, 207

HOMINY
california rancho-style hominy, 159
hominy and red chile stew, 100

Hoppin' john, crab, 123

Huevos rancheros new york style, 99

I-J-K

Ice cream, rice pudding, 256
INDONESIAN CUISINE
   achar kuning, 220
   indonesian fried tofu and bean sprout salad (tahu goreng), 39
   indonesian mixed salad with pineapple, 220
   indonesian rice with raisins, 109
   indonesian sweet soy sauce (kecap manis), 39
   layered salad with spicy peanut dressing (gado-gado), 114
   spiced fruit salad, 221
   sweet-and-sour vegetable stir-fry, 107
   urab urab, 217
   yellow rice, 196
JALAPENO PEPPERS
   jalapeño risotto with sonoma dry jack cheese, 58
   pickled jalapeños, 269
Jasmine rice, thai, 108
Jerusalem artichoke, potato and, gratin, 176
Kasha with wild mushrooms and toasted walnuts, 187
Kecap manis, 39
KIDNEY BEANS
   french-style white kidney beans, 186
   mussel and red bean salad, 124
   three-bean casserole, 101
Kim chee, 267
KOHLRABI
   root vegetable cakes, 57

L

Lamb, marinated, salad with radishes and currants, 143
LEEKS
   corn-and-leek soup, 72
   flageolet and leek soup, 71
   papillote of leeks with parmesan cheese, 49
   roasted garlic and leek chowder, 82
Lentils, french, salad with glazed pearl onions, 236

LIMA BEANS
   baby lima beans, cherry tomatoes and pears, 185
   grilled fresh tuna and lima bean salad, 118
   lima beans with brown butter, 184
   simple stew of lima beans, corn and tomatoes, 102
   tangy corn and lima beans, 228
   three-bean casserole, 101
   white lima beans with sorrel and parsley, 184
Lobster salad with tarragon and sweet peppers, 44
Lyonnaise potato cake, 175

M

Macaroni salad, three-bean, 233
MILLET
   millet-and-eggplant casserole, 101
   saffroned millet pilaf with roasted red peppers, 188
MORELS
   cream of asparagus and morel soup, 70
   morel and fontina pizza, 84
   potato cases with shiitake and morel filling, 52
MUSHROOMS. *See also* Wild mushrooms
   artichoke hearts with mushrooms, 148
   eggplant and mushrooms with balsamic glaze, 48
   mushroom hazelnut soup, 71
   mushroom-barley soup, 70
   peas and mushrooms in croustades with chervil, 52
   pickled eggplant and mushrooms, 268
   polenta with peppers, mushrooms and onions, 106
   potato-mushroom cakes with cumin, 175
   spinach-stuffed mushrooms with cheddar and chervil, 17
   stuffed white mushrooms in phyllo, 50
   three-mushroom soup with port and tarragon, 69
   tomato and mushroom pie, 87
   triangles with two-mushroom filling, 16
MUSSELS
   mussel and potato salad, 124
   mussel and red bean salad, 124
   scallop, mussel and asparagus salad with orange-saffron dressing, 43
   warm mussel salad with lemon grass and fresh coriander, 125
   wild rice and mussel salad, 124
Mustard greens, braised, 163

N-O

NACHOS
   classic nachos, 23
   salad nachos vinaigrette, 23
NAPPA CABBAGE
   kim chee, 267
No-yolk caesar dressing, 259
NOODLES
   singapore noodle salad, 115
   thai rice noodle salad, 116
Nuoc cham dipping sauce, 133
ONIONS
   Au Pied de Cochon's onion soup, 68
   baked onions with balsamic vinaigrette, 165
   caramelized onion and bacon pizza, 86
   caramelized onion with salsa, 49
   cider-braised onions, 164
   potato and onion croquettes with deep-fried parsley, 168
   richly glazed pearl onions, 165
   tempura onion rings, 165
   three-onion tart, 89
   vermouth-pickled onions, 266
   vidalia onion tart, 90
Orange salad, strawberry, kiwi and, 240
Orange-rhubarb breakfast cake, 246
Oyster, artichoke, corn and, gratin, 47

P

PAPAYAS
   green papaya salad, 40
   papaya-tomatillo chutney, 263
   thai papaya scallop salad, 126
PAPILLOTE

INDEX

papillote of leeks with parmesan cheese, 49
potatoes and sun-dried tomatoes en papillote, 167
potatoes baked in parchment, 104
wild mushrooms en papillote, 51
Parmentier pie, 97
PARSLEY
deep-fried parsley, potato and onion croquettes with, 168
parsley-rice soufflé, 194
spiced tempura parsley, 166
PARSNIPS
lemon-glazed parsnip tea bread, 244
parsnip chips, 24
root vegetable cakes, 57
PASTA
composed salad of sausage and orzo with fresh peas and mixed greens, 143
noodle galettes with basil water and tomatoes, 57
pepper and spaghetti frittata, 98
singapore noodle salad, 115
three-bean macaroni salad, 233
tortellini salad, 141
PASTRY, PIE
caraway pie dough, 89
savory cheese crust, 92
PEACHES
peach and watercress salad, 204
peach sauce, 256
peaches-and-cream rice pudding, 254
russian salad with fresh peaches, 204
Peanut dressing, 262
PEARL ONIONS
fragrant pickled beets with pearl onions, 268
richly glazed pearl onions, 165
PEARS
braised pear, celery and endive with parmesan and basil, 49
salad of pears and mixed greens and chèvre, 37
salad of pears and stilton with sage leaves, 38

PEAS
chilled minted pea soup, 61
composed salad of sausage and orzo with fresh peas and mixed greens, 143
curried cauliflower with peas, 156
fresh pea salad with pancetta, 216
fresh pea soup garnie, 78
green pea soup, 78
greenport soup, 79
peas and mushrooms in croustades with chervil, 52
rich pea soup, 79
Pecans, peppered, 200
PEPPERS, BELL
black bean and bell pepper salad, 228
chickpea and baked bell pepper salad with cumin, 235
eggplant and red pepper pizza with rosemary, 85
grandma's potato, red pepper and zucchini gratin, 176
green beans and roasted peppers, 153
marinated red and green peppers, 21
pepper and spaghetti frittata, 98
polenta with peppers, mushrooms and onions, 106
potato frittata sandwiches with roasted red peppers, 98
potato salad à l'orange with sweet peppers, 226
red pepper and eggplant terrine, 48
red pepper slaw, 223
roasted peppers and celery on toasted croutons, 22
roasted peppers with anchovies and capers, 265
roasted peppers with olives and thyme, 264
rosemary-scented sweet red pepper bisque, 81
saffroned millet pilaf with roasted red peppers, 188
sardinian roasted pepper salad, 217
Persian rice, 197
PICKLES

abbreviated bread-and-butter pickles, 266
achar kuning, 220
cauliflower-carrot pickles in honey-mustard, 267
dilled carrot sticks, 266
fragrant pickled beets with pearl onions, 268
kim chee, 267
new dill slices, 265
old-fashioned bread-and-butter pickles, 266
olive oil pickled beans, 268
oriental pickled vegetables, 269
pickled beets, 267
pickled eggplant and mushrooms, 268
pickled jalapeños, 269
pickled vegetables with pink peppercorns, 269
roasted peppers with anchovies and capers, 265
roasted peppers with olives and thyme, 264
vermouth-pickled onions, 266
PIES AND TARTS, SAVORY
asparagus tart à la bicyclette, 89
deep-dish spinach pie, 88
Nancy Silverton's potato pie, 92
parmentier pie, 97
pipérade pie, 91
potato pie à l'alsacienne, 97
ratatouille pie with basil crust, 91
three-onion tart, 89
tomato and mushroom pie, 87
vidalia onion tart, 90
PIES AND TARTS, SWEET
apple and green tomato tart with calvados cream, 247
fresh rhubarb tart, 248
rhubarb pie with latticed walnut crust, 247
PILAF
coffee-flavored nutty pulau, 197
saffroned millet pilaf with roasted red peppers, 188
spiced potato pilaf, 195
wild mushroom and scallion pilaf, 195

wild rice and apple cider pilaf, 194
PINEAPPLE
   indonesian mixed salad with pineapple, 220
   pineapple and beet salad, 208
   tomato, cucumber and pineapple salad, 218
Pipérade pie, 91
PIZZA
   basic pizza dough, 84
   caramelized onion and bacon pizza, 86
   cornmeal pizza with greens and fontina, 86
   eggplant and red pepper pizza with rosemary, 85
   morel and fontina pizza, 84
   three mushroom and cheese pizza, 84
Plantain gratin, 166
POBLANO PEPPERS
   black bean, corn and poblano pepper salad with cumin vinaigrette, 233
   poblanos rellenos with coriander-corn sauce, 106
POLENTA
   fried polenta with gorgonzola, 18
   polenta with peppers, mushrooms and onions, 106
PORK
   cool and spicy salad with shrimp and pork, 128
   kentucky pork salad, 139
   pork tenderloin salad with apples and apricots, 141
   shrimp, pork and watercress salad, 133
   vietnamese salad with chile-mint dressing, 139
   wild south sea salad, 140
POTATO SALADS
   curried potato and yogurt salad, 224
   "green" potato salad, 225
   indian fruit and potato salad, 227
   lightened potato salad, 226
   mussel and potato salad, 124
   potato and white bean salad, 227
   potato salad à l'orange with sweet peppers, 226
   red potato salad with green beans and red pepper strips, 227
   tomato-potato salad, 225
   warm potato salad with salmon, 117
   wine-drenched potato salad with white wine vinaigrette, 225
POTATOES. *See also Potato salads*
   bacon, potato, white bean and red pepper soup, 72
   baked potato with avocado puree and chives, 167
   basil mashed potatoes, 178
   basilico potatoes, 167
   california scalloped potatoes, 177
   celery root, green bean and potato puree, 158
   cheese and onion rösti, 173
   chewy potato crisps, 173
   cold scalloped potatoes, 177
   crisp potato-scallion roast, 173
   curried oven fries, 166
   grandma's potato, red pepper and zucchini gratin, 176
   lyonnaise potato cake, 175
   mashed potatoes with romano cheese, 178
   Nancy Silverton's potato pie, 92
   pan-roasted potatoes with lemon and marjoram, 168
   parmentier pie, 97
   parslied potato cakes, 174
   potato and jerusalem artichoke gratin, 176
   potato and onion croquettes with deep-fried parsley, 168
   potato and wild mushroom gratin, 176
   potato cases with shiitake and morel filling, 52
   potato crisps, 173
   potato frittata sandwiches with roasted red peppers, 98
   potato pie à l'alsacienne, 97
   potato soup with greens and crisp potato skin croutons, 65
   potato-mushroom cakes with cumin, 175
   potatoes and sun-dried tomatoes en papillote, 167
   potatoes baked in parchment, 104
   red potatoes with gorgonzola cream, bacon and walnuts, 18
   rutabaga and potato puree, 178
   scalloped celery root and potatoes, 157
   scalloped potatoes with sweet marjoram and parmesan cheese, 177
   spiced potato pancakes, 174
   spiced potato pilaf, 195
   spicy potato-cucumber soup with dill, 60
   twice-baked potatoes with cream and wild mushrooms, 167
PUDDING, SWEET
   coconut-basmati rice pudding with satin chocolate sauce, 254
   cracked-wheat pudding, 253
   old-fashioned baked rice pudding, 253
   peaches-and-cream rice pudding, 254
   rice pudding with blueberry and bay compote, 255
PUMPKIN
   pumpkin bourbon pound cake, 245
   pumpkin-ginger cake, 245
Pumpkin seeds, spicy, 16

Q-R

QUINOA
   mixed grain trio, 193
   quinoa salad, 237
RADISHES
   double radish sandwiches, 16
   shredded radish slaw, 222
   zucchini-radish salad, 219
RAITA
   cauliflower raita, 224
   cucumber and tomato raita, 223
   cucumber, walnut and dill raita, 223
RATATOUILLE
   ratatouille pie with basil crust, 91
   skillet ratatouille, 162
RELISH/CHUTNEY
   fresh coriander chutney, 264

INDEX

papaya-tomatillo chutney, 263
spiced fresh tomato chutney, 263
tomato chutney with leeks and carrots, 264
yogurt and grilled eggplant relish, 263

RHUBARB
buttermilk rhubarb cobbler, 253
fresh rhubarb tart, 248
orange-rhubarb breakfast cake, 246
rhubarb pie with latticed walnut crust, 247

RICE. *See also* Rice pudding; Risotto
basic and beautiful fried rice, 110
basic cooked rice, 109
bean, corn and rice salad with chili vinaigrette, 234
chinese almond rice, 195
coffee-flavored nutty pulau, 197
curried basmati rice, 196
curried rice salad, 237
flagstaff green rice, 197
fried ginger rice, 198
fried-rice patties, 194
indonesian rice with raisins, 109
parsley-rice soufflé, 194
persian rice, 197
rice carlotta, 255
rice salad with mint and tomatillos, 238
spiced potato pilaf, 195
spinach-rice timbales with lemon cream, 193
sweet rice with apricot croquettes, 242
taiwanese rice with pineapple, 110
thai jasmine rice, 108
warm egg and rice salad, 117
wild mushroom and scallion pilaf, 195
yellow rice, 196

RICE PUDDING
coconut-basmati rice pudding with satin chocolate sauce, 254
old-fashioned baked rice pudding, 253
peaches-and-cream rice pudding, 254
rice pudding ice cream, 256
rice pudding with blueberry and bay compote, 255

RISOTTO
amarone risotto with pancetta and prunes, 58
jalapeño risotto with sonoma dry jack cheese, 58
wild mushroom risotto, 109

ROQUEFORT
Chardenoux's salad of blue cheese, nuts and belgian endive, 203
mixed leaf salad with roasted walnuts and roquefort, 203
Rösti, cheese and onion, 173

RUTABAGA
root vegetable cakes, 57
rutabaga and potato puree, 178
rutabaga chips, 25
rutabaga soup with curried couscous dumplings, 67

S

SALAD DRESSING
beet vinaigrette, 259
buttermilk herb dressing, 261
caramelized shallot vinaigrette, 259
chunky-creamy blue cheese dressing, 261
creamy roasted garlic dressing, 260
creamy tarragon vinaigrette, 261
dijon vinaigrette, 258
exotic oil vinaigrette, 258
green peppercorn teriyaki dressing, 260
no-yolk caesar dressing, 259
oil-and-egg vinaigrette, 258
peanut dressing, 262
port dressing, 258
red wine vinaigrette, 258
salsa vinaigrette, 262
tomato-cumin dressing, 260
vinaigrette dressing with garlic and fresh herbs, 261
virgin mary vinaigrette, 260
white wine vinaigrette, 258

SALADS, FIRST-COURSE
apricot-shrimp salad, 41
arugula and scallop salad, 43
arugula and watercress salad with swiss cheese beignets, 37
avocado salad with grapefruit and bacon, 32
black-eyed pea and mint marigold salad, 39
Citrus's crab coleslaw, 44
classic squid salad, 42
green papaya salad, 40
indonesian fried tofu and bean sprout salad (tahu goreng), 39
lobster salad with tarragon and sweet peppers, 44
mixed greens with polenta croutons and gorgonzola, 32
salad of flageolet beans with fennel and walnuts, 38
salad of pears and mixed greens and chèvre, 37
salad of pears and stilton with sage leaves, 38
scallop and orange salad, 42
scallop, mussel and asparagus salad with orange-saffron dressing, 43
shrimp salad hilary, 42
spicy cucumber and fruit salad, 40
swedish west coast salad, 45
warm artichoke salad with bacon and mustard vinaigrette, 40
warm salad of feta cheese and tomatoes with garlic shrimp, 46
warm shrimp salad with champagne vinegar sauce, 41
wilted escarole salad with pancetta and garlic, 46
wilted salad with chicken livers, apples and mustard seed, 45

SALADS, MAIN-COURSE
avocado and veal salad with walnuts, 138
barley-ham salad with honey mustard dressing, 141
beefsteak tomatoes with minted cracked wheat salad, 116
broiled thai chicken salad, 137
cabbage salad with shrimp (achar udang), 128
chinese-style chicken salad, 136
composed salad of sausage and orzo with fresh peas and mixed greens, 143

INDEX

cooked beef salad with mustard-anchovy dressing, 144
cool and spicy salad with shrimp and pork, 128
coriander swordfish salad with jicama and thin green beans, 119
couscous salad, 135
crab and avocado salad with coriander-tomatillo vinaigrette, 122
crab hoppin' john, 123
crunchy chicken salad, 136
curried salmon salad with lemon grass vinaigrette, 120
curried turkey salad with grapes and almonds, 135
dandelion and bacon salad with hard cider vinaigrette, 117
dilled tabbouleh, 121
fresh tuna salad niçoise, 119
grilled fresh tuna and lima bean salad, 118
grilled swordfish and avocado salad baja-style, 120
ham and cheese salad, 142
ham salad with raspberries and pearl onions, 142
jade crab salad, 122
kentucky pork salad, 139
layered salad with spicy peanut dressing (gado-gado), 114
lemon shrimp salad, 126
madrigal salad, 144
marinated lamb salad with radishes and currants, 143
molded seafood salad with shiitake mushrooms and avocado sauce, 134
mussel and potato salad, 124
mussel and red bean salad, 124
oriental beef salad, 144
paillard of veal with bitter greens, 138
pork tenderloin salad with apples and apricots, 141
seafood salad with watercress, red pepper and fresh herbs, 134
shrimp and endive salad, 127
shrimp salad with cauliflower, 127
shrimp, pork and watercress salad, 133
singapore noodle salad, 115
smoked seafood salad with roasted peppers and capers, 121
smoked trout niçoise, 122
soft-shell crab and hazelnut salad, 123
south sea beef salad, 146
spicy southeast asian shrimp salad, 127
squid salad with lemon and fresh mint, 133
stir-fried scallop and red pepper salad, 125
thai beef salad, 146
thai papaya scallop salad, 126
thai rice noodle salad, 116
thai-style beef and lettuce salad with chile-lime dressing, 145
tortellini salad, 141
triple-mustard chicken salad, 137
tuna and white bean salad, 118
tuscan-style white bean salad, 115
veal tonnato salad, 138
vietnamese salad with chile-mint dressing, 139
warm caesar salad, 114
warm egg and rice salad, 117
warm mussel salad with lemon grass and fresh coriander, 125
warm potato salad with salmon, 117
wild rice and mussel salad, 124
wild south sea salad, 140
SALADS, SIDE-DISH. *See also* Bean salads; Potato salads; Raita; Slaws
achar kuning, 220
asparagus and beet salad with tarragon, 207
baby carrot and turnip salad, 214
bagna cauda salad, 219
baked beet and endive salad, 208
beet and avocado salad with tarragon vinaigrette, 208
beet and chicory salad, 207
carrot and orange salad with black mustard seeds, 214
Chardenoux's salad of blue cheese, nuts and belgian endive, 203
cold cucumber fettuccine salad, 214
cracked wheat-spinach salad, 236
cucumber and chive salad, 214
curried rice salad, 237
fennel and eggplant salad, 215
fennel salad, 215
field salad with shallots and chives, 201
fresh and sun-dried tomatoes with zucchini and balsamic vinegar, 218
fresh pea salad with pancetta, 216
fresh tomato salad, 217
garden green bean and tomato salad with warm bacon and pine nut dressing, 206
grapefruit and avocado salad with tomato-cumin dressing, 221
green bean and benne salad, 206
green beans mimosa, 205
green gulch lettuces with sonoma goat cheese, 202
green salad with lemon and fennel, 205
green beans in aquavit, 206
hearts of palm, beet and endive salad, 207
indonesian mixed salad with pineapple, 220
lebanese eggplant and yogurt salad, 224
marinated cucumber salad, 215
marinated green beans, 21
marinated red and green peppers, 21
minted zucchini and apple salad, 219
mixed green salad with soured cream and blue cheese dressing, 201
mixed leaf salad with roasted walnuts and roquefort, 203
oriental-style tomato salad, 218
peach and watercress salad, 204
pineapple and beet salad, 208
port and stilton salad, 203
Quatorze's chicory and bacon salad, 202
quinoa salad, 237

INDEX

rice salad with mint and tomatillos, 238
russian salad with fresh peaches, 204
salad of mixed greens and herbs with oranges, 205
salad of spinach crowns, 200
salade bressane, 201
sardinian roasted pepper salad, 217
shredded raw beet and celery salad, 213
southern greens salad with peppered pecans, 200
spiced fruit salad, 221
spinach salad with feta, mint and olives, 202
sugar snap pea and summer squash salad with rye croutons, 216
sugar snap peas in toasted sesame seed vinaigrette, 216
sweet potato salad, 226
tabbouleh, 236
tender greens with prosciutto, croutons and walnut oil vinaigrette, 204
tomato, cucumber and pineapple salad, 218
tomato-potato salad, 225
tomatoes on basil chiffonade, 218
urab urab, 217
vegetable salad à la grecque, 220
warm broccoli salad with toasted pine nuts, 213
warm string bean salad, 213
zucchini-radish salad, 219

SALMON
curried salmon salad with lemon grass vinaigrette, 120
warm potato salad with salmon, 117

SALSA
fresh tomato salsa, 100
red tomato salsa, 262
tomatillo salsa, 262
tomato-coriander salsa, 262
yellow tomato salsa, 263

Salsify, scalloped, with parmesan and pecans, 179

SANDWICHES
double radish sandwiches, 16
potato frittata sandwiches with roasted red peppers, 98

SAUCES, SAVORY
ancho chile cream, 77
chili soy sauce, 20
cilantro cream, 77
fresh tomato salsa, 100
garlic-olive mayonnaise, 99
herb coulis, 193
indonesian sweet soy sauce (kecap manis), 39
nuoc cham dipping sauce, 133
sweet and sour sauce, 19

SAUCES, SWEET
apricot sauce, 242
blueberry and bay compote, 255
peach sauce, 256
satin chocolate sauce, 254

SCALLOPED DISHES
california scalloped potatoes, 177
cold scalloped potatoes, 177
scalloped celery root and potatoes, 157
scalloped potatoes with sweet marjoram and parmesan cheese, 177
scalloped salsify with parmesan and pecans, 179

SCALLOPS
arugula and scallop salad, 43
scallop and orange salad, 42
scallop, mussel and asparagus salad with orange-saffron dressing, 43
stir-fried scallop and red pepper salad, 125
thai papaya scallop salad, 126

SHALLOTS
caramelized shallot vinaigrette, 259
crispy shallot dip, 28
roasted shallots, 179

SHIITAKE
braised wild mushrooms, 51
potato cases with shiitake and morel filling, 52
three mushroom and cheese pizza, 84
three-mushroom soup with port and tarragon, 69

SHRIMP
apricot-shrimp salad, 41
cabbage salad with shrimp (achar udang), 128
cool and spicy salad with shrimp and pork, 128
lemon shrimp salad, 126
shrimp salad hilary, 42
shrimp salad with cauliflower, 127
shrimp, pork and watercress salad, 133
spicy southeast asian shrimp salad, 127
warm salad of feta cheese and tomatoes with garlic shrimp, 46
warm shrimp salad with champagne vinegar sauce, 41

SLAWS
cabbage salad with garlic, 213
Citrus's crab coleslaw, 44
coleslaw with bulgur, 222
creamy garlic coleslaw, 222
five-vegetable slaw, 222
red pepper slaw, 223
sassy slaw, 223
shredded radish slaw, 222
warm cabbage slaw with bacon, 221

Soft-shell crab and hazelnut salad, 123
Som tam (green papaya salad), 40
Soufflé, parsley-rice, 194

SOUPS, COLD. See also Soups, hot
avocado soup with papaya-pepper relish, 62
chilled cream of summer squash soup, 62
chilled minted pea soup, 61
cool asparagus soup, 61
corn soup with spicy pumpkin seeds, 61
german summer salad soup, 60
spicy potato-cucumber soup with dill, 60
sweet potato vichyssoise, 60

SOUPS, HOT. See also Soups, cold
acorn squash and turnip soup, 67
asparagus soup, 64
Au Pied de Cochon's onion soup, 68
bacon, potato, white bean and red pepper soup, 72
broccoli, onion and cheese soup, 69
butternut squash consommé with leek ravioli, 63

cauliflower cress soup with corn wafers, 68
charred carrot soup, 80
collard greens and black-eyed-pea soup with cornmeal croustades, 65
corn-and-leek soup, 72
cream of asparagus and morel soup, 70
cream of carrot soup with brandy and chervil, 80
cream of celery soup, 77
creamy carrot soup, 80
flageolet and leek soup, 71
fresh pea soup garnie, 78
great lakes corn chowder, 81
green pea soup, 78
greenport soup, 79
grilled corn soup with cilantro and ancho chile creams, 77
mushroom hazelnut soup, 71
mushroom-barley soup, 70
new jersey tomato and sweet potato chowder, 82
potato soup with greens and crisp potato skin croutons, 65
red cabbage soup, 64
rich pea soup, 79
roasted garlic and leek chowder, 82
rosemary-scented sweet red pepper bisque, 81
rutabaga soup with curried couscous dumplings, 67
shrimp and endive salad, 127
southwestern black bean soup, 71
spiced tomato soup, 63
three-mushroom soup with port and tarragon, 69

SPINACH
braised artichokes with spinach, 148
cracked wheat-spinach salad, 236
deep-dish spinach pie, 88
eggplant and spinach timbales, 161
provençale spinach dip with herb toasts, 27
salad of spinach crowns, 200
spinach salad with feta, mint and olives, 202
spinach-rice timbales with lemon cream, 193

SQUASH. *See specific squashes*

SQUID
classic squid salad, 42
squid salad with lemon and fresh mint, 133

STEWS
chickpea stew with greens and spices, 102
hominy and red chile stew, 100
ragout of fall vegetables, 104
santa fe farmers' market stew, 103
simple stew of lima beans, corn and tomatoes, 102

STILTON
port and stilton salad, 203
salad of pears and stilton with sage leaves, 38

Stock, vegetable, 80

STRAWBERRIES
strawberry and fresh fig salad, 240
strawberry, kiwi and orange salad, 240

SUGAR SNAP PEAS
sugar snap pea and summer squash salad with rye croutons, 216
sugar snap peas in toasted sesame seed vinaigrette, 216

SUMMER SQUASH
chilled cream of summer squash soup, 62
sugar snap pea and summer squash salad with rye croutons, 216

SUN-DRIED TOMATOES
fresh and sun-dried tomatoes with zucchini and balsamic vinegar, 218
potatoes and sun-dried tomatoes en papillote, 167

SWEET POTATOES
baked buttery sweet potato chips, 180
new jersey tomato and sweet potato chowder, 82
sweet potato salad, 226
sweet potato vichyssoise, 60

SWORDFISH
coriander swordfish salad with jicama and thin green beans, 119
grilled swordfish and avocado salad baja-style, 120

T

TABBOULEH
dilled tabbouleh, 121
tabbouleh, 236

TEMPURA. *See also Fritters*
avocado tempura with coriander and lime, 20
spiced tempura parsley, 166
tempura onion rings, 165

Terrine, red pepper and eggplant, 48

THAI CUISINE
broiled thai chicken salad, 137
green papaya salad, 40
shrimp, pork and watercress salad, 133
thai beef salad, 146
thai jasmine rice, 108
thai papaya scallop salad, 126
thai rice noodle salad, 116
thai-style beef and lettuce salad with chile-lime dressing, 145

TIMBALES
eggplant and spinach timbales, 161
spinach-rice, with lemon cream timbales, 193

TOFU
bulgur wheat, peppers and bean curd, 104
indonesian fried tofu and bean sprout salad (tahu goreng), 39

TOMATILLOS
papaya-tomatillo chutney, 263
rice salad with mint and tomatillos, 238
tomatillo salsa, 262

TOMATOES
apple and green tomato tart with calvados cream, 247
baked cherry tomatoes with herbs, 180
baked stuffed tomatoes with eggplant and zucchini, 181
beefsteak tomatoes with minted cracked wheat salad, 116
braised eggplant and tomato with sage, 161
calamata-stuffed tomatoes, 182

INDEX

couscous-stuffed tomatoes, 181
cucumber and tomato raita, 223
fresh and sun-dried tomatoes with zucchini and balsamic vinegar, 218
fresh tomato salad, 217
fresh tomato salsa, 100
fresh tomatoes stuffed with rice and cheese, 108
garden green bean and tomato salad with warm bacon and pine nut dressing, 206
new jersey tomato and sweet potato chowder, 82
oriental-style tomato salad, 218
red tomato salsa, 262
savory tomato charlottes, 180
simple stew of lima beans, corn and tomatoes, 102
spiced fresh tomato chutney, 263
spiced tomato soup, 63
tomato and mushroom pie, 87
tomato bread crisp, 22
tomato chutney with leeks and carrots, 264
tomato, cucumber and pineapple salad, 218
tomato-coriander salsa, 262
tomato-potato salad, 225
tomatoes on basil chiffonade, 218
yellow tomato salsa, 263
Tortellini salad, 141
Trout, smoked, niçoise, 122
TUNA
fresh tuna salad niçoise, 119
grilled fresh tuna and lima bean salad, 118
tuna and white bean salad, 118
Turkey, curried, salad with grapes and almonds, 135
TURNIPS. *See also Rutabaga*
acorn squash and turnip soup, 67
baby carrot and turnip salad, 214
orange-glazed turnips, 182
turnips and turnip greens with ginger and garlic, 183
turnips anna, 182

V

VEAL
avocado and veal salad with walnuts, 138
paillard of veal with bitter greens, 138
veal tonnato salad, 138
VEGETABLES, MAIN-COURSE. *See also Pies and tarts, savory; Pizza*
black-eyed-pea cakes with salsa, 107
bulgur wheat, peppers and bean curd, 104
chickpea stew with greens and spices, 102
eggplant crêpes, 105
fresh tomatoes stuffed with rice and cheese, 108
garlic-braised eggplant and chickpea casserole, 101
hominy and red chile stew, 100
millet-and-eggplant casserole, 101
poblanos rellenos with coriander-corn sauce, 106
polenta with peppers, mushrooms and onions, 106
potatoes baked in parchment, 104
ragout of fall vegetables, 104
santa fe farmers' market stew, 103
simple stew of lima beans, corn and tomatoes, 102
sweet-and-sour vegetable stir-fry, 107
three-bean casserole, 101
VEGETABLES, SIDE-DISH. *See also the names of specific vegetables*
miniature vegetables with lime and basil, 184
roasted vegetables with vinaigrette, 183
skillet ratatouille, 162
Vichyssoise, sweet potato, 60
VIETNAMESE CUISINE
cool and spicy salad with shrimp and pork, 128
nuoc cham dipping sauce, 133
vietnamese salad with chile-mint dressing, 139

W-Z

WATERCRESS
cauliflower cress soup with corn wafers, 68
peach and watercress salad, 204
WHITE BEANS
bacon, potato, white bean and red pepper soup, 72
potato and white bean salad, 227
tuna and white bean salad, 118
tuscan-style white bean salad, 115
WILD MUSHROOMS
braised wild mushrooms, 51
cèpes bordelaise, 164
cream of asparagus and morel soup, 70
kasha with wild mushrooms and toasted walnuts, 187
morel and fontina pizza, 84
potato and wild mushroom gratin, 176
three mushroom and cheese pizza, 84
three-mushroom soup with port and tarragon, 69
twice-baked potatoes with cream and wild mushrooms, 167
wild mushroom and scallion pilaf, 195
wild mushroom risotto, 109
wild mushrooms en papillote, 51
WILD RICE
wild rice and apple cider pilaf, 194
wild rice and mussel salad, 124
ZUCCHINI
baked stuffed tomatoes with eggplant and zucchini, 181
fresh and sun-dried tomatoes with zucchini and balsamic vinegar, 218
grandma's potato, red pepper and zucchini gratin, 176
La Mère Poulard's zucchini crêpes, 183
minted zucchini and apple salad, 219
zucchini ribbons with arugula and creamy goat cheese sauce, 50
zucchini-radish salad, 219

CONTRIBUTORS

JEFFREY ALFORD is a cookbook author, food writer, photographer and the author (with Naomi Duguid) of *Flatbreads & Flavors* (Morrow).

ANN CHANTAL ALTMAN is a chef, food writer and cooking teacher (Peter Kump's New York Cooking School).

JEAN ANDERSON is a food/travel writer and the author of numerous cookbooks, including *The Doubleday Cookbook* (with Elaine Hanna) and (with Hedy Würz) the upcoming *The New German Cookbook*.

LEE BAILEY is a food/travel writer and the author of thirteen cookbooks, including the recent *Lee Bailey's New Orleans* and *Lee Bailey's Corn* (Clarkson Potter).

MELANIE BARNARD is a restaurant critic, food writer and co-author (with Brooke Dojny) of four cookbooks, including *Parties!* and *Cheap Eats* (both from HarperCollins).

SARAH BELK is a senior editor at *Bon Appetit* and the author of *Around the Southern Table* (Simon and Schuster). She is currently working on a second southern cookbook.

MARK BITTMAN is a food writer and the author of the upcoming *Commonsense Guide to Today's Fish* (Macmillan).

LINDA BURUM is a cooking teacher, food journalist and the author of a number of cookbooks, including *Asian Pasta* (Aris) and *A Guide to Ethnic Food in Los Angeles* (HarperCollins).

HUGH CARPENTER is a chef, cooking teacher, food writer and the co-author (with Teri Sandison) of *Chopstix* and *Pacific Flavors* (both from Stewart, Tabori & Chang).

KATHY CASEY is a restaurant consultant and the author of *Pacific Northwest the Beautiful Cookbook* (Collins).

BOB CHAMBERS is a New York-based chef and food stylist.

LAURA CHENEL is the co-author (with Linda Siegfried) of *Chèvre! The Goat Cheese Cookbook* and *American Country Cheese* (both Aris Books).

BRUCE COST is a restaurateur and the author of *Bruce Cost's Asian Ingredients* (Morrow), *Ginger East to West* (Addison-Wesley) and the upcoming *How to Steam a Bear* (Morrow).

SUSAN COSTNER is the editor of *Tables* magazine (for Beringer Vineyards) and the author of *Great Sandwiches* and *Good Friends, Great Dinners* (both from Crown).

BEVERLY COX is the author of *Spirit of the Harvest: North American Indian Cooking* (Stewart, Tabori & Chang).

PEGGY CULLEN is a New York-based baker, candy maker and food writer.

ROBERT DEL GRANDE is chef/owner of Cafe Annie in Houston.

BROOKE DOJNY is a food writer and the co-author (with Melanie Barnard) of four cookbooks, including *Parties!* and *Cheap Eats* (both from HarperCollins).

RICK ELLIS is a New York-based food writer and stylist.

MARY EVELY is chef at Simi Winery in Healdsburg, California.

CAROL FIELD is a food writer and the author of *The Italian Baker* (HarperCollins), *Celebrating Italy* (Morrow) and the forthcoming *Italy in Small Bites* (Morrow).

JIM FOBEL is the author of numerous books, including *Jim Fobel's Old-Fashioned Baking Book* and *The Whole Chicken Cookbook* (both from Ballantine).

ANNIE GILBAR is the author of *Recipex*.

MARCY GOLDMAN is a Montreal-based food writer, pastry chef and cooking teacher.

JOYCE GOLDSTEIN is chef/owner of Square One in San Francisco and the author of numerous cookbooks, including *The Mediterranean Kitchen* (Morrow) and the upcoming *Mediterranean the Beautiful* (HarperCollins).

DORIE GREENSPAN is the author of *Sweet Times* and *Waffles from Morning to Midnight* (both from Morrow).

KEN HAEDRICH is the author of *Home for the Holidays* and *Ken Haedrich's Country Baking* (both from Bantam).

JESSICA B. HARRIS is a culinary historian and the author of *Tasting Brazil* (Macmillan), *Sky Juice and Flying Fish* (Fireside) and *Iron Pots and Wooden Spoon* (Atheneum).

MADHUR JAFFREY is a cooking teacher and the author of numerous cookbooks, including *Madhur Jaffrey's World-of-the-East Vegetarian Cooking* (Knopf).

SUSAN SHAPIRO JASLOVE is a food writer and recipe developer.

NANCY HARMON JENKINS is a food writer, culinary historian.

MIREILLE JOHNSTON is a cookbook author currently working on a series for BBC called "Mireille Johnston's Cook's Tour of France."

BABA S. KHALSA is the author of *Great Vegetables from Great Chefs* (Chronicle).

CHRISTER LARSSON is executive chef of Aquavit in New York City.

KAREN LEE and ALAXANDRA BRANYON are the co-authors of several books, including *Nouvelle Chinese Cooking* (Macmillan) and the upcoming *Sometimes Vegetarian* (Warner).

SALLY GUTIERREZ and JONATHAN LOCKE are Minneapolis-based food writers.

CONTRIBUTORS

SUSAN HERRMANN LOOMIS is a food writer and the author of *Farmhouse Cookbook, The Great American Seafood Cookbook* and *The Clambake Book* (all Workman).

DEBORAH MADISON is the author of *The Savory Way* and *The Greens Cookbook* (both from Bantam).

COPELAND MARKS is a culinary anthropologist and a cookbook author specializing in exotic cuisines whose most recent work is *The Exotic Kitchens of Korea* (Chronicle).

LYDIE MARSHALL is a cooking teacher and the author of, most recently, *A Passion for Potatoes* (HarperCollins). She is currently working on a book of Provençal cooking.

ELIN McCOY and JOHN FREDERICK WALKER are contributing wines and spirits editors for *Food & Wine* and the authors of *Thinking About Wine* (Simon and Schuster).

MICHAEL McLAUGHLIN is a food writer and the author of numerous cookbooks, including most recently *52 Meat Loaves* and *Cooking for the Weekend* (Simon and Schuster).

LINDA MERINOFF is a cooking teacher and the author of, most recently, *Gingerbread* (Fireside).

LESLIE NEWMAN is the author of *Feasts* (HarperCollins) and the upcoming *The Family Table*.

MOLLY O'NEILL is the food columnist for *The New York Times* and the author of *New York Cookbook* (Workman).

CARL PARISI is a food writer, chef and restaurant consultant.

JACQUES PEPIN is a cooking teacher and the author of twelve cookbooks, the most recent of which are *Today's Gourmet* and *Today's Gourmet II* (KQED, Inc.).

STEPHAN PYLES is a chef/restaurateur, cooking teacher and the author of *The New Texas Cuisine* (Doubleday).

KEVIN RATHBUN is executive chef of Baby Routh in Dallas.

MICHEL RICHARD is chef/owner of Citrus in Los Angeles and the author of the forthcoming *Michel Richard's Home Cooking with a French Accent* (Morrow).

RICK RODGERS is a cooking teacher and the author of *The Turkey Cookbook* (HarperCollins).

DAVID ROSENGARTEN is a cooking teacher, food writer, wine columnist and the author of *Food & Wine*'s "Food and Drink" column.

JULIE SAHNI is a chef, cooking teacher and the author of *Classic Indian Cooking* and *Classic Indian Vegetarian and Grain Cooking* (both from Morrow).

RICHARD SAX is a cookbook author currently working on a dessert cookbook and a healthful cooking cookbook.

PHILLIP STEPHEN SCHULZ is the author of several cookbooks, including the upcoming *American Days Cookbook* (Simon and Schuster).

NANCY SILVERTON is pastry chef and co-owner of Campanile in Los Angeles.

ANDRE SOLTNER is chef/owner of Lutéce in New York.

ANNIE SOMERVILLE is executive chef of Greens in San Francisco.

JOHN MARTIN TAYLOR is a food writer, culinary historian and the author of the forthcoming *Hoppin' John's Lowcountry Cooking* (Bantam).

DORIS TOBIAS is a New York-based food and wine writer.

BARBARA TROPP is a cooking teacher, chef/owner of China Moon Cafe in San Francisco and the author of several cookbooks including *China Moon Cookbook* (Workman) and the upcoming *The China Diet*.

MICHELE URVATER is the author of *The Monday-to-Friday Cookbook* and an upcoming pasta cookbook in the same series (both from Workman).

JEAN-GEORGES VONGERICHTEN is chef and co-owner of Vong and JoJo in New York City.

PATRICIA WELLS is a Paris-based food journalist and the author of numerous books including most recently *Patricia Wells' Trattoria* and *Simply French* (both Morrow) and the upcoming *Patricia Wells' French Country Kitchen*.

BRILL WILLIAMS is executive chef at The Inn at Sawmill Farm in West Dover, Vermont.

EILEEN YIN-FEI LO is a cooking teacher and the author of several cookbooks, including *Eileen Yin-Fei Lo's New Cantonese Cooking* (Viking) and an upcoming book titled *Chinese Vegetarian Cooking*.

We would also like to thank the following individuals and restaurants for their contributions to *Food & Wine* and to this book: Carolina's, Charleston, South Carolina; Henri Charvet, Le Lafayette, Fort-de-France, Martinique; Adam Esman; Gaylord India; Michael Kalajian, Quatorze, New York City; Clara Lesueur, Chez Clara, Ste. Rose, Guadeloupe; Robert McGrath; Steve Mellina; Joseph Phelps Vineyards, St. Helena, California.

And the members of the *Food & Wine* staff, past and present: Mimi Ruth Brodeur, Anne Disrude, Rosalee Harris, Marcia Kiesel, John Robert Massie, W. Peter Prestcott, Tracey Seaman, Diana Sturgis, Elizabeth Woodson, Susan Wyler.

PHOTO CREDITS

Cover: Jerry Simpson. Pages 33-36: Jerry Simpson. Pages 53-55: Maria Robledo. Page 56: Elizabeth Watt. Pages 73-76 and 93-96: Mark Thomas. Pages 129-132: Dennis Galante. Pages 149-152: Steven Mark Needham. Pages 169-172: Dennis Galante. Pages 189-192: Mark Thomas. Pages 209-212: Cynthia Brown. Pages 229-232: Robert Jacobs. Pages 249-251: Lisa Charles Watson. Page 252: Mark Thomas.